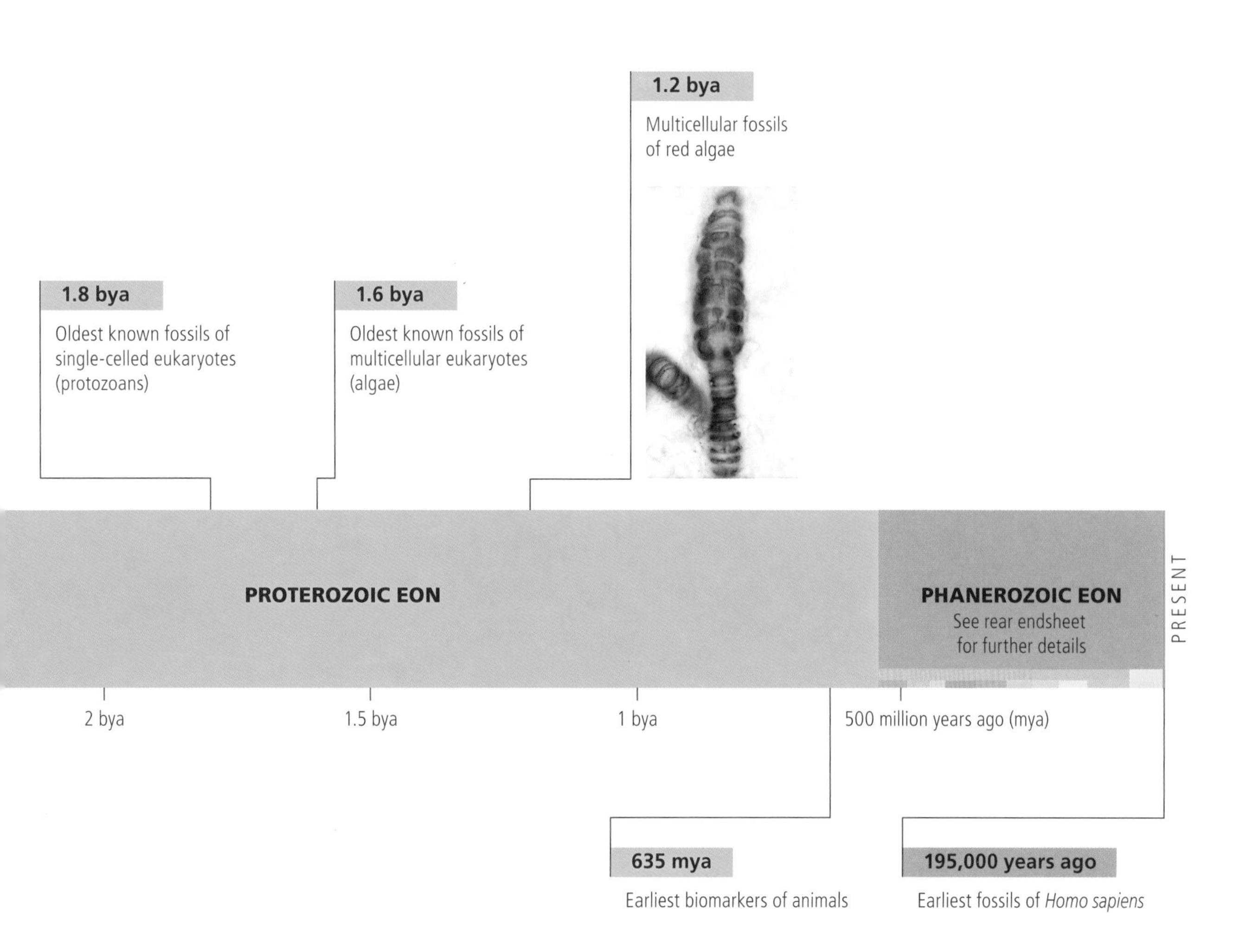

For a detailed geologic time scale, visit http://geosociety.org/science/timescale/

The Tangled Bank

The Tangled

An Introduction to

Bank Evolution

Carl Zimmer

SCIENTIFIC ADVISORS

Joel Kingsolver
University of North Carolina

Kevin Padian
University of California, Berkeley

Gregory Wray
Duke University

Marlene Zuk
University of California, Riverside

ROBERTS AND COMPANY
Greenwood Village, Colorado

THE TANGLED BANK: AN INTRODUCTION TO EVOLUTION
Roberts and Company Publishers

4950 South Yosemite Street, F2 #197
Greenwood Village, Colorado 80111 USA
Internet: www.roberts-publishers.com
Telephone: (303) 221-3325
Facsimile: (303) 221-3326

ORDER INFORMATION
Telephone: (800) 351-1161 or (516) 422-4050
Facsimile: (516) 422-4097
Internet: www.roberts-publishers.com

Publisher:	Ben Roberts
Copyeditor and Indexer:	Gunder Hefta
Proofreaders:	Lynn Golbetz, Michael Zierler
Artists and Art Studios:	Carl Buell, Emiko Rose Paul at Echo Medical Media, Quade Paul at Echo Medical Media, Lineworks, Inc.
Production Editor:	Betty Gee at Side By Side Studios
Interior Designer:	Mark Ong at Side By Side Studios
Cover Image:	Carl Buell
Cover Designer:	Mark Ong at Side By Side Studios
Photo Permissions Coordinator:	Laura Gabbard Roberts
Compositor:	Side By Side Studios

ISBN: 978-0981519470

Library of Congress Cataloging-in-Publication Data

Zimmer, Carl, 1966-
The tangled bank : an introduction to evolution / Carl Zimmer.
p. cm.
Includes bibliographical references and index.
ISBN 978-0-9815194-7-0
1. Evolution (Biology)--Popular works. I. Title.
QH367.Z56 2010
576.8--dc22
2009021802

10 9 8 7 6 5 4 3 2 1

It is interesting to contemplate an entangled bank, clothed with many plants of many kinds, with birds singing on the bushes, with various insects flitting about, and with worms crawling through the damp earth, and to reflect that these elaborately constructed forms, so different from each other, and dependent on each other in so complex a manner, have all been produced by laws acting around us. These laws, taken in the largest sense, being Growth with Reproduction; inheritance which is almost implied by reproduction; Variability from the indirect and direct action of the external conditions of life, and from use and disuse; a Ratio of Increase so high as to lead to a Struggle for Life, and as a consequence to Natural Selection, entailing Divergence of Character and the Extinction of less-improved forms. Thus, from the war of nature, from famine and death, the most exalted object which we are capable of conceiving, namely, the production of the higher animals, directly follows. There is grandeur in this view of life, with its several powers, having been originally breathed into a few forms or into one; and that, whilst this planet has gone cycling on according to the fixed law of gravity, from so simple a beginning endless forms most beautiful and most wonderful have been, and are being, evolved.

—Charles Darwin, *The Origin of Species* (1859)

Contents

Evolution

1

An Introduction

One of the best feelings paleontologists can ever have is to realize that they have just found a fossil that will fill an empty space in our understanding of the history of life. Hans Thewissen got to enjoy that feeling one day as he dug a 49-million-year-old fossil out of a hillside in Pakistan. As he picked away the rocks surrounding the bones of a strange mammal, he suddenly realized what he had found: a whale with legs.

Left: The earliest whales, such as the 49-million-year-old *Ambulocetus*, still had legs. Above: Paleontologist Hans Thewissen has discovered many of the bones of *Ambulocetus*.

There were no whales three billion years ago. There were no mushrooms or trees, either, not to mention people. Paleontologists have scoured ancient rocks for signs of life, and, as far as they can tell, the Earth three billion years ago was home only to single-celled microbes. Some of those tiny organisms were tossed by ocean waves. Others formed slimy films on the seafloor. Others thrived in undersea chambers of boiling water heated by volcanoes. The whales, the mushrooms, the trees, and all the rest came later. That transformation of life occurred through the process known as evolution.

This book is an introduction to evolution, both to the process by which life evolves and to the pattern evolution has produced over life's history. It is also about how scientists study evolution. When Charles Darwin formulated the theory of evolution in the mid-1800s, the most sophisticated tool he could use was a crude light microscope. Today, scientists can study evolution by analyzing our DNA. They can probe the atoms of ancient rocks to determine the age of fossils. They can use powerful computers to apply new statistical equations to the diversity of life. They can observe evolution unfolding in their laboratories.

Evolution is the foundation of modern biology. But that doesn't mean that only biologists should know about it. Evolution underlies many of the most important issues society faces. We are witnessing a wave of extinctions the likes of which the Earth may not have seen for millions of years. Doctors are battling rapidly evolving bacteria and viruses. Evolution is also part of the answer to some of the biggest questions we all ask. How did we get here? What does it mean to be human? This book is intended for those who are not planning to be biologists—in other words, most people. It does not deal at length with the mathematics and the advanced experimental techniques evolutionary biologists use. But it does describe the key concepts in evolution with the help of graphs, diagrams, and other illustrations.

What Is Evolution?

The short answer is *descent with modification*. The patterns of this modification, and the mechanisms by which it unfolds, are what evolutionary biologists study. They have much left to learn—but, at this point in the history of science, there is no doubt that life has evolved and is still evolving.

The fundamental principles that make evolution possible are pretty straightforward. Organisms inherit traits from their ancestors because they receive a molecule called DNA from them (page 87). Cells use DNA as a guide to building biological molecules, and, when organisms reproduce, they make new copies of DNA for their offspring. Living things do not replicate their DNA perfectly, sometimes introducing errors in its sequence. Such errors are referred to as mutations (Chapter 6). A mutation may be lethal; it may be harmless; or it may be beneficial in some way. A beneficial mutation may help an organism to fight off diseases, to thrive in its environment, or to improve its ability to find mates.

Evolution takes place because mutated genes become more or less common over the course of generations. Many mutations eventually disappear, while others spread widely. Some mutations spread simply by chance. Others spread because they allow organisms to produce more offspring. This nonrandom spread of beneficial genes is known as natural selection. The effect of a mutation depends on more than just the mutation itself. It is also influenced by all the other genes an organism carries. The environment in which an organism lives can also have a huge effect. As a result, the same mutation to the same gene may be devastating in one individual and harmless in another. Depending on the particular circumstances, natural selection may favor a mutation or drive it to oblivion.

These processes are taking place all around us every day, and they have been transforming life ever since it began, some 3.5 billion years ago (Chapter 3). Charles Darwin argued that over such vast stretches of time, natural selection could have produced very complex organs, from the wings of birds to human eyes. Today, the weight of evidence overwhelmingly supports that conclusion (Chapter 8). Changes in organs are not the only adaptations that have emerged through evolution; behavior and even language have evolved (Chapter 14).

To trace the origin of these traits, evolutionary biologists reconstruct the tree of life (Chapter 4). Natural selection and other processes can make populations genetically distinct from one another. Over time, the populations become so different from one another they can be considered separate species (Chapter 9). One way to picture this process is to think of the populations as branches on a tree. When two populations diverge, a branch splits in two. As the branches diverged over and over again over billions of years (and sometimes joined back together), the tree of life emerged.

To reconstruct the tree, evolutionary biologists identify which species are closely related to each other. They do so by comparing anatomical traits and DNA among many different species. Close relatives share more traits inherited from their common ancestor. We humans have a bony skull, for example, as do all other mammals, as well as birds, reptiles, amphibians, and fishes (Chapter 4). We are more closely related to these animals than to animals without a skull, such as earthworms and ladybugs.

Biologists today not only understand what evolution is, but what it is not. Evolution is not a steady progress towards some particular goal. Our apelike ancestors did not evolve big brains because they "needed" them. The conditions in which they lived—searching for food on the African savanna in big social groups—favored genetic changes that led to bigger brains. The long-term process of evolution emerges from the way life works on a generation-to-generation scale.

Biologists also recognize that evolution does not make life perfect. All adaptations have shortcomings. Humans have evolved very large brains, which have allowed us to become nature's great thinkers (Chapter 14), but those big brains also make childbirth much more dangerous for human mothers than for other female primates. Evolution is imperfect because it does not invent things from scratch: it only modifies what already exists. There are only a limited number of beneficial mutations any particular organism can acquire, and so evolution can only produce new forms under tight constraints. Because mutations can have several different effects at once, evolution also faces trade-offs. This means that evolution, unfortunately, has left us susceptible to many diseases. But it also means that studying evolution can help researchers better understand—and perhaps even treat—those disorders (Chapter 13).

It's also a mistake to think that evolution produces a peaceful harmony in nature. There are many helpful partnerships in nature, such as the one between flowering plants and the insects that pollinate them (Chapter 11). But the same process by which species adapt to one another can also give rise to what looks to us like cruelty. Predators are exquisitely adapted to finding and killing prey. Parasites can devour their hosts from the inside out. Their adaptations are finely honed, allowing them to manipulate individual molecules within their hosts. Yet parasites and predators are not evil. They are just part of a dynamic balance that is constantly shifting, driving the diversification of life but also leading to extinctions.

The diversity of nature, in other words, is not eternally stable. More than 99 percent of all species that ever existed have become extinct, and, at some points in the Earth's history, millions of species have disappeared over a relatively short span of time. We may be at the start of another period of mass extinctions, this time caused by our own actions (Chapter 10).

Evidence and Evolution

Evolution is a process that takes place over time—from months to millions of years. It's not possible to track evolution from one millisecond to the next, even in a carefully designed experiment in the confines of a laboratory. Scientists have no choice but to reconstruct parts of the history of life by analyzing a wealth of clues.

In this respect, evolutionary biology is like many other sciences. Ecologists cannot keep track of every fish on a coral reef as they track population booms and busts. Instead, they have to make small measurements and then extrapolate to an estimate. Astronomers cannot track the movement of a single photon from the center of the Sun out into space. Instead, they must use what they know about photons, about hydrogen and the other gases that make up the Sun, to generate a hypothesis that they can test. In all these cases, scientists analyze evidence, form hypotheses, and then test those hypotheses with fresh evidence.

The evidence that evolutionary biologists gather comes in many forms. In laboratory experiments, for example, scientists can measure the effects of natural selection as it alters populations of bacteria, insects, or other fast-breeding organisms (page 118). It's much harder to estimate the strength of natural selection in the wild. Nevertheless, there are now thousands of examples of carefully documented cases of natural selection in our own lifetime. By comparing the genomes of different species, scientists can also observe how genetic changes make natural selection possible by producing new variations.

Over longer time scales, fossils illuminate the dark corners of life's history. Paleontologists have unearthed a record of fossil life that reaches back about 3.5 billion years. Life's many forms did not appear at once but rather emerged gradually. Single-celled microbes are the earliest known forms of life; the oldest known fossils of multicellular organisms are almost two billion years younger. The oldest known fossil traces of animals are only 635 million years old; the oldest fossils of mammals are only 200 million years old (Chapter 3).

Evolutionary biologists also study the similar traits that closely related species share. Humans have arms and bats have wings, for example, but a close look reveals that their bones form the same basic pattern. Both fossils and living organisms share traits inherited from their common ancestors. From these sorts of clues, scientists build hypotheses about how different species are related to one another. These hypotheses allow them to make new predictions about when

Many books about evolution include a separate chapter about the evolution of our own species, *Homo sapiens*. This book takes a different approach. Human evolution is not separate from the evolution of other species, and so humans are not isolated in a chapter of their own here. Instead, the story of human evolution is woven into most of the chapters in this book.

Here's a guide to the sections in which human evolution is discussed:

certain groups of species first arose in the history of life. As new evidence arises—new fossils, new comparisons of living species—biologists can see whether their hypotheses pass or fail new tests.

The best way to illustrate how scientists learn about evolution—by gathering many lines of evidence, from fossils to DNA; by understanding how evolution shapes everything about an organism, from how it moves to how it ages to how it behaves—is with a case study. So let's take a look at one of the most interesting of those cases: the origin of whales.

A Case of Evolution: Why Do Whales Have Blowholes?

Today, whales and dolphins have no legs. Their bodies are sculpted with the same sleek curves you can find on tunas and sharks, allowing them to use relatively little energy to shoot through the water. The tails of whales and dolphins narrow down to a small neck and then expand into flattened flukes, which they lift and lower to generate thrust. Sharks and tunas have similar tails, except that they move theirs from side to side.

Yet, unlike tunas and sharks, whales and dolphins must rise to the surface of the ocean in order to breathe. They do so by opening up a blowhole on the top of the head, which allows air to enter a passageway that leads to the lungs. Fishes, on the other hand, can usually get all the oxygen they need from the water. They pump water through their gills, and some of the dissolved oxygen passes into their blood vessels.

The differences keep piling up. Whales and dolphins have tiny bones embedded in their flesh just where the hips would be on a land vertebrate, but fishes have none. Fishes have relatively simple sets of muscles that form vertical blocks from head to tail, whereas whales and dolphins have long muscles that run horizontally down the length of their bodies. Whales and dolphins give birth to live young that cannot get their own food; instead, the young must drink milk produced by their mothers. Together, these differences make whales and dolphins starkly unlike the fishes. Those traits—and many others—are found only in mammals.

How did whales evolve? Clues come from everything from fossils to whale DNA.

Charles Darwin hypothesized that whales and dolphins descended from mammals on land and that their ancestors had gradually evolved into ocean-going animals without legs. But no one at the time had found a fossil from that transition, and it would take a very long time before someone did. In 1979, almost a century after Darwin died, a paleontologist from the University of Michigan named Philip Gingerich discovered in Pakistan a 50-million-year-old skull of a whale that appeared to be adapted to life on land. He found the fossil in rocks that had formed on a continent, rather than on the floor of an ocean, and the skull looked more like a dog's than a dolphin's. Gingerich knew that the animal was related to whales, however, because it shared several traits that are found today only in whales, such as a distinctive bony wall surrounding its ear. That whale is now known as *Pakicetus* ("whale of Pakistan" in Latin).

Thirteen years later, Hans Thewissen (a student of Gingerich's) traveled to a different part of Pakistan to look for mammals that lived a few million years after *Pakicetus*. One day he and his Pakistani colleagues happened across the fossil of a strange beast, and they slowly excavated its bones from the tail to the head. Its tail was massive, its legs were stubby, and its rear feet were shaped like paddles. Thewissen discovered that its head was long like an alligator's, but it had teeth with distinctive shapes that are found today only in the teeth of mammals. In fact, the teeth looked like those of *Pakicetus*. And when he excavated the bones around the ear, he discovered a distinctive bony wall around the ear—evidence that he had actually found a whale. He dubbed it *Ambulocetus*: the walking whale.

Thewissen's discovery was historic—it was the sort of animal that Darwin had predicted, and that many skeptics claimed could not have existed. But paleontologists do not stop with a single fossil. They have continued to dig up new fossils to understand the evolution of whales, and other scientists have looked for other kinds of evidence as well. In the mid-1990s geneticists began to sequence whale DNA and to compare it to the DNA of other animals. If whales were indeed mammals, then they should have DNA like the mammals that live on land. They do, in fact: all whales carry genetic markers found only in one group of land

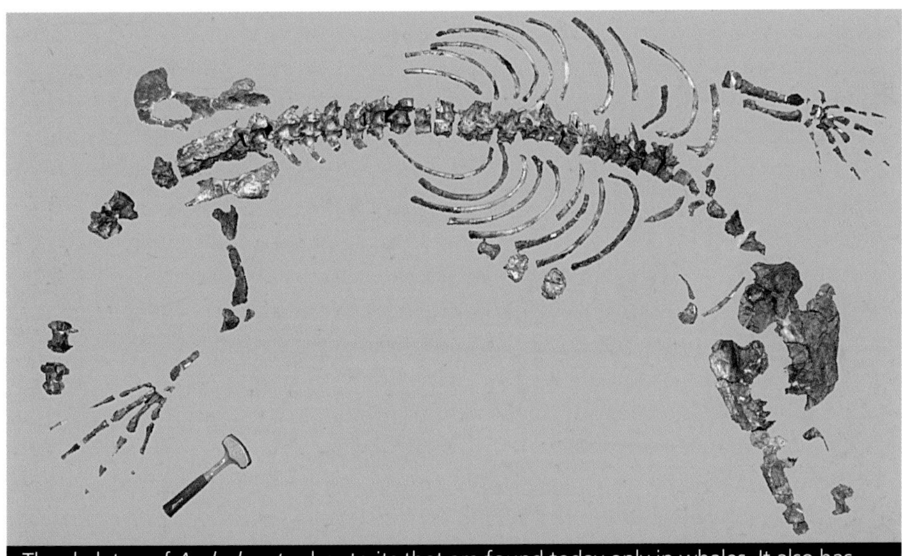

The skeleton of *Ambulocetus* has traits that are found today only in whales. It also has legs and other adaptations found in land mammals.

mammals, known as the artiodactyls. This is a group that includes camels, cattle, hippopotamuses, and goats.

So how did an artiodactyl ancestor of whales go into the sea and lose its legs? To shed light on that question, paleontologists have searched for new fossils and have reanalyzed fossils discovered many years ago. In 2007, Thewissen and his colleagues published a study in which they argued that the closest relative of *Pakicetus* and other whales was a 47-million-year-old artiodactyl called *Indohyus.* They pointed to a set of traits in its skeleton that link it to whales, such as the distinctive ear bone. Yet, if you could see a living *Indohyus,* you might never guess at the kinship. *Indohyus* was a small, slender-legged creature that bore a striking resemblance to an African mammal known as a mouse deer.

Figure 1.1 is an evolutionary tree that illustrates how *Indohyus* and other fossil species are related to living whales. To draw this kind of tree, scientists identify key traits that are shared by groups of species and determine the most likely course of evolution that could have produced them. These trees help scientists to understand major evolutionary transformations, such as the transition of the ancestors of whales from land to sea. (For more on evolutionary trees, see Chapter 4.)

Indohyus and *Pakicetus* evolved when the ancestors of whales still had four long legs. They may have been adept swimmers, but probably no more so than mouse deer and many other land mammals are today. After their ancestors branched off, new whales evolved that were more adapted to life in water. *Ambulocetus,* which lived 49 million years ago along the coastline of what is now Pakistan, had short legs and massive feet. It probably swam like an otter, kicking its large feet and bending its tail. Another species from around the same age, *Rodhocetus,* looked more like a seal, being able only to drag itself around on land. All told, scientists have found more than 30 fossil species that mark the transition from land mammals to the living groups of whales and dolphins.

As whales adapted to the ocean, their legs gradually dwindled. Their forelimbs changed gradually from hooved limbs to flat flippers they used for steering. But evolution did not completely retool whales from the ground up. To breathe underwater, they did not reinvent gills (which had been lost long before by the early vertebrates that moved onto land). Instead, their nostrils shifted up along their skulls until they passed over the eyes.

It's almost certain that scientists will never be able to read the genes of 40-million-year-old whales, because DNA is too fragile to last for more than a few hundred thousand years. But it is possible to study the genes of living whales to learn about some of the genetic changes that turned four-legged land mammals into fishlike whales. Some of the most important genetic changes that take place during major evolutionary transitions change the timing of gene activity in embryos (Chapter 8).

When legs begin to develop on the embryos of humans or other land vertebrates, a distinctive set of genes become active. Thewissen and a team of embryologists discovered that these leg-building genes also become active in dolphin

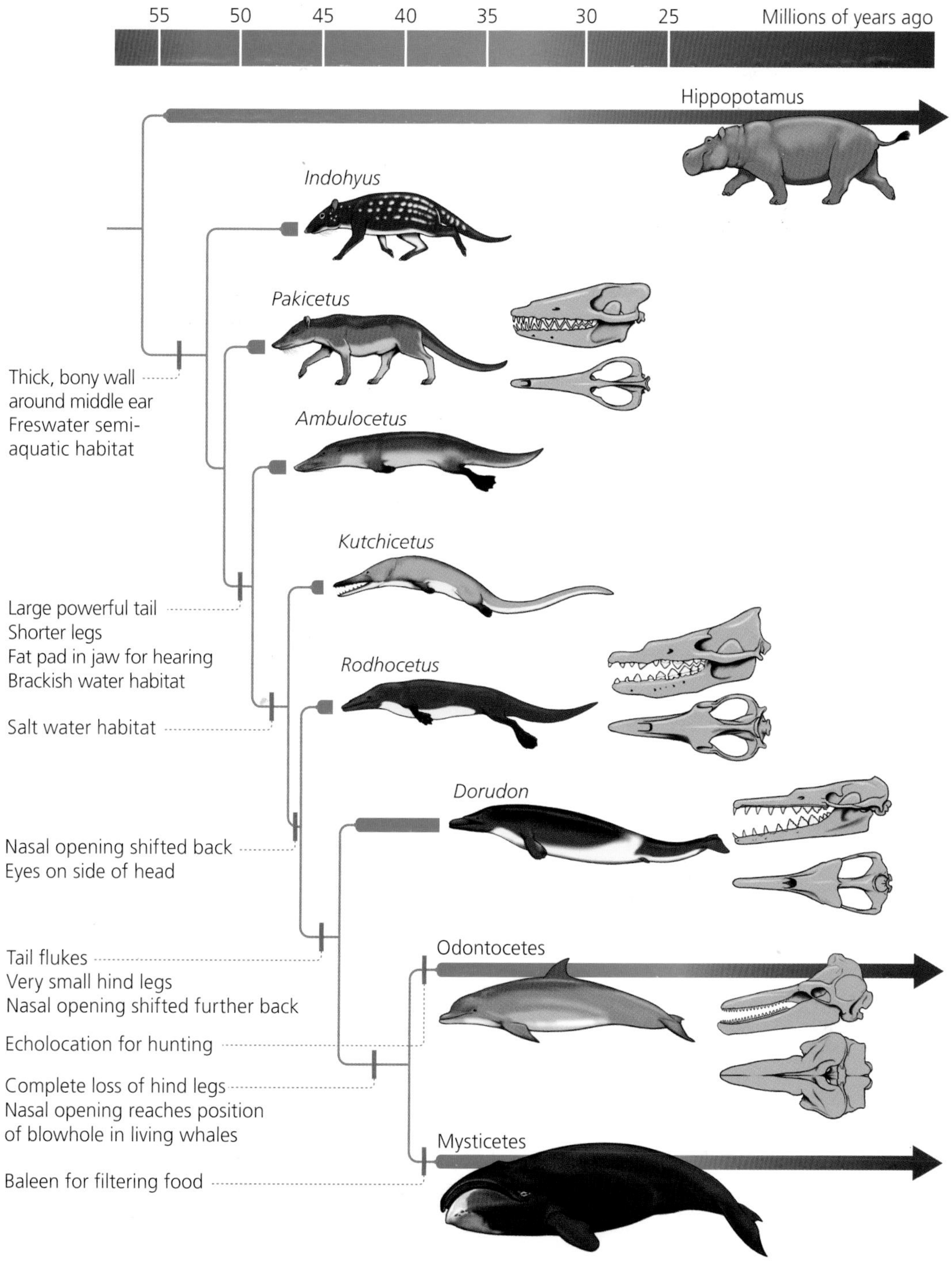

55 50 45 40 35 30 25 Millions of years ago

Hippopotamus

Indohyus

Pakicetus

Ambulocetus

Kutchicetus

Rodhocetus

Dorudon

Odontocetes

Mysticetes

Thick, bony wall around middle ear
Freswater semi-aquatic habitat

Large powerful tail
Shorter legs
Fat pad in jaw for hearing
Brackish water habitat

Salt water habitat

Nasal opening shifted back
Eyes on side of head

Tail flukes
Very small hind legs
Nasal opening shifted further back

Echolocation for hunting

Complete loss of hind legs
Nasal opening reaches position of blowhole in living whales

Baleen for filtering food

Figure 1.1 This diagram shows how some extinct species of early whales are related to living species. The animals illustrated here are only a fraction of the fossil whales paleontologists have discovered in recent decades. By studying fossils, paleontologists have been able to show how the traits found in living whales emerged gradually, not all at once.

embryos. They help build tiny buds of tissue, but these buds stop growing after a few weeks and then die back.

Fossils show that this change in the timing of gene activity took millions of years to complete. Some 40 million years ago, fully aquatic species had evolved. One species, called *Basilosaurus*, grew to be 50 feet long. But Gingerich and his colleagues have discovered a well-preserved *Basilosaurus* fossil that still had hind legs. Its legs were only as big as those of a human child, but they still retained ankles and toes.

Scientists can get clues about the origins of whales and dolpins not just from the shapes of their fossils, but from the individual atoms inside them. Living whales and dolphins can drink seawater, while land mammals can drink only freshwater. The two kinds of water are different in several ways, and not just because seawater is salty and freshwater is not. Both kinds of water contain oxygen atoms, but the oxygen atoms are slightly different. Like other elements, oxygen atoms are made up of a combination of negatively charged electrons, positively charged protons, and neutral neutrons. All oxygen atoms have eight protons, and most have eight neutrons. But a fraction of oxygen atoms on the Earth have extra neutrons. Scientists have observed that seawater has more oxygen atoms with 10 neutrons than freshwater. And animals that live on land and at sea reflect this difference in the oxygen atoms incorporated in their bones. Living whales and dolpins have a larger percentage of heavy oxygen in their bones than mammals that live on land.

Thewissen wondered if the oxygen atoms in ancient whale fossils might indicate where they lived. So he and his colleagues ground up tiny samples of ancient whale teeth and measured the ratio of light and heavy oxygen. They discovered that *Pakicetus* still drank freshwater. *Ambulocetus*, which belongs to a younger branch of the whale tree, had an intermediate ratio, suggesting that it was drinking brackish water near the shore, or a mix of freshwater and seawater. Younger fossil whales had the ratios you would expect in animals that drank seawater. The chemistry of these whales documents a transition from land to estuaries to the open ocean—the same transition documented in the changing shape of their skeletons.

The earliest lineages of whales are long extinct. The two lineages of whales alive today evolved from a common ancestor that lived about 35 million years ago. One lineage, known as the toothed whales, evolved muscles and special organs that they used to produce high-pitched sounds with their blowholes and to hear the echoes that bounced off animals and objects around them in the water. Today, dolphins and other toothed whales use these echoes to hunt for their prey.

The other living lineage, the baleen whales, lost their teeth and evolved huge, stiff pleats in their mouths that allowed them to swallow huge amounts of water, and to push it back out, straining out any fish or shrimp that the water contained. Scientists are now beginning to find important new clues to these two transitions. Fossils from about 25 million years ago show that the toothed ancestors of baleen whales probably grew small patches of baleen. Only later did their

Dolphins live in large groups and can communicate with each other. Their social life may have favored the evolution of big, powerful brains.

teeth disappear, much like hind legs of their ancestors. Baleen whales still carry genes for building teeth, but all of those genes have been disabled by mutations.

It's not just organs such as baleen and flippers that evolve. Every aspect of an organism may be sculpted by evolution. Consider, for example, the fact that whales can live to be very old. No one knows exactly how long they can live, but scientists have found some astonishing clues. In 2007, Eskimo hunters hunting off the coast of Alaska killed a bowhead whale that still had the tip of an antique harpoon lodged in its blubber. It was a kind of harpoon that had been used for only a few years around 1890, which means that the whale was about 130 years old when it was killed. In 1999, scientists examining the growth patterns in the teeth of a dead bowhead whale estimated that it was 211 years old. By contrast, the closest living relatives of whales on land have much shorter lives. Hippopotamuses are known to live up to 61 years; camels can live up to 35 years.

The lifespans of all species, ours included, are shaped by evolution (page 312). Animals face a trade-off between living long and having lots of offspring. Small animals that are commonly preyed on may grow up fast and die young. Thanks to their large size, whales don't face a lot of attacks by other animals. Bowhead whales, the longest-lived species, may also enjoy the extra luxury of little competition for food as they swim the frigid depths of the Arctic Ocean.

Aging is not the only threat to the health of whales, however. They are beset by parasites, including viruses, bacteria, single-celled protozoans, fungi, flukes, and sea lice. Not all of these parasites are harmful, though. Take an intestinal worm called *Anisakis*. It lives first in fishes. If a whale eats the fish, the worm becomes a parasite of the whale. It doesn't make the whale sick, however. It feeds harmlessly on the food the whale eats and lays its eggs in the whale's droppings. Some studies suggest that *Anisakis* have evolved molecules that they can use to hide from their host's immune system, so that they can stay safe and their hosts don't get sick. Scientists have discovered that the most closely related species of *Anisakis* often live inside the most closely related species of whales. This mirror-like pattern suggests that the evolution of *Anisakis* has been closely tracking the evolution of its hosts for millions of years (Chapter 11).

We don't have this long history with *Anisakis*. And so, if we happen to eat a fish infected with *Anisakis* (in a bad piece of sushi, for example), things sometimes go badly. The worm does not get the normal signals it would use in a whale to guide its journey to the intestines. Instead, it drills a hole in the stomach and crawls out into the abdominal cavity. It wanders around aimlessly, causing excruciating pain and dangerous infections. (Chapter 13 explores other insights that evolution offers about diseases.)

Biologists have long been impressed with the size and complexity of whale brains. Aside from humans, dolphins have the biggest brains in proportion to their body of any animals. Dolphins can also use their oversized brains to solve remarkably complicated puzzles that scientists make for them. A number of studies suggest that big brains evolved in dolphins as a way to solve a particular kind of natural puzzle: figuring out how to thrive in a large social group. Dozens of dolphins live together, forming alliances and competing for mates. They communicate with each other with high-frequency squeaks, and each dolphin can tell all the other dolphins apart by their whistles. Natural selection appears to have favored dolphins with extra brainpower for processing social information. The same process appears to have driven the expansion of brain size in primates as well, culminating in our own extraordinarily oversized brains (page 344).

Unfortunately, 50 million years of evolution has not prepared whales and dolphins very well for life in a human-dominated world. In the 1800s, sailors crisscrossed the world to hunt big whales for their oil and baleen. (The oil was used for lamps and the baleen for corset stays.) Many species of whales came perilously close to extinction before the whale-oil industry collapsed and laws were passed to protect the surviving animals. Because whales are so long-lived, they reproduce slowly, and so their populations have not expanded very much toward their previous levels over the past 100 years. Animals that have only small populations are at greater risk of extinction, in part because diseases can spread more effectively through small populations, and in part because small populations have little genetic variability, making them more susceptible to genetic disorders. Scientists are analyzing the DNA of whales to understand the risks they face from over-

Whale hunting in the nineteenth century nearly drove many species of whales extinct.

hunting and to determine how best to preserve them from extinction (page 235). Unfortunately, whales and dolphins also face other risks. Pollution and heavy fishing in the Yangtze River drove the Chinese river dolphin to extinction in 2006.

Learning about the evolution of whales makes them all the more fascinating. We can discover a dense tapestry of history on display in the living things with which we share the planet. As Charles Darwin wrote at the end of *The Origin of Species*, "There is grandeur in this view of life, with its several powers, having been breathed into a few forms or into one; and that, whilst this planet has gone cycling on according to the fixed law of gravity, from so simple a beginning endless forms most beautiful and most wonderful have been, and are being evolved."

TO SUM UP...

- **Evolution is descent with modification.**
- **Evolution produces complex adaptations, but it does not move towards a particular goal.**
- **Evolutionary biologists test hypotheses about evolution with many different lines of evidence.**
- **Whales evolved from land mammals about 50 million years ago. Evolution has shaped many aspects of their biology, from their behavior to their aging.**

Biology

2

From Natural Philosophy to Darwin

In the Pacific Ocean seven hundred miles west of Ecuador lies an isolated cluster of extinct volcanoes known as the Galápagos Islands. On these strange outcrops are strange kinds of life. There are big birds with bright blue feet. There are scaly iguanas that leap into the ocean to eat seaweed and then wade back out to bask on the rocks. Giant tortoises chew peacefully on cactuses. Finches are so tame that they will let you pick them up by hand.

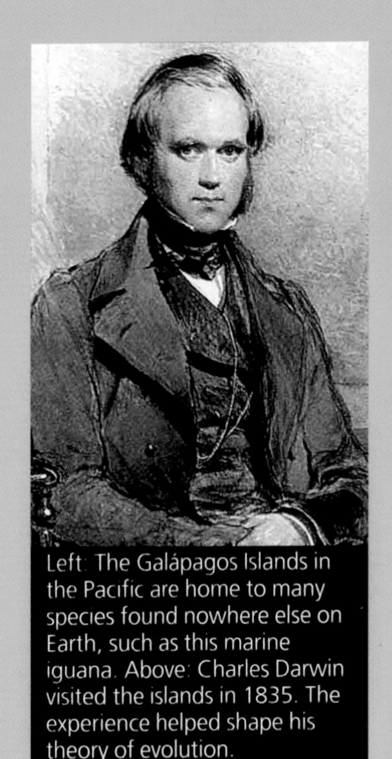

Left: The Galápagos Islands in the Pacific are home to many species found nowhere else on Earth, such as this marine iguana. Above: Charles Darwin visited the islands in 1835. The experience helped shape his theory of evolution.

Every year, dozens of scientists come from across the world to the Galápagos Islands in order to study these species, which exist nowhere else in the world. It is like a laboratory of evolution, where scientists can study an isolated example of how life changes over millions of years. It takes those scientists a long time to get to the Galápagos Islands—but not as long as the journeys by steamer that scientists took a hundred years ago. And those steamer trips were much faster than the voyage of a British surveying ship that sailed to the Galápagos Islands in 1835. On board the HMS *Beagle* was a young British naturalist named Charles Darwin. He had been traveling aboard the *Beagle* for almost four years, during which time he had studied the marine life of the Atlantic, trekked in the jungles of Brazil, and climbed the Andes. But even after all that, Darwin was astonished by the Galápagos Islands. "The natural history of this archipelago is very remarkable: it seems to be a little world within itself," he wrote later in his book *The Voyage of the Beagle.* Darwin spent five weeks on the islands, clambering over jagged volcanic rocks and gathering plants and animals. The experience would later lead him to a scientific revolution.

Darwin was born in 1809, at a time when everyone—including the world's leading naturalists—thought that the world was only thousands of years old, not billions. They generally believed that species had been specially created, either at the beginning of the world, or from time to time over the Earth's history. But after Darwin returned from his voyage around the world in 1836, his experiences in places like the Galápagos Islands caused him to question those beliefs. He opened up a notebook and began jotting down ideas for a new theory of life, one in which life evolved.

To understand evolutionary biology today, we have to start with Darwin, the naturalist who first recognized its fundamental elements. But it would be a mistake to reduce evolutionary biology to Darwin alone. Charles Darwin was not the first person to contemplate evolution, and he did not uncover all of the evidence on which he built his argument. That evidence came instead from centuries of investigations into the natural world. Darwin reassessed all that information, combined it with all of his own observations and inferences, and came up with a new theory.

But that does not mean that Darwin discovered all there was to know about evolution, leaving later generations of scientists with nothing left to do. Only after Darwin's death did scientists discover how genes work, for example. He never got to see some of the most spectacular fossils that have been discovered over the past century, which paleontologists have used to document evolution's great transitions. Since Darwin's day, scientists have not just found new evidence for evolution, they've developed new concepts about how it works. The study of evolution has, in a sense, evolved.

Nature Before Darwin

Charles Darwin first began to learn about nature as a teenager in the 1820s. The concepts he was taught had emerged over the previous two centuries, as naturalists pondered two questions: what were the patterns in nature's diversity, and how did those patterns come to be?

Understanding the diversity of life had been a practical necessity. People needed names for different kinds of plants and animals, for example, so that they could pass on their wisdom about which kinds were safe to eat or useful as medicines. For thousands of years, people had been well aware that some kinds of animals and plants were similar to other kinds. Cats and cows and humans all nourished their young with milk, for example. In the 1600s, naturalists became more systematic in the way they sorted the diversity of life. They came up with rules for naming species and schemes for classifying species into different groups. This urge to compare and classify reached its most glorious form in the mid-1700s with the work of the Swedish botanist Carl Linnaeus.

Linnaeus organized all life into a single hierarchy of groups. Humans belonged to the mammal order, for example, and within that order, the primate family; and within that family, the genus *Homo*; and within that genus, the species *Homo sapiens*. Each species could be assigned to a particular genus, family, or order according to the traits it shared with other species. His system was so useful that biologists continue to use it today.

Carl Linnaeus (1707–1778) invented a way to classify species into larger groups.

Linnaeus believed that the pattern of his system reflected a divine plan. "There are as many species as the Infinite Being produced diverse forms in the beginning," he wrote. In some cases, Linnaeus believed that species had later changed. He believed that two species of plants could sometimes interbreed, producing a new hybrid species. For the most part, though, Linnaeus believed that the overall patterns of life's diversity had not changed since the biblical creation of the world.

While Linnaeus studied the diversity of life in its present forms, other naturalists were looking back over its history. They discovered that, when animals and plants die, their remains are sometimes preserved, transformed into stone. One of the first naturalists to realize this was Nicolaus Steno, a seventeenth-century Dutch anatomist and bishop. In 1666, some fishermen brought to him a giant shark that they had caught. As Steno studied the shark's teeth, it occurred to him that they looked just like triangular rocks that were then known as tongue stones. Steno proposed that tongue stones had started out as shark teeth. After the sharks had died, their teeth had gradually been transformed into stone.

But if fossils really were the remains of once-living things, Steno would have to explain how it was that stones shaped like sea shells had come to be found on top of certain mountains. How could animals that lived in the ocean end up so far from home? Steno argued that a sea must have originally covered the mountains. Shelled animals died and fell to the ocean floor, where they were covered over in sediment. As sediments accumulated, they turned to rock. The layers of rocks exposed on the sides of mountains, Steno recognized, had been laid down in succession, with the oldest layers at the bottom and the youngest ones at the top.

What Is Science?

Evolutionary biologists want to understand things that are hidden from view. When they see a bee feeding on nectar from a flower—and picking up pollen to fertilize other flowers—they want to know how that partnership came to be over millions of years. They want to know why there are no 50-foot-long marine reptiles in the ocean today, despite the fact that those giants swam the seas for tens of millions of years. They want to turn back time—to know, for example, what life on Earth was like three billion years ago.

How can scientists possibly know anything about something that happened an unimaginably long time ago, something with which they've had no direct experience? That's a reasonable question, but it could be asked about anything that scientists study. Physicists cannot walk alongside electrons, making personal observations of how they behave. They must run experiments to gather indirect clues to how electrons act, and then use that evidence to come up with an explanation. Geologists cannot lay their hands on an earthquake, but they can eavesdrop on the reverberations that earthquakes send out across the planet, and they can use those aftereffects to discover why the Earth shudders. Epidemiologists track epidemics without ever seeing the vast majority of the viruses or bacteria that are making people sick. All scientists, in other words, seek to understand the invisible. Their mission is to find explanations that can account for indirect clues gathered through experiment and observation.

If scientists must rely on indirect clues, how can they ever know that their explanations are correct? They must find a way to test their explanations against more evidence. The better job that an explanation does in predicting the new evidence, the more confidence the scientists have in it. If the prediction fails, the scientists give their explanations a critical look and alter them in order to create a better one.

This cycle of evidence, explanation, and testing can only work effectively on phenomena that follow reliable rules. Today, for example, the bonds that join together atoms in a water molecule have the same energy as they did yesterday. You could imagine a world in which some supernatural force could alter the bonds from moment to moment based on some mysterious whim. It would not be possible to use science to learn about how such a world works, because you would have no idea whether new evidence would be produced by the same processes that you were observing before.

Science is a complicated way of learning about the world, but, in 2008, the National Academy of Sciences, the leading scientific body of the United States, came up with a concise definition: "the use of evidence to construct testable explanations and predictions of natural phenomena, as well as the knowledge generated through this process."

If all that certain scientists have to go on are indirect clues, how can they ever know that their explanations are correct?

Paul Turner, a biologist at Yale, studies the evolution of viruses.

Nicholas Steno (1638–1686) recognized that triangular rocks known as "tongue stones" were in fact fossils of teeth from sharks.

Steno was still a traditional believer in a biblical Earth that was just a few thousand years old. But he introduced a radically new idea: life and the planet that supported it had a history filled with change, and the Earth itself kept a record of that history.

Evolution Before Darwin

The concept that life changes over the course of vast stretches of time—what would come to be known as evolution—was already being vigorously debated before Darwin was born. One of the earliest evolutionary thinkers was the eighteenth-century French nobleman Georges Louis LeClerc de Buffon (1707–1788). Buffon was the director of the King's Garden in Paris, and he also owned a huge estate in Burgundy, where he harvested timber for the French navy and carried out research on the strength of different kinds of wood. Buffon also spent years writing an encyclopedia in which he intended to include everything known about the natural world.

Georges Buffon (1707–1788) was one of the earliest naturalists to argue that life had changed over time.

Like other thinkers in the mid-1700s, Buffon recognized that the new sciences of physics and chemistry offered a radically different way of thinking about the universe. It had become clear that the world was made up of minuscule particles, which we now call atoms and molecules. These particles reacted with each other according to certain laws, and when they came together into larger objects, the objects obeyed certain laws as well. They were attracted to one another by gravity, for example, and repelled by electric charge. Following these laws, the particles moved about, and the complexity of the universe emerged spontaneously as a result.

Buffon proposed that the Earth had formed according to the laws of physics. A comet struck the Sun, he argued, breaking off debris that formed into a planet. The scorching Earth cooled down and hardened, and oceans formed and dry land emerged. The entire process took more than 70,000 years, Buffon calculated—a span of time too vast for most people in Buffon's day to imagine.

The fact that living things were made from the same kinds of particles found in rocks and water struck Buffon as profoundly important. He argued that each species had a supply of organic particles that somehow transformed an egg or a seed into its adult form. He envisioned that these organic particles had first come together in the hot oceans of the early Earth. Animals and plants sprang into existence in the process, and, as the planet cooled, they retreated to the warm tropics. Those migrations could explain the stunning discovery in the mid-1700s of fossil elephants in Siberia and North America, far from the tropics where elephants live today.

According to Buffon, when life first emerged, it was already divided into a number of distinct types—an "internal mould," as he called it, that organized the organic particles that made up any individual creature. But life could be transformed. As a species moved to a new habitat, its organic particles changed, and its mould changed as well. Buffon was, in other words, proposing a sort of proto-evolution.

Fossils and Extinctions

Steno's realization that fossils were the remains of living things helped open up a new science that came to be known as paleontology. During the eighteenth century, naturalists discovered more fossils and studied how groups of fossil species gave way to new ones over time. Some fossils were left by species of animals and plants that exist today. But some fossils belonged to species that did not live in the same place any longer. Elephant-like animals left fossils in Italy, France, and Siberia, where no elephants now live.

Georges Cuvier, a French paleontologist (1769–1832), compared these fossils to the skeletons of living elephants from Africa and India. He demonstrated that some of the fossils were distinct from living elephants in some crucial ways, such as the shapes of their teeth. These fossil animals, which he called mammoths and mastodons, were species that no longer existed. They had, in other words, become extinct.

Cuvier and others went on to document the extinction of many species. Paleontologists began to investigate how these species could have died out. The answer turned out to be hidden in the rocks themselves—or, more precisely, their geography. Over the course of the eighteenth century, a debate raged about whether the Earth's features had formed from volcanic eruptions or floods. An important step forward came when James Hutton, a Scottish farmer, realized that rocks formed through imperceptibly slow changes—many of which we can see around us today. Rain erodes mountains, while molten rock pushes up to create new ones. The eroded sediments form into layers of rock that can later be lifted above sea level, tilted by the force of the uprising rock, and eroded away again. Some of these changes can be tiny, but over enough time, Hutton argued, they could produce vast changes. The Earth must therefore be vastly old—Hutton envisioned it as a sort of perpetual-motion machine passing through regular cycles of destruction and rebuilding that made the planet suitable for mankind.

Hutton's vision of a slowly transformed Earth came to be accepted by most geologists in the 1800s. They looked closely at layers of exposed rock and began to determine how they were formed by volcanoes and settling sediments. And they also began to figure out the order in which those layers had formed. Some of the most important clues to the history of the Earth came from fossils. William Smith, a British canal surveyor, came to this realization as he inspected rocks around England to decide where to dig canals. He noticed that the same kinds of fossils tended to appear in old rocks, and different ones appeared in younger layers. Smith could find the same sets of fossils in rocks separated by hundreds of miles.

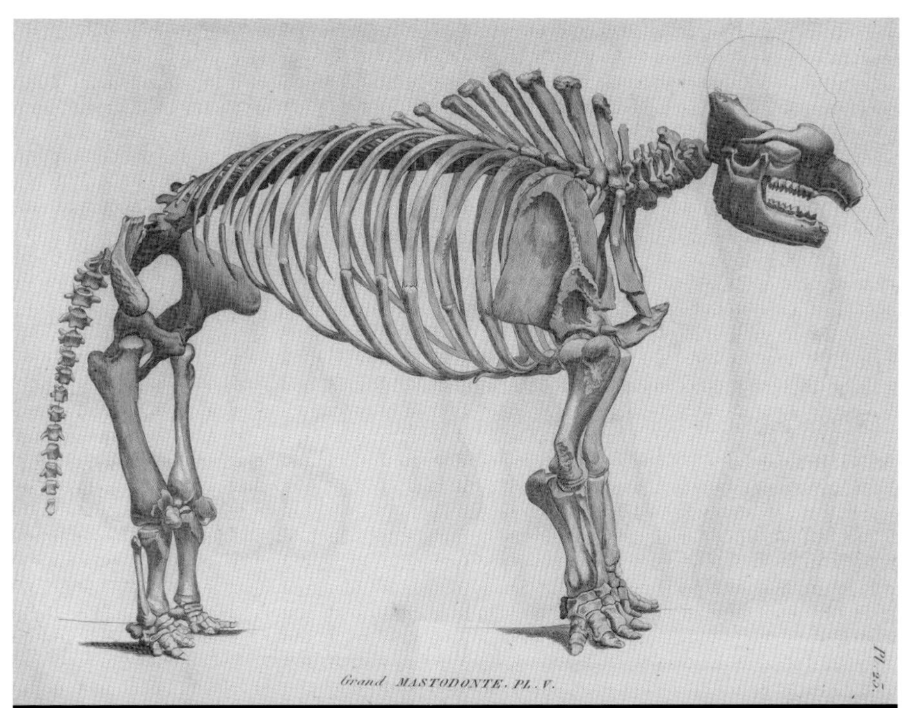

In the late 1700s, paleontologists recognized that some fossils belonged to species that no longer exist, such as this mastodon, a relative of elephants that became extinct 11,000 years ago.

William Smith (1769–1839) discovered that layers of rocks contain distinctive groups of fossils. He did not know how old species disappeared and new ones emerged.

William Smith learned how to recognize the same layers of rocks in different parts of England by looking at the fossils they contained.

In the early 1800s, geologists came to agree that the surface of the planet had been gradually sculpted over vast spans of time. Smith realized that each type of animal had lived across a wide geographical range for a certain period of time, and so the rocks that formed during that time preserved their fossils. As those animals became extinct and new ones emerged, younger rocks contained their own sets of fossils. By marking the places where he found certain fossils, Smith was able to organize strata into a geological history, from oldest to youngest.

Other researchers, including Cuvier, later used the same method to map the geology of other parts of the world. They discovered that formations of rock exposed in one country could often be found in others. They began to give names to the sequences of these far-flung rock formations (see the geological chart on the endpapers). Many fossil species were restricted to just a few layers of rock. Larger groups of species spanned more geological history, but they had their own beginnings and endings as well. In the early 1800s, for example, fossil hunters discovered the bones of gigantic reptiles, some of which had lived on land and some of which had lived in the sea. These fossils came only from rocks dating back to the Mesozoic Era, disappearing abruptly at its end.

In the early 1800s, geologists established that exposed rocks in different regions were parts of the same formations. These formations were arranged in layers, with the youngest at the top. Paleontologists found distinctive groups of fossils in rocks of different ages.

Why species emerged and disappeared over the history of life was a subject of fierce debate. Cuvier, for example, rejected Buffon's earlier suggestion that life had evolved. He believed that life's history had been punctuated by revolutions that had wiped out many species and brought many new ones to take their place. But one of his colleagues at the National Museum of Natural History in Paris was about to make a new case for evolution.

Evolution as Striving

In the early 1800s, a new voice for evolution emerged: a colleague of Cuvier's, named Jean-Baptiste Pierre Antoine de Monet, Chevalier de Lamarck (1744–1829). Lamarck was an expert on plants and invertebrates, and he was struck by the similarities between some of the species he studied. He was also impressed by the fossil record, which, at the time, was becoming detailed enough to reveal a dynamic history of life.

Jean-Baptiste Lamarck (1744–1829) argued that complex species had evolved from simple ones.

Lamarck combined these two lines of thought into a single argument. He proposed that life was driven inexorably from simplicity to complexity, with humans and other large species descending from microbes. To explain why there are microbes today, Lamarck argued that primitive life was being spontaneously generated all the time. Today's bacteria are just the newest arrivals.

Lamarck also believed animals and plants could adapt to their environment. If an animal began to use an organ more than its ancestors had, the organ would increase in its lifetime. If a giraffe stretched its neck for leaves, for example, a "nervous fluid" would flow into its neck and make it longer. Lamarck claimed that these changes could be passed down from an animal to its offspring. A giraffe could inherit a longer neck; if it continued stretching for leaves, it would pass on an even longer neck to its descendants.

Things ended badly for Lamarck. He was criticized by Cuvier and many other naturalists of the day for his elaborate speculations, and he died in poverty and obscurity in 1829. But just eight years after his death, a young British naturalist, newly returned from a voyage around the world, would quietly embrace the notion that life had evolved. And three decades after Lamarck's death, that naturalist—Charles Darwin—would publish *The Origin of Species*, changing the science of biology forever.

The Unofficial Naturalist

Today, Darwin is practically synonymous with evolution, but he was hardly the first naturalist to wonder about nature's patterns. By the time Darwin was born in 1809, Lamarck was already famous (and infamous) for arguing that life had changed over a long history. When Darwin finally presented his own theory of evolution at age 50, however, he could not be so easily dismissed. He had assembled a towering edifice of evidence and argument for evolution. As later chapters of this book will show, a vast amount of subsequent research supports most of his argument. Like any scientist, Darwin got some things wrong, but his errors were not critical to his major ideas. The theory of evolution has matured and grown enormously since his day. That's how science works, and it does nothing to diminish Darwin's achievements, which stand among the most significant in the history of science.

Darwin's biography makes his breakthrough all the more remarkable. He was born into comfortable wealth, thanks to the fortune his mother's family made manufacturing china and pottery. Darwin's father, a physician, expected Charles and his brother Erasmus to follow him into medicine, and he sent them to Edinburgh for training. There Charles also learned about geology, chemistry, and natural history, and soon he realized that he would much rather spend his life studying nature than practicing medicine. It was common in Darwin's day for well-to-do young men interested in nature to train in theology and become clergymen, using their spare time to pursue their investigations. Darwin started down that path, leaving Edinburgh to study theology at the University of Cambridge.

But Darwin grew restless. He reveled in a journey to Wales, where he was able to study geological formations. He devoured books about the travels of great naturalists to distant tropical countries. And then, out of the blue, Darwin got a chance to go on a voyage of his own.

In 1831, Darwin was invited to join the company of a small British navy ship, HMS *Beagle*, on its voyage around the world. The captain, Robert Fitzroy, feared that the long journey might drive him to suicide, a fate one of his uncles had met as a ship's captain. So Fitzroy began to search for a gentleman who might act as an unofficial naturalist for the voyage and whose companionship might keep him from succumbing to depression. He eventually settled on the 22-year-old Darwin. Darwin joined the *Beagle's* company and would not return home for five years.

The *Beagle* traveled from England to South America. Along the way, Darwin gathered fossils of extinct mammals. He trapped birds and collected barnacles. He observed the ecological complexity of the jungles of Brazil. Darwin also learned a great deal about geology in South America. He recognized the layers of rock that had gradually formed and had then been reworked into mountains and valleys. He experienced an earthquake in Chile, and he observed that the shoreline had been lifted a few centimeters as a result. When Darwin had boarded the *Beagle,* one of the books he had brought with him was the first volume of *The Principles of Geology,* which had just been published by the Scottish lawyer and scholar Charles Lyell. Lyell made the provocative argument that the Earth's landscapes had been created not by catastrophes but by a series of many small changes. During the earthquake in Chile, Darwin had seen one of these changes take place as the coastline rose in an instant. Darwin's travels turned him into a passionate Lyellian.

Charles Darwin spent five years aboard HMS *Beagle*, traveling the world and gathering clues that he would later use to develop his theory of evolution.

Darwin did not realize the full importance of his observations until he returned to England in 1836. At the Galápagos Islands, for example, Darwin had collected a number of birds that had dramatically different beaks. Some had massive beaks good for crushing seeds, while others had slender needle-like beaks for feeding on cactus plants. Darwin assumed he had found species of blackbirds, wrens, and finches. But when he gave his birds to a London ornithologist named James Gould, Gould made a surprising discovery: the birds were all finches. Despite their radically different beaks, they shared a number of telltale traits only found in finches.

Darwin was puzzled. Some naturalists of his day argued that species had been created where we now find them, well suited to their climate. But if the finches had all been created on the Galápagos Islands, why were they so different from one another? Darwin began to wonder if instead the birds had not been unchanging since creation. Perhaps they had evolved into their current forms.

Darwin's finches helped lead him to conclude that all of life had evolved. Only evolution—the fact that all living things share a common ancestry—could explain the patterns in nature today. As for *how* life evolved, Darwin rejected mechanisms previous naturalists had proposed, such as Lamarck's escalator of progress. Instead, Darwin envisioned a simpler process based on variation and selection.

Darwin spent years painstakingly gathering evidence for his theory. He wanted to answer every possible objection that a critic might have, knowing how fierce the opposition to the very idea of evolution was at that time. Darwin was finally spurred to publish his ideas in 1858, when he received a letter from Indonesia.

The letter was from another English naturalist named Alfred Russel Wallace. Wallace, 14 years Darwin's junior, had patterned his own life after Darwin's famous travels. He had spent years in the jungles of Southeast Asia, gathering plants and animals that he sold to museums and wealthy collectors in Europe. Wallace also kept careful records of the diversity of life he saw, and, as he reflected on his observations, he concluded that life had indeed evolved. He even came up with a mechanism for evolution very much like Darwin's idea of natural selection. Wallace wrote to Darwin to describe his new ideas, and he asked Darwin to present them to the Linnean Society, one of England's major scientific organizations.

If Wallace were to publish first, Darwin knew, his own years of work could be cast into shadow. Darwin also knew that he had worked out his own argument in far more detail than Wallace. On the advice of Lyell and others, Darwin decided to turn the matter over to the Linnean Society. In July 1858, letters from both Wallace and Darwin were read at a meeting of the Linnean Society and later published in the society's scientific journal.

Strangely, though, neither the letters nor the article made much of an impression. It was not until Darwin wrote a book about his theory and published it in 1859 that the world sat up and took notice.

Darwin was surprised to discover that finches on the Galápagos Islands have dramatically different beaks.

On the Origin of Species by Means of Natural Selection, or the Preservation of Favoured Races in the Struggle for Life was an immediate sensation, both in scientific circles and among the public at large. Scientists who embraced "Darwinism" engaged in fierce public debates with those who rejected it. Darwin himself, however, did not personally enter the fray. He went on working quietly and patiently at his rural home, known as Down House. There he continued to carry out experiments to investigate his theory, studying everything from orchids to earthworms.

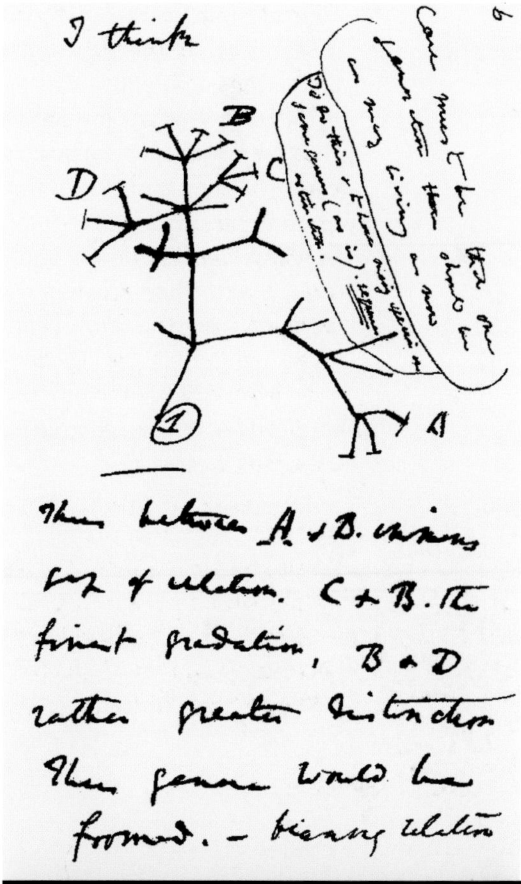

At the same time, Darwin also cultivated a global network of contacts who could supply him with information about the natural world from its remotest corners. He went on to write more books about evolution and other aspects of nature—including human nature, which was the subject of his 1871 book *The Descent of Man, or Selection in Relation to Sex*. When Darwin first argued for evolution, he shocked many readers. But, over the years, much of the public came to accept much of what he had to say. When *The Descent of Man* came out 12 years after *The Origin of Species*, it generated far less controversy. As the botanist Joseph Hooker, Charles Darwin's friend, wrote in a letter, "I dined out three days last week, and at every table heard evolution talked about as accepted fact, and the descent of man with calmness."

After his return to England, Darwin began to develop his theory of evolution in notebooks. Here is a tree he drew in 1837 to illustrate how different species evolve from a common ancestor.

As for scientists, some were skeptical of some of the ways in which Darwin thought evolution occurred, but few disagreed that life had indeed evolved. Perhaps most importantly, Darwin had established evolution as a subject that could be studied scientifically: by running experiments, by comparing species, and by thinking of processes that could explain the patterns of nature. When Darwin died, in 1882, he was buried in Westminster Abbey, in the company of kings and queens, great writers and prime ministers, and other great scientists, including Isaac Newton.

Common Descent

One of Darwin's great achievements was to show the world that we and all other species on Earth are related, like cousins in a family tree. For evidence, he used the patterns of nature that had puzzled naturalists for so long.

In the mid-1800s, anatomists had become keenly aware that underneath the diversity of life there were many common themes. Consider a seal's flippers, a bat's wings, and your arms (See **Figure 2.1**). The seal uses its flippers to swim through the ocean; the bat uses its wings to fly; and people use their arms to cook, sew, write, perform surgery, and steer spacecraft. These appendages serve very different functions, and yet they have a deep similarity. The bones, for example, are arranged in the same way. A long bone (the humerus) extends from the shoulder. On its far end, it meets two thin parallel bones (the radius and ulna), and the bones can bend at the elbow. At the end of the radius and ulna are a cluster of wrist bones. The same set of bones can be found in each species' wrist. Extending from the wrist are five digits. Of course, any given bone in one species is different from the corresponding bone in other species. A seal's humerus is short and stout, for example, while a bat's looks more like a chop-

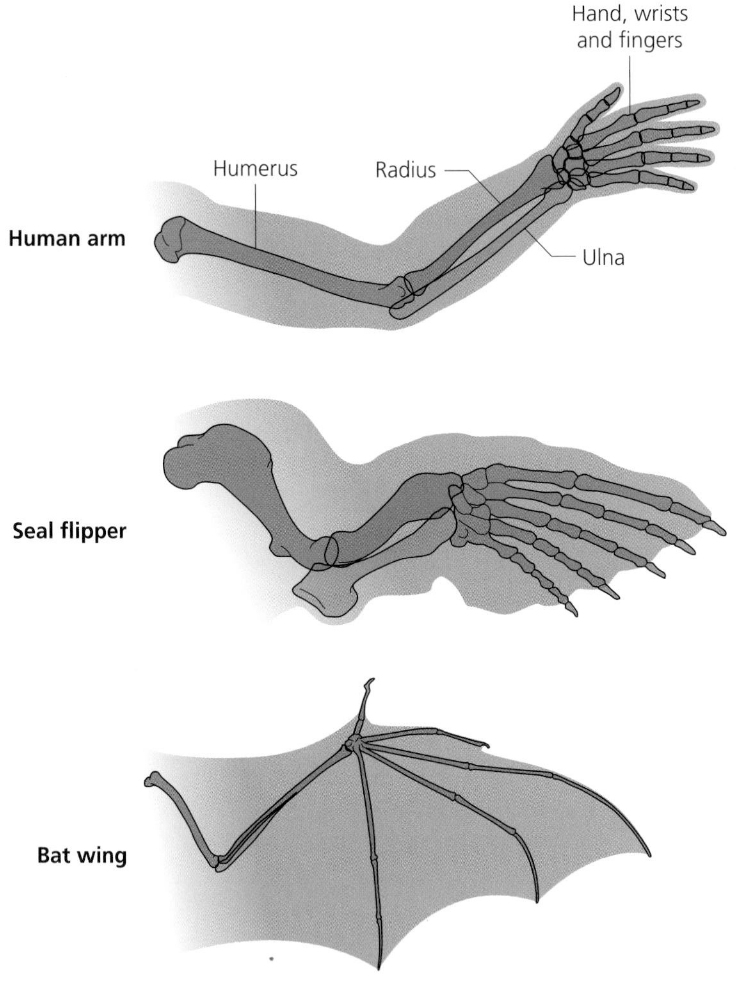

Figure 2.1 Bats, humans, and seals have seemingly different limbs, which they use for different functions. But the bones in one species correspond to bones in the others. Darwin argued that this similarity was a sign of common ancestry.

stick. But those differences don't obscure the arrangement that all of those limbs share. Naturalists called this similarity homology.

What accounts for this combination of differences and similarities? Some anatomists in the mid-1800s argued that each species was created according to an "archetype"—a fundamental plan to which some variations could be added. Darwin preferred a simpler, less transcendental explanation: seals, bats, and humans all shared a common ancestor that had limbs with wrists and digits. That ancestor gave rise to many lineages. In each of them, the limbs evolved, yet the underlying legacy of our common ancestor survived.

Darwin's case for descent with modification was strengthened by the fact that many homologies are found together in the same groups of species. Bats, humans, and seals don't just share limbs, for example. They also have hair, and the females of each species secrete milk to nurture their young. Taxonomists used these common traits to classify humans, bats, and seals into the same category: all three species are mammals. Darwin argued that the very fact that we can classify species this way is consistent with the notion that they evolved from a common ancestor. While each species may evolve new traits of its own (we can't fly like bats can, for example), they all descend from an ancestral mammal. And that mammal, Darwin argued, shared an even older ancestry with other animals. For example, we humans share many homologies with fish. We have

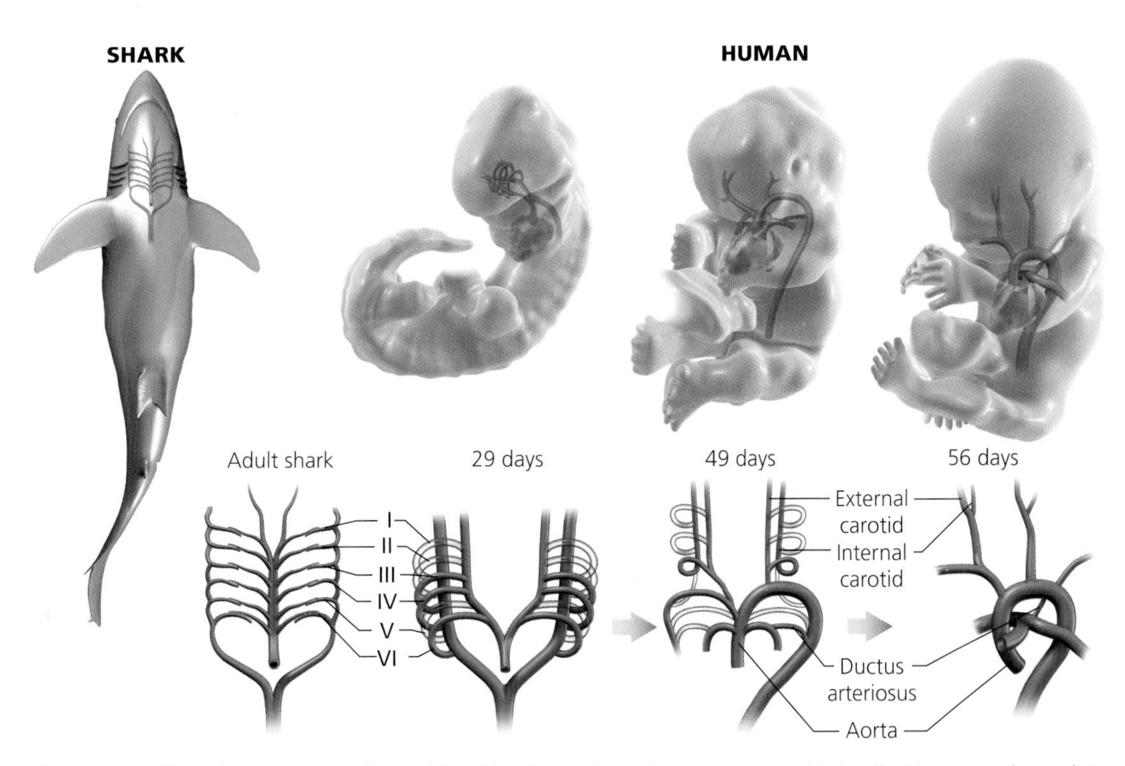

Figure 2.2 Fishes have a series of branching blood vessels to absorb oxygen in their gills. Human embryos (at 29 days) grow blood vessels in the same arrangement, but later the vessels change for absorbing oxygen through our lungs.

eyes with the same arrangement of lenses, retinas, and nerves. We have skulls, livers, and many other organs in common.

Of course, we are different in some important ways. Just about all vertebrates on land have lungs. So do vertebrates that have gone back to the ocean, such as whales and seals. Some fishes have lung-like structures for breathing. But they all have gills, which let them draw in dissolved oxygen from water. No land vertebrate has true gills.

Darwin argued that these differences might not actually be as profound as they first appear. In some cases, homologies are clear only when animals are still embryos, not when they are adults. While fish and land vertebrates are still embryos, for example, they all develop the same set of arches near their heads. In fishes, those arches go on to become gills. In land vertebrates like us, they go on to form a number of different structures in the head and neck, including the lower jaw. A human embryo initially develops blood vessels in the same pattern seen in fish gills. But later the blood vessels are modified (**Figure 2.2**). Darwin argued that those arches are homologies inherited from a common ancestor. In our ancestors, the arches that once supported gills evolved to take on a new function in adulthood.

Natural Selection

Darwin argued that patterns in biology—in homologies, in the fossil record, and so on—could be explained by the inheritance of these features from common ancestors: in other words, evolution. He also argued for a new kind of mechanism that drove much of that change. To account for evolution, Darwin's predecessors usually proposed mysterious, long-term drives. Lamarck, for example, claimed that the history of life followed a trend towards "higher" forms. Many German biologists in the early 1800s argued that life evolved much as an embryo develops in the womb, from simplicity to complexity. What made the new ideas of Darwin and Wallace important was that they depended instead on a process that was not just natural but also observable. Darwin called it natural selection.

Darwin and Wallace both got their inspiration from an English clergyman named Thomas Malthus. In 1798, Malthus published a book called *An Essay on the Principle of Population*, in which he warned that most policies designed to help the poor were doomed because population growth would always outstrip the ability of a nation to produce more food for its people. A nation could easily double its population in a few decades, but its food production would increase far more slowly. The result would be famine and misery for all. Malthus claimed that only those who could adapt to society's needs to produce useful work would be able to survive and reproduce.

When Darwin and Wallace read Malthus, they both realized that animals and plants should experience just this sort of pressure. A fly can take just a few weeks to go from egg to maturity, which meant that its population could explode far more quickly than our own. But the world is not buried in a thick layer of flies.

No species can reproduce to its full potential. Many individuals die before they become adults. They are vulnerable to droughts and cold winters and other environmental assaults, and their food supply is not infinite. Individuals must compete—albeit unconsciously—for the limited resources necessary for survival.

Survival and reproduction do not come down to pure chance. If an individual animal or plant has some trait that helps it to thrive in its environment, it may leave more offspring behind than other individuals of its species. Its traits would therefore become more common over the course of generations (**Figure 2.3**).

As Darwin wrestled with natural selection, he spent a great deal of time with pigeon breeders, learning their methods. In order to produce new breeds—pigeons with ruffles of feathers around the neck, for example, or brilliant white plumage—breeders would select a few birds from each generation with the traits they desired. Over many generations, this selective breeding would cause the traits to get more and more exaggerated.

Darwin saw in this process an analogy for what was happening in nature. Pigeon breeders artificially select certain individ-

Darwin recognized that some traits evolve not because they help organisms survive, but because they help organisms mate more often. In some species, male beetles grow huge horns that they use to joust with other males and to attract females.

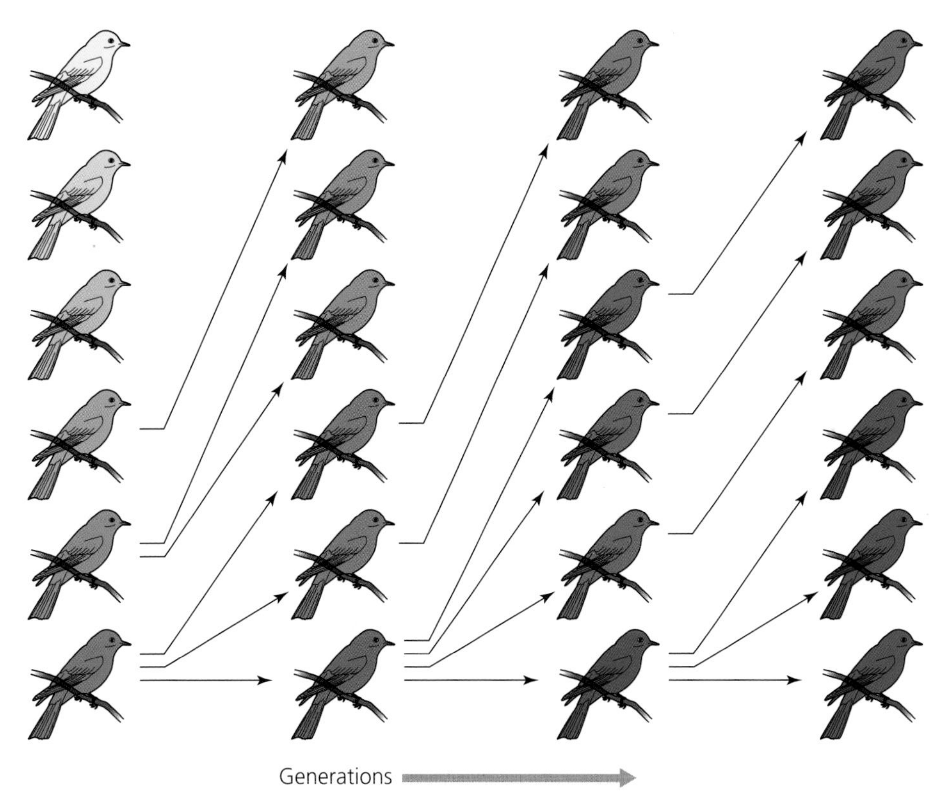

Figure 2.3 Natural selection occurs because some individuals in a species are better adapted to their environment than others. Thanks to their survival, they can produce more offspring. Over generations, their traits become more common. In this diagram, dark birds are better adapted to avoiding predators than light ones.

ual animals to reproduce. Nature, on the other hand, unconsciously selects individuals that are best suited to surviving in their local conditions. Given enough time, Darwin and Wallace argued, natural selection could produce new types of body parts, from wings to eyes.

Beyond Natural Selection

It's been 150 years since Charles Darwin first offered his argument for evolution. His insights are now the foundation of modern biology.

After 150 years, biologists have established that natural selection is a real, powerful force in nature (Chapter 6). But Darwin himself stressed that it was not the full story of evolution. Today biologists recognize that much of our genetic material has been altered by a series of random flukes known as genetic drift. Many animal species are also shaped by a special form of selection Darwin called sexual selection. In many species, females are attracted to males with certain traits, such as showy feathers or big antlers. These attractive traits don't help the males survive the elements, and in some cases they can even make the animals easier for predators to kill. In Chapter 12, we'll see why such extravagant features can evolve.

It has been 150 years since Darwin published *The Origin of Species*, and in that time he has been spectacularly vindicated. Natural selection and sexual selection are real phenomena, and they can even be observed taking place in living organisms. But that does not mean that Darwin had figured out all there was to know about evolution by 1859. Inherited traits were central to his theory, for example, and yet he did not know anything about the molecules that make heredity possible. He speculated that living species evolved from transitional ancestors, such as whales with legs, but the fossils documenting many of those transitions would not be discovered until long after his death. But we think no less of Isaac Newton for not knowing about the hidden structure of atoms. Just as Newton opened a door for generations of physicists, Darwin opened a door for biologists.

TO SUM UP...

- **In the seventeenth and eighteenth centuries, naturalists devised systems for classifying life and recognized fossils as the remains of living things.**
- **Science is the use of evidence to construct testable explanations and predictions of natural phenomena, as well as the organization of knowledge generated through this process.**
- **Georges Buffon proposed that the Earth was very old and that life had gradually changed during its history.**

- Georges Cuvier helped to establish that many fossils were the remains of extinct species.
- The geological record revealed a succession of different species that had lived on Earth.
- Jean-Baptiste Lamarck developed an early theory of evolution, based in part on the idea that acquired traits are passed down through the mechanism of heredity.
- Charles Darwin and Alfred Russel Wallace developed a theory of evolution by means of natural selection.
- Homology is the similarity in different species of characteristics inherited from common ancestors.
- Comparisons between embryos can reveal homologies not evident in adulthood.
- Natural selection is the process by which individuals with heritable traits are better able to succeed in their environments and to leave more offspring with those traits in later generations, while unfavorable heritable traits become less common because the organisms that have them have fewer offspring that are able to live long enough to reproduce.
- Darwin argued that natural selection is only one process by which life evolves.

3

What the Rocks Say

Abigail Allwood searches for clues to the evolution of life in one of the most remote, inhospitable places on Earth. Allwood, a paleontologist at NASA's Jet Propulsion Laboratory, travels with her colleagues deep into the Outback of Australia, where one can find plenty of blue-tongued goanna lizards and galah cockatoos but virtually no people. Water is scarce among the

Above: Geologist Abigail Allwood and her colleagues have discovered 3.4-billion-year-old rocks in Australia (bottom left) that appear to be fossils of mats of microbes called stromatolites (top left). They are among the oldest evidence of life on Earth.

bare outcrops and hills, and the days can be scaldingly hot. The name of the geological formation where Allwood works is a grim joke: North Pole.

Allwood hikes along the exposed rocks, taking photographs and sometimes hammering off pieces to take home to study further. There's nothing in the rocks that looks alive to the inexperienced eye. The most notable thing about them is that they are made up of fine, even layers, which curve and sag into strange shapes. Some look like upside-down ice cream cones, and others look like egg cartons.

What could anyone learn about life from this experience? It may be hard to believe, but Allwood's research indicates that these mysterious striped rocks were once swarming with life. They were formed by mats of bacteria that stretched far across the floor of a shallow sea. What is particularly striking about the rocks is their age. They are 3.43 billion years old, an age that makes them some of the oldest traces of life on Earth.

Allwood is one of thousands of scientists who wander the planet in search of traces of the history of life. Together, they are creating a record—from fossils, molecules, and even atoms—that chronicles how life first emerged on Earth, how it flourished, diversified, suffered extinctions, and continued to change for at least 3.5 billion years, giving rise only 200,000 years ago to a new species of upright apes: our own species, *Homo sapiens*. (The major milestones in the history of life are summarized on the endpapers of this book.)

The Ancient Earth

Charles Darwin is now best known as an evolutionary biologist, but it was as a geologist that he first came to fame after the voyage of the *Beagle*. What Darwin learned about rocks on his journey helped to shape his thoughts about life. Rocks preserved some of the most important evidence for evolution. They demonstrated that the Earth was very old, Darwin argued, far older than the few thousand years that Biblical scholars had proposed, and even older than many scientists of his day thought possible. Darwin held that evolution unfolded very slowly, and the geological record showed that there had been enough time for such a gradual process to produce the diversity of life today.

Unfortunately, Darwin and his fellow nineteenth-century geologists had no way to determine exactly how old a particular fossil or rock might be. They could only offer rough estimates of how long it had taken for a geological formation to emerge, based on the rate at which sediments accumulated on riverbanks or in coastal waters. Some scientists rejected this argument. The most famous of all these critics was the eminent physicist William Thomson (Lord Kelvin).

Thomson argued that the world could not be as old as some scientists, such as Darwin, proposed. His argument was based not on formations of rocks but on their temperature. Let's assume that the Earth began as a ball of molten rock, Thomson said. A hot rock cools at a steady rate, and so Thomson reasoned that you could use the current temperature of rocks to estimate how long they had been cooling. Rocks on the planet's surface would not give a reliable estimate, because they were heated by the Sun every day and cooled every night. The rocks deep underground in mine shafts, on the other hand, stayed at the same warm temperature year round. Based on those mine rocks, Thomson calculated that the Earth could only be 20 million years old at most—an age far younger than Darwin argued for.

Darwin was able to refute most of his critics quite effectively, but he could not find a way to respond to Thomson. As Darwin struggled to flesh out his theory of evolution, he wrote, "then comes Sir W. Thomson like an odious specter."

Thomson, it would later turn out, was wrong. Three decades after he estimated the age of the Earth, scientists discovered that the planet is warmed by radioactivity, a form of energy released by unstable atoms. In the 1900s, scientists discovered that they could use those unstable atoms to make much more precise estimates of the Earth's true age. The Earth, it turns out, is 4.55 billion years old. Scientists can also use radioactivity to calculate the ages of individual rocks and the fossils they contain. Darwin's estimates, they found, were a lot closer to reality than Thomson's. (For more on radioactive clocks, see the box on page 40.)

A Vast Museum

Darwin recognized that evolution made sense of the fossil record. But he also knew that some critics would try to use the fossil record to challenge him. Why, they might ask, hadn't paleontologists found fossils from every stage of evolution from one species to another? "I believe the answer mainly lies in the record being incomparably less perfect than is generally supposed," Darwin wrote in *The Origin of Species*. "The crust of the earth is a vast museum; but the natural collections have been imperfectly made, and only at long intervals of time."

Over the past 150 years, scientists have confirmed that the fossil record is far from complete. To understand why most living things don't turn to stone and a few do, researchers have studied the process of fossilization. They've observed how dead animals and other organisms decay over time, and they have replicated some of the chemistry that turns living tissues into rock.

Most organisms don't fossilize simply because other organisms eat them. Animals such as hyenas or vultures may scavenge the muscles and organs from a carcass, while insects, bacteria, and fungi work more slowly on what's left. Within a few months, most cadavers are so thoroughly devoured, trampled, sunbeaten, or rain-soaked that nothing is left to become a fossil. The same goes for marine animals. Soft-bodied creatures, such as jellyfish, generally just disinte-

Radioactive Clocks

All atoms are made of three kinds of particles: protons, neutrons, and electrons. The atoms of each element have a unique number of protons. Hydrogen has one proton, helium has two, and carbon has six. Alongside their protons, atoms have neutrally charged particles called neutrons. But while all atoms of a particular element have the same number of protons, they can have different numbers of neutrons. The most common form of carbon on Earth, carbon-12, has six protons and six neutrons, for example, but there are trace amounts of carbon-13 and carbon-14 as well.

The protons and neutrons in an atom are a bit like piles of fruit at a grocery store: in some arrangements, they're perfectly stable, but in other arrangementss they will sooner or later fall apart. When an unstable atom loses some of its protons, it becomes a different element. Uranium-238, for example, breaks down by releasing a pair of neutrons and a pair of protons, thereby turning into thorium-234. Thorium-234 is also unstable, and, in time, it decays into protactinium-234, which, in turn, decays again. It takes a chain of 13 intermediates for uranium-238 to settle into a stable form: lead-206.

Physicists and geologists in the early twentieth century realized that the half-lives of radioactive elements opened up many ways to estimate the ages of rocks.

The Earth, along with the other planets, formed out of a disk-shaped cloud of dust. Among the components of that dust were small amounts of radioactive elements. They steadily broke down after the Earth formed. Each time a radioactive atom decays, it releases some of the energy that was holding its particles together. The energy released from each decay helped to heat up the surrounding rock. As a result, radioactivity has slowed down the cooling of the Earth, tricking Thomson into thinking it is far younger than it really is.

Each radioactive isotope decays at a particular rate, which scientists call its half-life. The half-life of carbon-14 is 5,730 years, for example, which means that if you put a kilogram of carbon-14 in a time capsule and someone dug it up in 5,730 years, half the carbon-14 would have broken down. If the time capsule was buried for another 5,730 years, only a quarter of the original carbon-14 would be left.

Physicists and geologists in the early twentieth century realized that the half-lives of radioactive elements opened up many ways to estimate the ages of rocks. The older a rock gets, for example, the less of its original radioactive elements it contains. But you can't calculate its age simply by measuring its radioactive elements and their breakdown products. To see why, imagine lava pouring out of a volcano and forming a rock. The rock contains a radioactive isotope called rubidium-87, which breaks down to strontium-87 very slowly, with a half-life of 48.8 bllion years. But let's imaging that there was also a lot of strontium-87 in the lava, so that the rock had equal portions of rubidium-87 and strontium-87. The rock cools, and then a month later you measure the levels of the two isotopes. If you simply measured their proportions, you might conclude that the rock was 48.8 billion years old—about four times older than the universe itself.

Fortunately, rocks provide a way to get around this problem. Scientists can measure the age of a rock by measuring rubidium and strontium in *several spots* in the same rock. Here's how this method works. A newly formed rock is made up of different minerals. The chemistry of each type of mineral causes it to take up different amounts of rubidium and strontium. Some mineral grains will have a lot of rubidium and a little strontium, and some will have a lot of strontium and a little rubidium.

Let's focus now just on the strontium in those minerals. Let's imagine that it's a mixture of strontium—strontium-86 and strontium-87, the breakdown product of rubidium-87. Chemically speaking, these isotopes behave identically. That means that no matter how much strontium is in each mineral in a rock, the proportion of the two strontium isotopes will be the same when they form from the same lava.

Once the rock is formed, no more strontium can enter it. Now the only new source of strontium is the decay of rubidium-87. In each mineral grain, the rate of decay is the same. This decay causes two important changes in the proportions of atoms in the minerals. The proportion of strontium-87 to strontium-86 goes up. Meanwhile, the proportion of rubidium-87 to strontium-87 goes down (**Figure 3.1**).

In a mineral that started out with only a little rubidium, these two changes will be small. In a mineral that started out with a lot of rubidium, the changes will be bigger. Geochronologists can plot the change in these proportions in each mineral on a graph. As you can see in the figure below, they form a straight line (called an isochron). When a rock first forms, the isochron is horizontal. Over time, it moves like a clock hand turning counterclockwise. The angle at which it turns tells geochronologists how much time has passed since the rock formed.

In 1953, Claire Patterson, a geologist at the California Institute of Technology, used uranium

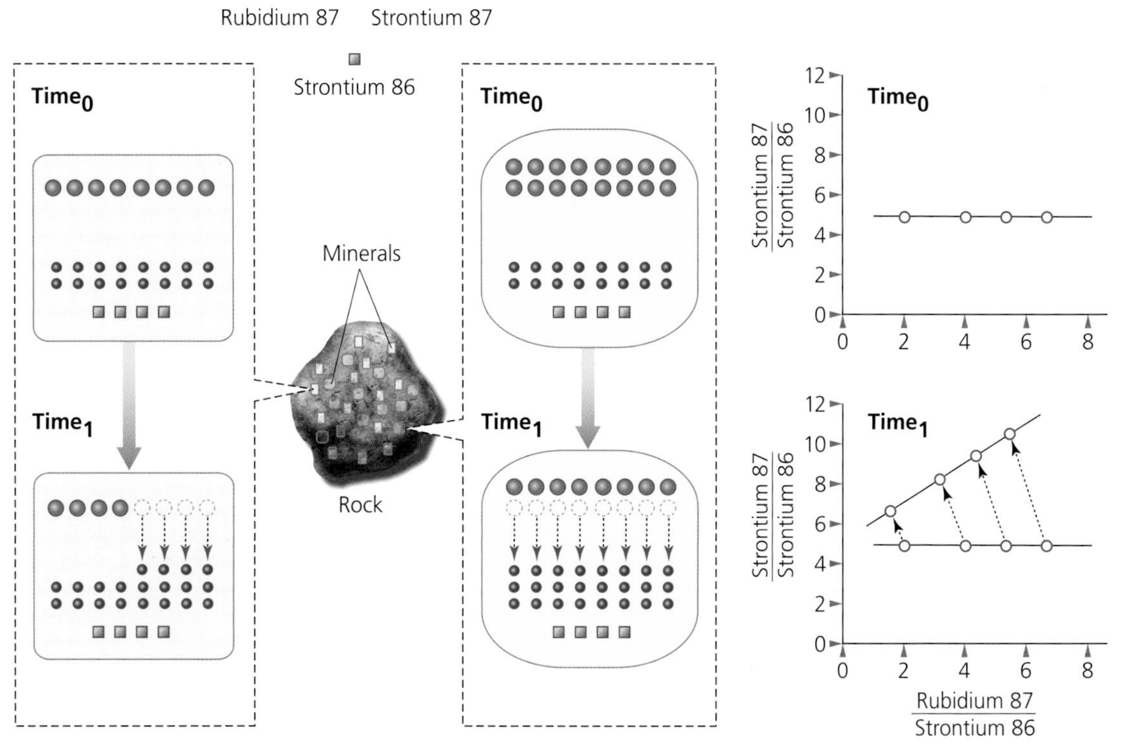

Figure 3.1 By comparing mineral grains in a single rock, geologists can estimate its age. Left: A rock contains a mix of rubidium and strontium atoms. Upper right: The proportions of atoms and isotopes are plotted on a graph. The ratio of rubidium to strontium varies from one region of the rock to another. But since strontium isotopes behave the same way in chemical reactions, the ratio of strontium isotopes is the same everywhere in the rock. Lower left: Over time, some of the rubidium atoms decay into strontium-87. The amount of strontium-86 stays the same. As a result the ratio of strontium-87 to strontium-86 increases everywhere in the rock. But the ratio increases more in parts of the rock that started out with a higher proportion of rubidium-87. Lower right: As rubidium decays to strontium-87, the line on the graph turns counterclockwise. By measuring the angle between the two lines, geologists can estimate the age of the rock. (Adapted from Miller, 2000)

and lead isotopes to figure out the age of the entire Earth. All the evidence gathered by astronomers points to Earth having formed from a dusty cloud swirling around the Sun. The other planets, along with comets and asteroids, formed from that same cloud. Sometimes meteorites fall to Earth, and scientists have found that their chemical makeup is very different from that of the rocks formed on our planet. However, since they all formed from the same primordial disk, they should record the same age with their isotopes.

Patterson compared the radioactive clocks in Earth rocks and meteorites. In fact, he consulted two different clocks at once. Uranium-238 decays into lead-206 with a half-life of 2.23 billion years. Uranium-235, meanwhile, decays into lead-207 with a much shorter half-life of 703 million years. Over time, rocks accumulate more lead-206 and lead-207 compared with a stable isotope of lead, such as lead-204. Patterson compared the proportions of the radioactively derived lead isotopes with stable lead-204 in meteorites and compared the results with what other geologists had found in rocks on Earth. Once again, the results fit along a remarkably straight line. Using the half-lives of the two uranium isotopes, Patterson calculated that the Earth is 4.55 billion years old.

Geologists use different isotopes to measure time on different scales. Potassium-40, for example, only takes 1.25 billion years to break down into argon-38. As a result, it provides a more accurate clock for dating the age of younger rocks.

Geologists can use isotopes not only to determine the ages of rocks but also to estimate the ages of fossils. Volcanic eruptions sometimes lay down thick layers of ash that is rich in potassium-40. If a fossil is sandwiched between layers of volcanic ash, the layers can create upper and lower bounds for the age of the fossil itself. In 1967, for example, scientists discovered fossils of humans at a site near the village of Omo, Ethiopia (**Figure 3.2**). The scientists knew the fossils were old, but it was hard to determine just how old they were. Almost three decades later, a team of scientists went back to the site to take a closer look. They discovered two layers of volcanic ash, one above the rocks where the fossils had been found, and another right below them.

Geologists can use isotopes not only to determine the ages of rocks but also to estimate the ages of fossils.

The argon in the upper layer yielded an age of 104,000 years, with a margin of error of 7,000 years. The lower layer was 196,000 years old, with a margin of error of 2,000 years. A careful study of the sediments in between those two layers indicated that the fossils were closer to the older boundary than to the younger one—perhaps as old as 195,000 years. Thanks to this research, the Omo fossils became the oldest known fossils of members of our own species.

Elements such as uranium and argon can only help geologists estimate the ages of rocks. But carbon-14 can, in some cases, allow scientists to determine the ages of fossils themselves. Carbon-14 is continually being generated in the atmosphere, as charged particles from space rain

grate. Hard-shelled animals, such as lobsters and clams, get drilled and cracked open by scavengers, their remains left to fall apart and perhaps to wash up on some distant shore.

A tiny fraction of the organisms that die each year are protected from this oblivion. In some cases, they happen to fall into a still lake, where they are rapidly covered in sediment before they can be torn apart by scavengers. Water percolating through the sediment can fill the tiny spaces between the latticework of

down toward the Earth and collide with atoms in the air. The collisions turn some stable nitrogen-14 atoms into unstable carbon-14. As plants take up carbon dioxide from the atmosphere, they accumulate carbon-14 in their tissues. Animals build up a supply of carbon-14 as well when they eat plants, or when they eat other animals that have eaten plants. At every meal, you pick up a little extra carbon-14 too. Once an organism dies, the carbon-14 in its remains steadily breaks down for thousands of years. Scientists can use radiocarbon dating to estimate the age of biological material that's up to about 40,000 years old.

By measuring carbon-14, geologists can date not just bones, but any material with some organic carbon in it. They can date the age of wooden tools, or the age of ash from a fire. In 1994, cave explorers in France discovered hidden chambers filled with beautiful paintings of horses, lions, and other animals. Scientists came to this cave, known as Chauvet, and scraped tiny samples of charcoal from the walls. Back at their lab, they isolated carbon isotopes and used them to estimate that the paintings were made between 26,000 and 32,000 years ago. If other scientists replicate these dates, the Chauvet cave contains the oldest known examples of painting in the world.

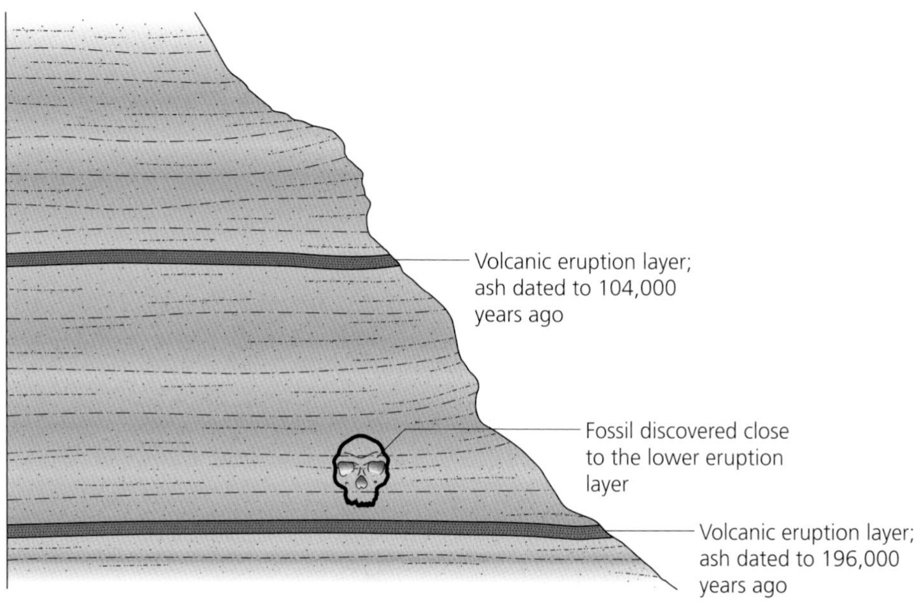

Figure 3.2 Paleontologists use as many lines of evidence as possible to estimate the age of fossils. By calculating the ages of layers of volcanic ash above and below a fossil, they can establish upper and lower bounds for when it formed. Dating volcanic layers allowed scientists to estimate the age of the oldest known human fossils: about 195,000 years old.

bone and shells with new minerals. Over thousands of years, the sediment surrounding the organism turns to rock. A fossil is formed (**Figure 3.3**).

Most of the fossils in the world's museums are the remains of bones, teeth, or shells. It takes an especially rare set of circumstances for soft tissues, such as muscles, to mineralize. Soft-tissue fossils are exquisitely important to scientists, because they preserve an incredible amount of detail. Some of the most impressive soft-tissue fossils in the world first came to light in 1998, when scientists from Harvard

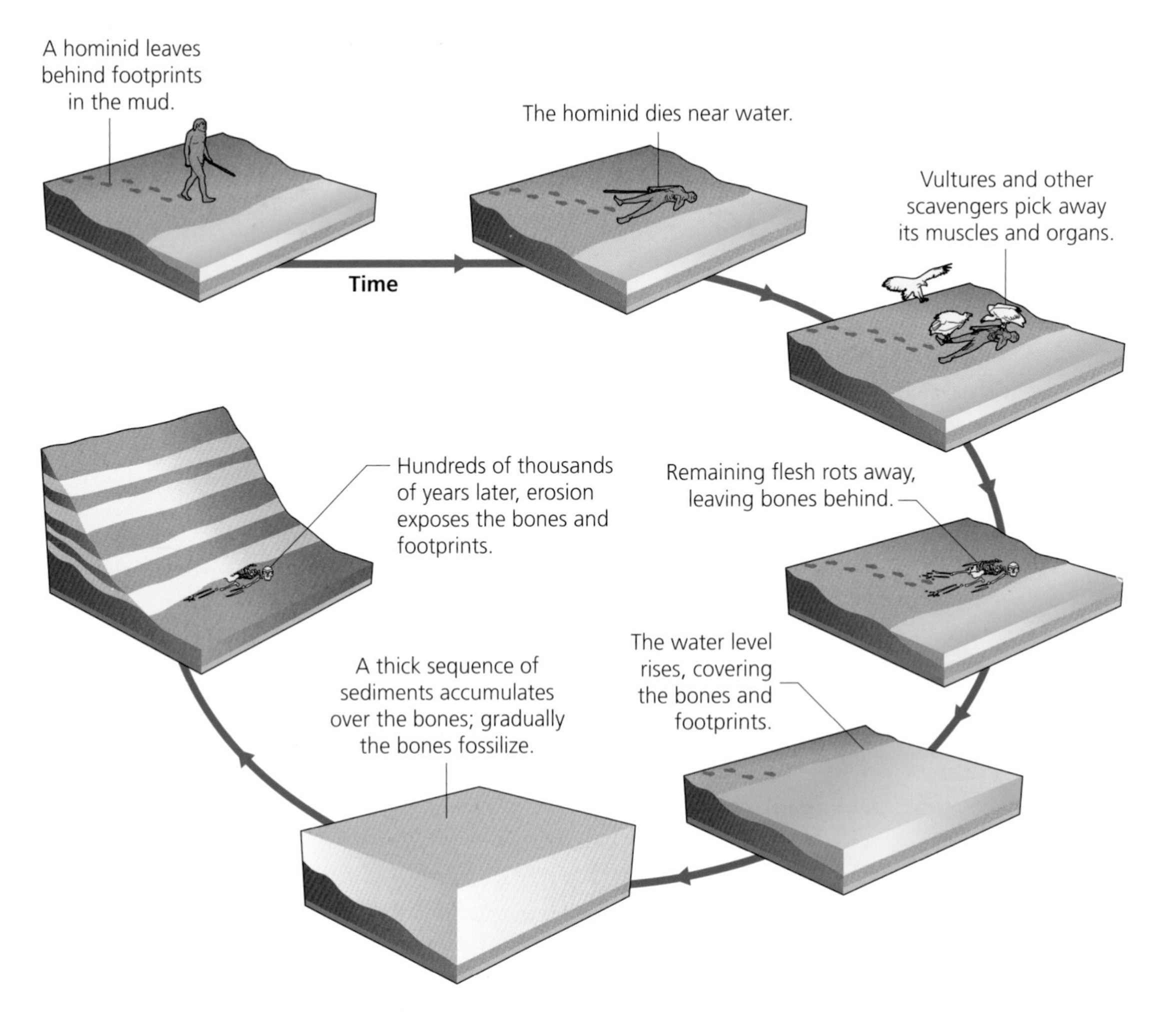

Figure 3.3 Fossils form after organisms die. In some cases, their bodies are covered by sediment and then gradually turned to minerals. (Adapted from Prothero, 2007)

University and Beijing University published a report on 580-million-year-old remains from a formation in southern China called Doushantuo. They are microscopic clumps of cells, some of which bear the hallmarks of animal embryos in their earliest stages of division. Experiments on living embryos suggest that bacteria made the exquisite preservation of the Doushantuo fossils possible. After the ancient embryos died, films of bacteria coated their cells. The films gradually became precise replicas of the cells they devoured. Later, the bacteria began to secrete calcium phosphate, which then created fossilized replicas of the embryos.

The fossil record may be incomplete, as Darwin suggested, but paleontologists have many ways to squeeze knowledge from it. Once they have brought a fossil back to their lab, for instance, they can put it into a CT scanner to capture an image of its internal structure. The shapes of bones can reveal how fast an extinct animal ran or how hard it could bite. Paleontologists can cut off microscopically thin slices from fossils and count the rings of new tissue that grew on the bone each year. They can even analyze the atoms in the fossil to determine what sort of food it ate or what kind of environment it lived in.

Traces of Vanished Life

A fossil is not the only thing that an organism can leave behind. Some rocks themselves are made from the remains of dead organisms. About 300 million years ago, for example, giant swamps spread across many of the continents. When plants died there, they did not immediately decay. Instead, they fell into the swamps and were rapidly buried in sediment. Bacteria then began to break them down. These bacteria—a special kind that could survive in the oxygen-free swamp bottoms—transformed the plant material into a tarry substance known as lignin. Eventually, the swamps were drowned by rising oceans, and then they were buried under vast amounts of marine sediment. The plant material was transformed yet again, under tremendous pressure and heat, into a hard, rock-like substance. We call this substance coal. Some fossils—of leaves and branches, for example—can be preserved in protected pockets in these deposits, known as coal balls.

In recent decades, scientists have figured out how to identify individual molecules in rocks that once belonged to living things. The membranes that surround cells, for example, include oily molecules known as lipids. Each lipid is a short chain of carbon atoms studded with oxygen and hydrogen atoms. When an organism dies, its cells rupture and their lipids spread out in the sediment. Over millions of years, the lipids react with the sediments and lose some of their oxygen and hydrogen atoms. But even after all that time, the molecules can still be recognized as the product of living things. (The only place where these lipids are known to be made is inside cells.) Ironically, these unimaginably small molecules have proven to be more durable than bones or shells. They've been discovered in rocks dating back billions of years.

Even the individual atoms in rocks can offer scientists clues about ancient life. About 18% of the human body is made up of carbon, and that carbon is a mix of carbon isotopes. (See the box on page 40 for more on isotopes.) The precise ratio of carbon isotopes in any organism is determined by its biology. Plants, for example, absorb carbon dioxide from the atmosphere, but it's a little more difficult for them to take in the heavier carbon-13 than carbon-12. As a result, the fraction of carbon-13 in a plant is lower than the fraction in the air. Different plants have slightly different ratios of carbon isotopes—grasses, for example, have lower levels of carbon-13 than trees or shrubs. This difference extends to the animals that eat the plants, too. Once these animals and plants (and microbes) die, their fossils can sometimes preserve the same balance of carbon isotopes. By measuring that balance, scientists can discover clues about the diet of the organisms. Scientists have measured the carbon isotopes in the teeth of ancient humans and their extinct relatives, for example, to discover when our ancestors began to hunt animals and eat meat.

Life's Earliest Marks

After the Earth formed from the primordial solar disk 4.55 billion years ago, it cooled over millions of years until its crust turned hard. Lighter formations of rock rose and became continents, while the surrounding crust became vast basins. Gases and water vapor escaped from the rocks and formed the Earth's atmosphere. The water gradually rained down into the basins to form oceans. For hundreds of millions of years, the Earth continued to collide with the remaining debris from the original solar disk. One such collision was so big that the rocky rubble thrown up from the impact began to orbit the Earth, and then it eventually coalesced to form the Moon. The giant impacts began to taper away about 3.8 billion years ago. Over the next billion years or so, the crust of the planet broke into plates. Hot rock rose up in some of the cracks between the plates and added to their margins. Meanwhile, the opposite margins of the plates were driven down under the crust. As this rock sank, it became hotter, until it melted away.

Thanks to the early impacts and the relentless burial of crust, just about all the original surface of the Earth has been erased. Along with it, whatever fossils formed on the early Earth have also been destroyed. Paleontologists have not found a single recognizable fossil from the first billion years of Earth's history. It's possible, however, that early life left other signatures behind. Before life began, the only source of carbon on the surface of Earth would have come from lifeless sources, like volcanoes. But once life emerged on Earth, it would have produced abundant amounts of organic carbon, which would have gradually become incorporated into sedimentary rocks. And since organisms have less heavy carbon than volcanoes, scientists might find a shift in the ratio of isotopes once Earth became home to living things.

In 2003, a geologist named Minik Rosing from the University of Copenhagen and his colleagues announced they had found this shift. They extracted bits of

Tiny specks of carbon can be preserved for billions of years in minerals known as zircons. The balance of carbon isotopes can provide clues to what life was like when they were trapped in the mineral.

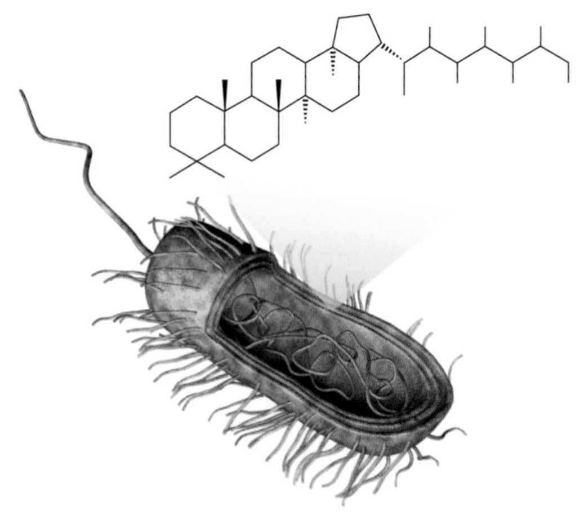

Some of the molecules in cells can survive for billions of years. These biomarkers are the earliest evidence for some groups of species, such as animals.

The Present and the Past in Science

Predictions are essential to science. When scientists find an explanation of a natural phenomenon, they can use it to generate predictions, and, in some cases, they can set up experiments to put those predictions to a test. If one scientist publishes an experiment that fulfills a prediction, other scientists try to come up with experiments of their own to see if they get the same result. Even a scrupulously honest scientist can get a misleading result from an experiment. If an experiment is set up badly, it may produce a "false positive"—in other words, a seemingly positive result that meets the scientist's predictions, but not for the reasons the scientist thinks. Experiments can also produce false negatives. A prediction may seem to fail, but only because the experiment wasn't set up carefully enough to test it.

Evolutionary biologists study events that happened millions or even billions of years ago. In these cases, they make predictions about the past.

Evolutionary biologists have run many experiments to test predictions. They have set up experiments in which natural selection has reshaped organisms in their laboratories, for example (page 118). But evolutionary biologists also study events that happened millions or even billions of years ago. In these cases, they must make predictions about the past.

If paleontologists find half of a fossil that shows key traits found only in one lineage of dinosaurs, they can predict that if anyone ever finds a fossil of the other half, they'll find other unique dinosaur traits, and not traits of clams or mammals. By comparing the genes of different species, scientists can draw an evolutionary tree showing how they are related to one another. This tree is a hypothesis, which scientists can test with new evidence. They can compare other species, or analyze a different set of genes to see if they end up with the same tree.

Many paleontologists are investigating how almost all species became extinct 252 million years ago (page 232). Based on the chemistry of rocks that formed at that time, some researchers proposed that much of the ocean's oxygen disappeared, starting in deep water and spreading out to the coasts. That hypothesis leads to a prediction: if paleontologists found a good fossil record, they'd find that deep-water species would become extinct first, and then shallow-water ones.

Catherine Powers and David Bottjer at the University of California found just such a record, in the fossils of bryozoans, coral-like animals that anchor themselves to the seafloor. Many deep-water bryozoan species went extinct first during the mass die-offs, and then shallow ones disappeared afterwards, just as predicted.

3.7-billion-year-old carbon from rocks in Greenland and discovered a biological ratio of carbon in them. Rosing and his colleagues concluded that this was the earliest sign of life, produced most likely by photosynthetic bacteria. But other researchers have challenged Rosing's conclusion. They argue that geological processes could have created the ratio of carbon isotopes in the rocks, without any need for life. Such uncertainty hovers over much of the earliest evidence for life on Earth. In the 1980s, in Australia, J. William Schopf of UCLA discovered what he proposed were 3.5-billion-year-old fossils of bacteria. Martin Brasier of the University of Oxford has challenged Schopf's results, arguing that the fossils were actually formed by tiny blobs of mineral-rich fluids.

To better understand the early history of life, scientists are continuing to scour ancient rocks. Abigail Allwood and her colleagues discovered their strange, egg-carton-like rocks in some of the oldest geological formations on Earth. The researchers then found striking microscopic similarities between the rocks and large mounds built today by colonies of bacteria. These mounds, known as stromatolites, grow on the floors of lakes and shallow seas. Sediments and minerals accumulate on the bacteria in thin layers, and more bacteria grow on top of the sediments and minerals. In 2006, Allwood and her collegues published their discovery, arguing that they had discovered stromatolite fossils. If they're right (and many of their colleagues think they are), they may have found some of the earliest life on Earth.

The fossil record slowly improves in younger rocks, offering scientists more information about life on the early Earth. If you could travel back in time 2.5 billion years, the world would have looked like a fairly desolate place. On land there were no trees, no flowers, not even moss. In some spots, a thin varnish of single-celled organisms grew. In the ocean, there were no fish or lobsters or coral reefs. Yet the ocean was full of microbial life, from hydrothermal vents on the sea floor to free-floating bacteria catching sunlight at the ocean's surface. Today Earth is home to millions of species of multicellular organisms—animals, plants, algae, and fungi. But microbes have never stopped being a hugely important form of life—some would argue the most important. They live in far more habitats than animals or plants. You can find bacteria inside undersea volcanic chambers or at the bottom of acid-soaked mine shafts, for example. By weight, microbes make up the bulk of Earth's biomass. Their genetic variability is enormous as well. Most genes on the planet belong to microbes or their viruses. It's a microbial world, in other words, and we just happen to live in it.

Life Gets Big

Early in the history of life, three main branches of life emerged. The members of two of those branches, known as Bacteria and Archaea, have remained single-celled up until today. The third branch (our own) is called Eukaryota. The most obvious trait that distinguishes eukaryotes from bacteria and archaea is a sac, called the nucleus, in which a eukaryote stores its DNA. Living eukaryotes include animals, plants, fungi, various kinds of algae, and many single-celled organisms called protozoans. The oldest fossils of single-celled eukaryotes date back 1.8 billion years. These fossils measure about 100 micrometers across (that's one hundredth of a centimeter, or roughly four thousandths of an inch). While they would have been invisible to the naked eye, they marked a giant leap in size, measuring about 100 times bigger than a typical bacterium. These protozoans also had ridges, plates, and other structures that are similar to those of living protozoans.

Over the next billion years, the diversity of protozoans increased. Along the way, a few multicellular forms of eukaryotes appeared as well. The oldest fossils of multicellular life, filaments of algae, are 1.6 billion years old. The oldest known fossils of red algae, are 1.2 billion years old. Green algae appeared 750 million years ago. The oldest traces of animals are 635 million years old. Gordon Love, a geochemist at the University of California at Riverside, and his colleagues drilled into oil-rich rock in Oman and discovered cholesterol-like molecules known to be made only by sponges.

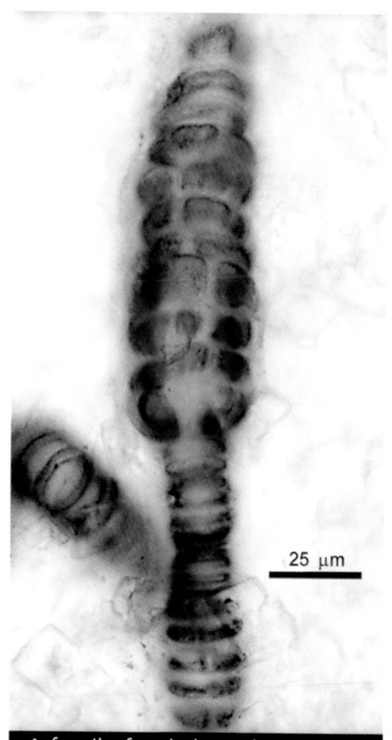

A fossil of red algae, known as *Bangiomorpha*, dating back 1.2 billion years, is among the oldest known fossils of multicellular life.

The earliest actual fossils of animals yet found come from the Doushantuo Formation in China. The oldest of the Doushantuo fossils, dating back to 580 million years ago, appear to be animal embryos encased in cysts. Fossils of sponges appear a few million years later in the Doushantuo Formation, along with the remains of tiny wormlike animals with bodies shaped like flattened slugs—that is, if slugs were only a fifth of a millimeter across. (It should be pointed out that some researchers aren't yet convinced that all of these fossils are from animals.) Paleontologists have also found lines, squiggles, and other marks in Doushantuo rocks that may be tracks made by some kinds of crawling wormlike creatures.

A few other sites around the world document a number of new kinds of animals. Paleontologists have found shells measuring only one or two millimeters across. Animals began to leave more complex tracks and even started plunging into the seafloor. The most striking fossils from this time were made by animals measuring up to a few feet long that didn't look very much like anything alive today. They had strange shapes: some looked like fronds, geometrical disks, or blobs covered with tire tracks.

Collectively, these weird species are known as the Ediacaran fauna (named for a region in Australia where paleontologists first recognized that these kinds of fossils dated back to before the Cambrian Period). Paleontologists have compared Ediacaran fossils to living species to figure out their place in the tree of life. Some fossils share many traits with living groups of animals. *Kimberella*, for instance, may belong to the mollusks, a group that includes clams and snails. Most Ediacarans were more mysterious, however. They were big and had complicated structures, like marine animals, but they didn't look like any particular group of animals alive today. It's possible that they belonged to lineages of animals that branched off early from other animals and later became extinct.

Many Ediacarans had disappeared by the beginning of the Cambrian Period, 542 million years ago, and the last were gone by about 535 million years ago. Meanwhile, new groups of animals were evolving. The early Cambrian, from 542 million to 511 million years ago, is divided into four stages, and each stage saw more of these first appearances of living groups than the

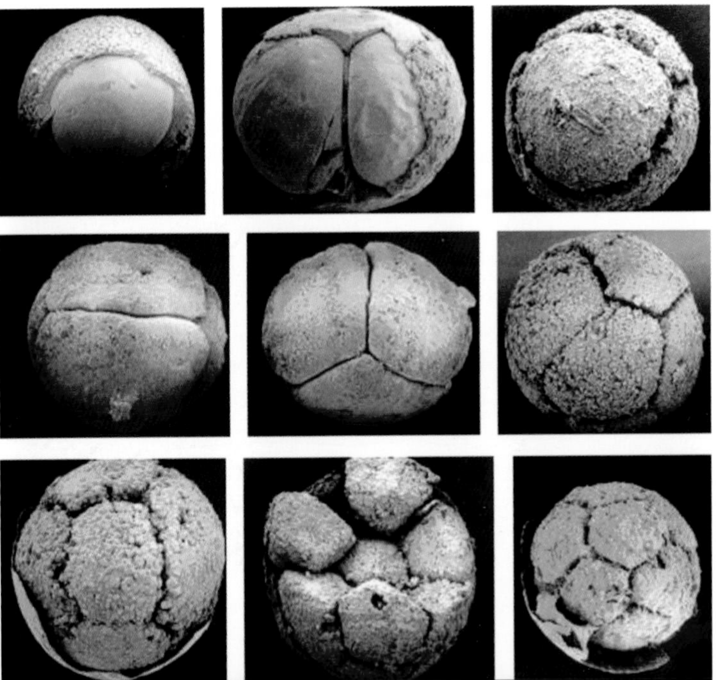

These microscopic fossils of clusters of cells date back about 580 million years. Based on studies of living cells, paleontologists have concluded they are probably animal embryos.

Between about 575 and 535 million years ago, a bizarre community of multicellular organisms known as the Ediacaran fauna dominated the world's oceans. Some of them belonged to living lineages of animals, while many others had become entirely extinct by 535 million years ago.

Arthropods, a lineage that includes insects and crustaceans, first emerged in the fossil record during the Cambrian Period. Left: *Opabinia*, which lived 505 million years ago, had a bizarre tentacle-like jaw and five eyes. Right: Trilobites were armored arthropods related to horseshoe crabs. Emerging 530 million years ago, they were common in the oceans until their extinction 250 million years ago.

We humans belong to the vertebrate lineage. Close relatives of vertebrates first appear in the fossil record during the Cambrian Period. Left: *Haikouichthys* was a small fish-like animal with some traits found only in vertebrates, such as a brain, a stiffening rod running next to its spinal cord, and arches that may have supported gills. By 380 million years ago, vertebrates had evolved into large predators, such as *Dunkleosteus*, (right) which grew up to 6 meters (18 feet) long.

previous one. We belong to the chordates, for example, a group that makes its first appearance in fossil-rich rocks in China called the Chenjiang Formation, dating back 530 million years ago.

Not all of the major groups of animals that emerged during the Cambrian Period can be found on Earth today. One of the most common groups of fossils from the Cambrian was the trilobites—relatives of horseshoe crabs that were covered by ribbed shields. After trilobites emerged during the Cambrian Period, they endured until 252 million years ago. The last trilobite species disappeared at around the same time that about 90% of all other species vanished. (For more on the causes and effects of mass extinctions, see page 232.)

Climbing Ashore

Along with the rise of multicellular life, another major transition documented in the fossil record is the transition of life from the ocean to land. As life evolved in the sea, dry land remained bare. The earliest hints of terrestrial life come from microbes. In South African rocks dating to 2.6 billion years ago, scientists have found remains of microbial mats that grew on land. Animals, plants, and fungi did not arrive on land until much later. In Oman, scientists have found 475-million-year-old fossils of spores embedded in plant tissues—the oldest plant fossils found so far. The earliest land plants resemble mosses and liverworts. Over the next 100 million years, fossils show that plants began to establish larger and larger ecosystems on land, until full-blown forests were growing.

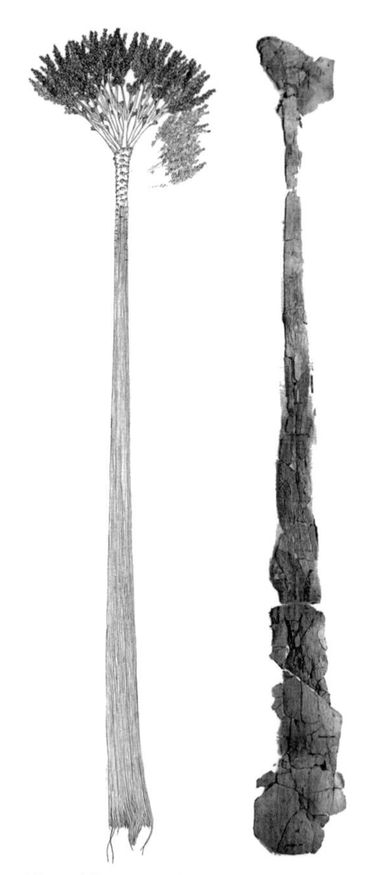

The oldest tree, known as *Wattieza*, was a 385-million-year-old plant that stood 8 meters (26 feet) tall.

Today, plants live in intimate association with fungi. Some fungi feed on dead plants, helping to convert them into soil. Others cause diseases in plants, such as chestnut blight, which wiped out almost all American chestnut trees in the twentieth century. Still others help plants, supplying nutrients to their roots in exchange for organic carbon that the plants create in photosynthesis. The oldest fungus fossils, which date back to 400 million years ago, belong to this last category. Their fossils are mingled with the fossils of plants. It appears that fungi and plants helped each other move from water onto land. (See Chapter 11 for more on how different species form intimate partnerships.)

Animals left only tentative marks on the land at first. In rocks dating back to about 480 million years ago, there are tracks that appear to have been made by invertebrate animals—probably ancient relatives of insects and spiders. The tracks were made on a beach dune; whether the animal that made them could actually have lived full-time on land is a mystery. The fossil of the oldest known truly terrestrial animal is more than 50 million years younger than those trackways. It's a 428-million-year-old relative of today's millipedes, found in Scotland in 2004 by a bus driver who hunts for fossils in his free time. Younger fossils mark the oldest evidence of other invertebrates on land. But it was not until about 350 million years ago that vertebrates moved ashore as well. The oldest land vertebrates (known as tetrapods, meaning "having four feet") were similar in some ways to living salamanders.

The oldest fossil of a land animal belonged to a 428-million-year-old millipede known as *Pneumodesmus newmani*.

Between about 370 and 350 million years ago, one lineage of vertebrates made the transition from water to land. *Silvanerpeton*, shown here, was one of the oldest known terrestrial vertebrates (known as tetrapods).

Recent Arrivals

One of the most important lessons from the fossil record is that some of the most familiar kinds of life today only emerged relatively recently. Most species of fish on Earth today, for example, belong to a group known as the teleosts. They

Mammals are descended from sprawling, reptile-like vertebrates called synapsids that first emerged 320 million years ago.

Dinosaurs evolved about 230 million years ago and were the dominant land vertebrates until 66 million years ago. One lineage of dinosaurs, the birds, still exists today.

include many of the most familiar fish, including tuna, salmon, and goldfish. But 350 million years ago there were no teleosts at all. Likewise, 350 million years ago there were no mammals, which today are the dominant vertebrates on land. Some 15,000 species of birds fly overhead, but not a single bird existed 350 million years ago.

The oldest insect fossils are 400 million years old. But many of the largest groups of living insect species evolved much later. The first flies, for example, evolved about 250 million years ago. This fly (a gall midge) was trapped in amber about 30 million years ago.

Before today's most common groups of species emerged, the planet was dominated by other groups. Before the rise of the fishes, for example, some of the ocean's top predators were giant sea scorpions, which measured up to six feet long. On land, 280 million years ago, the dominant vertebrates were relatives of today's mammals: ungainly, sprawling creatures called synapsids. The first synapsids that evolved into something that looked even remotely like today's mammals emerged about 200 million years ago. It was not until about 150 million years ago that the first members of the living groups of mammals evolved.

Meanwhile, the new lineges of reptiles were also evolving. One of the most successful was the dinosaur branch. Dinosaurs emerged about 230 million years ago and steadily grew more diverse. Their ranks included giant long-necked sauropods that were the largest animals ever to walk the Earth, as well as fearsome predators. Dinosaurs dominated ecosystems on land until they suffered a series of extinctions and finally disappeared about 66 million years ago. The only

The oldest known fossil that's more like humans than like other apes is *Sahelanthropus*. Discovered in the Sahara desert in 2001, it is estimated to have lived seven million years ago.

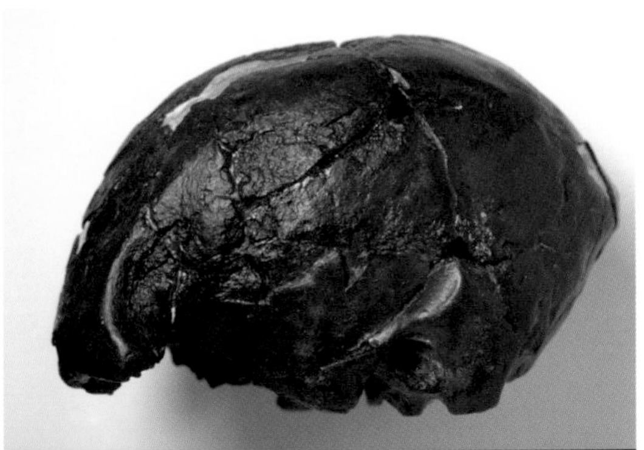
The oldest known fossil of our own species, discovered in Ethiopia, is estimated to be under 200,000 years old.

survivors of this lineage today are the birds, which branched off from other dinosaurs about 150 million years ago (page 71).

Most of the plants we see around us today are also relatively new in the history of life. There were no flowering plants 200 million years ago, nor were there any grasses. At the time, the dominant plants included relatives of living ferns and gingko trees. The oldest fossils of flowering plants date back to only 132 million years ago. The oldest fossils of grasses are tiny bits of tissue found in the 70-million-year-old droppings of dinosaurs. For the next 50 million years, the grasses remained rare. Only about 20 million years ago did extensive grasslands begin to evolve.

Sometime before the big dinosaurs became extinct 66 million years ago, the modern groups of mammals emerged. But it was only after the dinosaurs were gone that mammals began to evolve into dramatically new forms. Starting around 50 million years ago, whales evolved from land mammals into the ocean's top predators. At about the same time, bats evolved into the only flying mammals. And from rocks of the same age, paleontologists find the oldest fossils of small mammals with forward facing eyes and dexterous hands—members of our own lineage, the primates. Our closest living primate relatives are chimpanzees and the other great apes, with which we share a common ancestor. More closely related apes—now extinct—are known as hominids. The oldest fossils of hominids include *Sahelanthropus*, which was discovered in 2001 in the African country of Chad and dates back about seven million years. The first hominids were a long way from human—they had tiny brains, for one thing, and their bodies were about the size of those of chimpanzees. The oldest fossils of hominids matching our own stature only emerged about two million years ago; and the oldest fossils that clearly belong to our own species, found in Ethiopia, are estimated to be nearly 200,000 years old (page 42).

Two hundred thousand years may be an inconceivable span of time to us, but it occupies a tiny stretch of the 3.5 billion or more years that life has existed on Earth. If you shrank that time down to a year, our species would have emerged at about 11:30 p.m. on New Year's Eve.

The geological record of fossils, biomarkers, isotopes, and other traces of past life is clear evidence that life on Earth is immensely old. It also documents profound transformations from a planet inhabited solely by microbes to one they share with millions of species of animals and plants. To study the history of life, scientists do not simply catalogue lists of bones and stromatolites. They also figure out how different species—either alive today or long extinct—are related to one another. By determining their relationships, scientists can form hypotheses about the processes and patterns of evolution. How they discover life's kinship is the subject of the next chapter.

TO SUM UP...

- Geologists use the breakdown of radioactive isotopes to estimate the age of rocks.
- Organisms only rarely become fossils.
- Biomarkers, such as molecules from cell walls, can be preserved for hundreds of millions of years.
- The ratio of isotopes in fossils can give hints about the diets and ecology of extinct species.
- The oldest proposed evidence of life is about 3.7 billion years old.
- Stromatolites and other fossils of microbes date back about 3.5 billion years.
- The three main branches of life are bacteria, archaea, and eukaryotes.
- Multicellular eukaryote fossils date back as far as 1.6 billion years ago.
- Biomarkers of animals date back as far as 650 million years ago.
- The Ediacaran fauna is a puzzling collection of animals that existed between about 575 and 535 million years ago.
- Some of the first members of living groups of animals appeared during the Ediacaran Period, and more appeared during the Cambrian Period.
- Plant fossils date back 475 million years. Invertebrate animals may have walked on land by then.
- The oldest fossils of vertebrates with four legs (tetrapods) date back about 360 million years.
- The oldest known fossils of animals that looked much like living mammals are 200 million years old.
- The oldest fossils of our own species are about 200,000 years old.
- Paleontologists can test their predictions about fossils against new evidence as it is discovered.

The Tree of Life

4

Neil Shubin spends the school year at the University of Chicago, where he teaches paleontology and anatomy. But his summers have frequently taken him north of the Arctic Circle, to a barren patch of land called Ellesmere Island. It's a harsh, dangerous place, with so many hungry polar bears that Shubin and his colleagues all carry shotguns wherever they go. They spend so much time scanning the horizon for bears that their eyes play tricks on them. The scientists once saw what looked like a

Left: *Tiktaalik*, which lived 375 million years ago, illuminates the transition of our ancestors from water to land. Although it still lived underwater, it had evolved some of the traits found in all land vertebrates, such as wrists and a neck. Above: Paleontologists Neil Shubin (top right) and Ted Daeschler are part of a team that traveled to the Arctic and discovered a remarkable fossil they named *Tiktaalik*.

distant polar bear and scrambled for their guns, flares, and whistles. It took them a while to realize that the moving white blob was actually an Arctic hare, hopping along just a couple hundred yards away.

Shubin and his colleagues were drawn to this harsh, dangerous place because they wanted to find clues to one of life's major transitions. The polar bears, the Arctic hares, and Shubin himself are all four-limbed vertebrates, or tetrapods. Shubin wanted to learn about how tetrapods evolved from marine ancestors, and how they went from living underwater to living on land. He decided to search for fossils of extinct species that evolved during this transition—species that might reveal details that scientists could not find in living animals. So Shubin and his colleagues read the scientific literature to figure out the likeliest locations for such fossils. Their research pointed them to a stretch of northern Canada, including Ellesmere Island.

Shubin and his colleagues got to the island for the first time in 1999 and then returned summer after summer. They found fossils, but none of early tetrapods. By 2004, Shubin was wondering if it was time to bring the hunt to an end. But then, while cracking ice off of rocks in a lonely valley, he saw the outline of jaws that looked more like those of a tetrapod than those of a fish. The next day, his colleague Stephen Gatesy, a paleontologist from Brown University, found a second set of similar jaws. Gatesy could see that these bones were connected to a well-preserved skull. The skull was flattened, like the skulls of early tetrapods, and quite unlike the conical heads of fishes.

The scientists spent much of the summer of 2004 slowly excavating the rock in which the bones were lodged. The fossil was flown by helicopter, and then by airplane, to Chicago, where expert fossil preparators carefully picked away the rock, leaving the fossilized bones. The scientists had discovered a large portion of a skeleton of a truly remarkable creature that had lived 375 million years ago. It looked something like a fish with arms, with the flattened head of a salamander. It measured about three feet from its flat head to its swimming tail. It had gills and scales, like a fish, but its front pair of appendages could bend at the elbow and could support the weight of its body.

Shubin and his colleagues dubbed the creature *Tiktaalik roseae* (*Tiktaalik* is the name of a fish in the Inuktitut language of northern Canada, and *roseae* honored one of the people who funded the expeditions to Ellesmere). They announced their discovery in a pair of detailed scientific papers in 2006, and it quickly ended up on the front pages of newspapers around the world.

Shubin and his colleagues have found enough bones of *Tiktaalik* to make a detailed reconstruction of its skeleton.

Tiktaalik is just one of many fossils paleontologists have found in recent decades that are helping them understand evolution's great transitions. To glean clues from these fossils, biologists take advantage of one of Darwin's central insights: evolution can produce new lineages much as a tree grows branches. It was the tree of life that pointed Shubin and his colleagues to Ellesmere, and it is only by understanding *Tiktaalik*'s place on the tree of life that we can appreciate its truly remarkable significance. The methods that Shubin used to find *Tiktaalik*'s place on the tree of life are not unique to the evolution of tetrapods. Scientists use them to understand the evolution of major new adaptations, like eyes and brains. They even use them to figure out where new diseases come from, and what human genes do.

How to Build a Tree

The evolutionary relationships of a group of organisms is known as their phylogeny. Darwin envisioned phylogenies as being like branches on a tree. As a species splits into new lineages, its descendants inherit its traits. In each lineage, those traits may evolve, and as those lineages branch yet again, the traits may evolve even more. If you pick out any three of those descendant species, two will be more closely related to each other than either is to the third. That close relationship is reflected in the traits the two species share that the third one lacks.

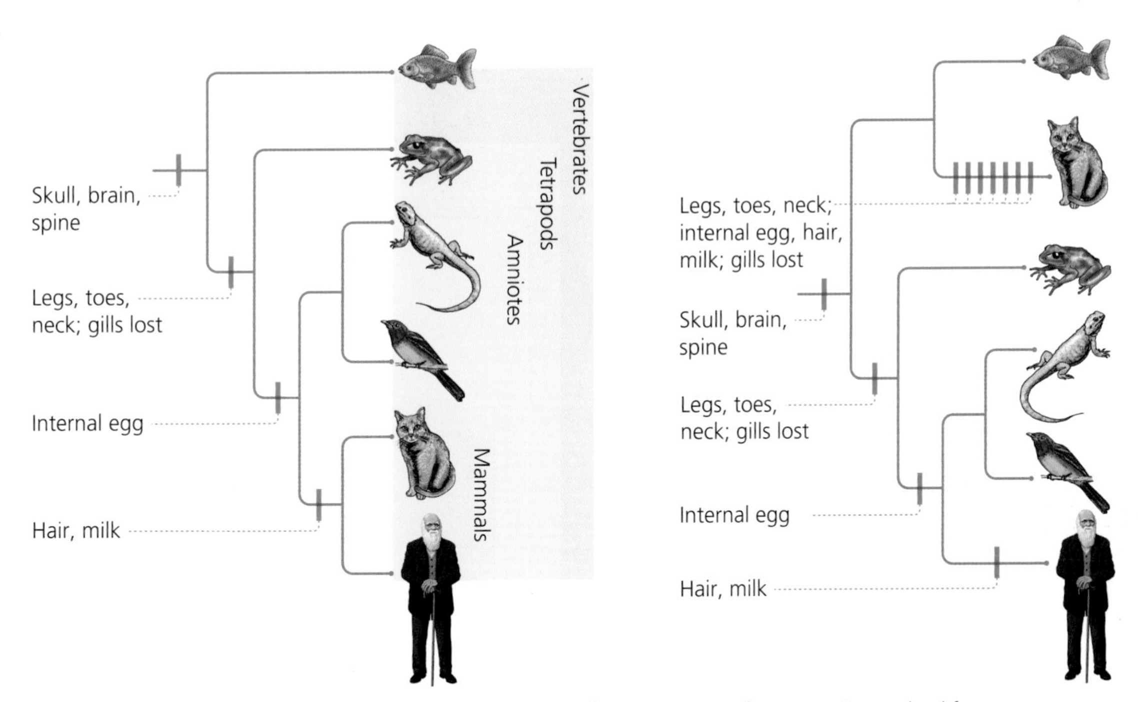

Figure 4.1 Left: This tree is a well-supported hypothesis for how six vertebrate species evolved from a common ancestor. The pink bars show when the ancestors of some of these species first evolved unique traits they share. Biologists use these shared traits to identify related groups of species (such as mammals). While the tree in this figure only includes a few traits found in vertebrates, a vast number of other traits support this tree. Right: Biologists test alternative trees to see how well they can explain the evidence. If cats were more closely related to goldfish than to humans, for example, they would have to have independently evolved a large number of traits.

Facts and Theories in Science

Scientists have to use a strict set of terms to communicate with each other. If a word means different things to two scientists, they can't test each other's explanations for natural phenomena. Many of those terms are technical jargon that can mystify the rest of us—words like *paraphyletic* and *dorsorostral*. Scientists also use words that sound familiar, but they often use them with quite unfamiliar meanings. One of these tricky words is *theory*.

Many people think a theory is simply a hunch: a vague guess based on little evidence. When they hear scientists speak of "the theory of evolution," they assume that it's mere speculation, far less certain than a fact.

But that's not what scientists mean when they speak of a theory. A theory is an overarching set of mechanisms or principles that explain a major aspect of the natural world. A theory makes sense of what would otherwise seem like an arbitrary, mysterious collection of data. And a theory is supported by independent lines of evidence.

Modern science is dominated by theories, from Newton's theory of gravitation, to the theory of plate tectonics, to the germ theory of disease, to Darwin's theory of evolution. Each of these theories came about when scientists surveyed research from experiments and observations and proposed an explanation that accounted for them in a consistent way. Scientists can use theories to generate hypotheses, which they can test with new observations and experiments. The better a scientific theory holds up to this sort of scrutiny, the more likely it to become accepted. At the same time, however, many theories have been revised with the discovery of new evidence.

Theories make sweeping claims. Molecular biologists in the 1950s came up with a theory that all cellular organisms use a molecule called DNA to store genetic information, and

Isaac Newton (1643–1727) developed a theory to explain the motion of objects. Today engineers rely on that theory to build cars, ships, and spacecraft.

Those shared traits were passed down from a more recent ancestor than the common ancestor of all three species.

Scientists can put together hypotheses about how species are related by comparing their traits, and then they can test those hypotheses by analyzing more traits or more species. **Figure 4.1** is a simple illustration of this process, comparing six species of animals: a goldfish, a frog, an iguana, a pigeon, a cat, and a human.

All six species have many traits in common that many other animals lack. For example, they all have skulls made out of a distinctive tissue called bone. They have a spinal column made up of small bones called vertebrae. You won't find skulls or vertebrae in clams or jellyfish. This pattern suggests that all six species share a common ancestor that had already evolved a skull and a spine. It also suggests that these traits evolved after the common ancestor of these animals branched off from other animals that lacked a skull and a spine.

that the structure of other molecules, known as proteins, is encoded by pieces of DNA. (page 87). At the time, molecular biologists had tested this theory on just a handful of species. But, given how consistently DNA turned up in those species, scientists were reasonably confident that every animal, plant, fungus, protozoan, and bacterium on Earth also used DNA as its genetic material.

After five decades, scientists have found this to be the case in hundreds of thousands of species. But there may be more than 10 million species on Earth. The fact that scientists have not tested every possible ramification of this theory has not stopped them from accepting it as a good theory. Instead, they look to its predictive power. Every species that scientists have looked at so far has DNA, and the theory that DNA is the storehouse of genetic information has allowed them to understand important things about how those species work.

A theory makes sense of what would otherwise seem like an arbitrary, mysterious collection of data

The same rules apply to the theory of evolution. The modern theory of evolution still embraces many of Darwin's central insights, such as the mechanism of natural selection and the concept of the tree of life. But the theory has matured, and it now takes a form that Darwin might not entirely recognize. That change has come about as one generation of scientists after another has examined the theory of evolution and has tested aspects of it in specific ways, whether they are measuring natural selection in our species, say, or are running experiments to see how sexual selection influences mating success. Today the theory of evolution is as well supported as any of the other leading theories of modern science, but that does not mean that scientists know all there is to know about evolution. New fossils are discovered every year. The DNA of humans and other species is yielding profound secrets about how evolution works, which scientists are only now beginning to understand. The great expanses scientists have yet to explore do not diminish the importance of the theory of evolution, however. As with other theories, scientists value the theory of evolution for what it has helped them to understand so far. A good theory is like a powerful flashlight helping scientists make their way into the dark.

It's ironic that those who would reject evolution often call it a mere "theory," implying that it's inferior to facts. For scientists, just the opposite is true. A good theory is superior to facts, because it organizes them, changing them from a loose collection of details into a meaningful, well-supported picture.

But some of these species share traits that the others lack. Only cats and humans have hair, for example. In Figure 4.1 the species are grouped according to the unique traits they share.

This sort of diagram is known as a cladogram ("clade" means a group of organisms that share a common ancestor). By comparing the traits in different species, scientists can build hypotheses about how they are related and how their shared traits evolved. The left cladogram in Figure 4.1 is not the only hypothesis for how these six animals are related. There are alternatives. Perhaps, for example, cats are more closely related to goldfish than to humans. Perhaps they evolved hair independently from humans.

To test these two hypotheses, scientists can find new traits to compare. Hair, it turns out, is not the only trait found in cats and humans but not in the other four species. Cats and people also have nipples and are able to produce milk, for example, while the other animals do not. As embryos, cats and people develop a placenta to help them get nutrients from their mothers. If cats were closely related to goldfish, then a long list of other traits would have to have evolved twice. Cats also share certain traits with birds, turtles, and frogs that they don't share with goldfish. For example, they live on land and have legs with digits on them, and each of them has a neck. Goldfish, on the other hand, have no neck, no legs, and no digits. Goldfish have gills rather than lungs. If cats were closely related to goldfish, all these traits would have to have evolved independently as well.

Scientists can also test a hypothesis about phylogeny by comparing more species. If you were to add trout to the animals in Figure 4.1, the simplest cladogram would have trout sharing a close common ancestor with goldfish. All the other relationships shown in the cladogram would not be altered. Another way to test a phylogenetic hypothesis is to use a different kind of evidence. In Chapter 7, we'll see how scientists use DNA to discover the relationships of species—and how DNA confirms many of the hypotheses that scientists originally developed by looking at anatomy.

From Fins to Feet

The evolutionary tree in Figure 4.1 has five tetrapods and one fish. The difference between the two groups is obvious. All the tetrapods have limbs with digits that they use to move around on land. The goldfish, by contrast, has fins it uses to swim underwater. The tissues inside its fin are dramatically different from the ones inside the tetrapod limb as well. Tetrapod limbs have the same basic arrangement of skeletal bones—a long bone close to the body, two more rod-shaped bones farther out, a set of small roundish bones, and finally a set of digits. A goldfish fin consists of a small fan of skeletal bones surrounded by fin rays, an entirely different kind of tissue.

So where did the tetrapod "body plan" come from? That's the question that propelled Neil Shubin and his colleagues to the Arctic Circle. But evolutionary trees guided them there.

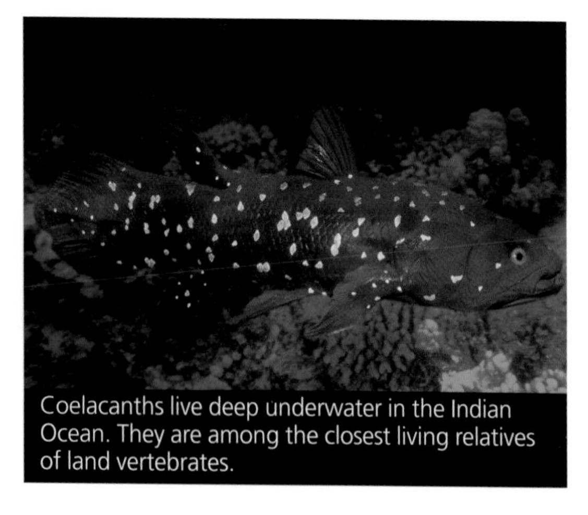

Coelacanths live deep underwater in the Indian Ocean. They are among the closest living relatives of land vertebrates.

Studies on fossils and on the anatomy of living vertebrates has revealed that the closest relatives of tetrapods are a handful of species of aquatic vertebrates. These close relatives include coelacanths, that live in the deep sea off the eastern coast of Africa as well as in the waters around Indonesia. Another group of closely related species, called lungfishes, live in rivers and ponds in Brazil, Africa, and Australia. Instead of the webbed fins of goldfish or salmon, coelacanths and lungfishes have fleshy lobes with stout bones inside. Many biologists refer to the group that includes tetrapods, coelacanths, and lungfishes as lobe-fins.

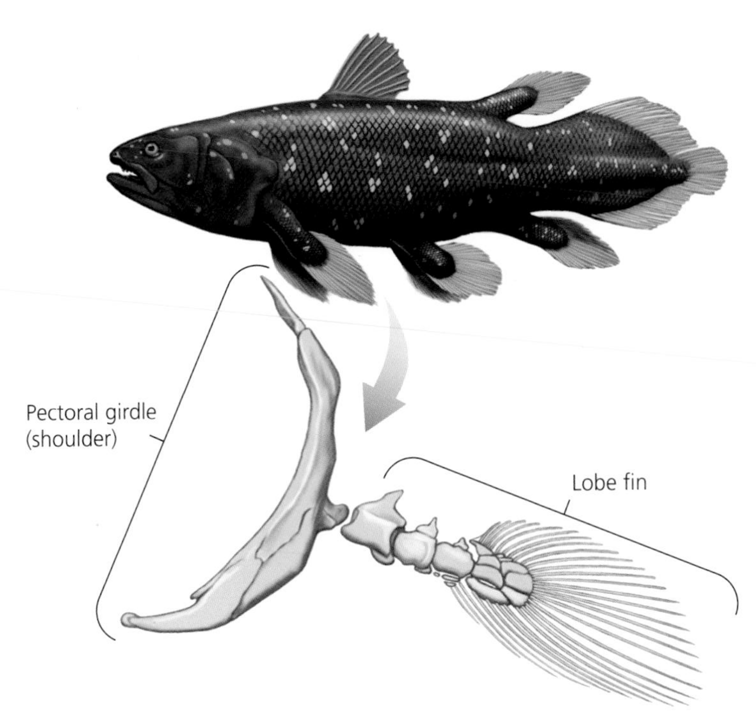

Goldfish and other ray-finned fishes have fins that are made up mostly of slender rays. Coelacanths, on the other hand, have a short chain of skeleton bones that anchor powerful muscles. Our own limbs evolved from a related "lobe fin."

Fossils are essential for understanding how an aquatic lobe-fin evolved into tetrapods like us. Paleontologists have found many fossils of different kinds of full-blown tetrapods in terrestrial rocks dating back over 300 million years. That means the transition must have taken place before then. In the late 1800s, paleontologists began to zero in on the transition, with the discovery of fossils of extinct lobe-fins that were more like tetrapods than they were like lungfishes or coelacanths. *Eusthenopteron*, for example, lived about 385 million years ago. Each of its lobe fins had a stout bone extending from its shoulder girdle, with two more bones extending further out.

Over the course of the 1900s, a few more transitional fossils emerged. In Greenland, for example, Jennifer Clack of the University of Cambridge found the remains of a 365-million-year-old tetrapod called *Acanthostega*. It had complete tetrapod limbs, including eight toes on each foot. But it also showed many signs of living in the water, not on land. Its skeleton was not adapted for supporting its body on land, for example, and it had bones for supporting gills. On a fossil hunt in Pennsylvania in 1996, Neil Shubin and his colleagues found an isolated shoulder bone from an early tetrapod that lived some 360 million years ago. That discovery led Shubin to investigate new places where he might find more complete fossils of early tetrapods.

In particular, Shubin wanted to find a species that was more closely related to humans than *Eusthenopteron*, but not as closely as *Acanthostega*. Such a discovery would shed more light on the transition of fins to limbs. Shubin knew that all of the fossils marking this transition dated from the mid-Devonian Period, from about 370 to 350 million years ago. So he narrowed down his search to rocks that had formed at the same time. Shubin also knew that early tetrapods and their closest lobe-fin relatives had lived in coastal wetlands and river deltas. So he narrowed his search further, to sedimentary rocks that had formed in those environments.

It turned out that northern Canada had a number of geological formations that fit Shubin's bill. Better still, no paleontological expedition had ever searched for tetrapod fossils in those formations before. So Shubin and his colleagues headed north.

Just as Shubin had hoped, those rocks revealed a species that was more tetrapod-like than *Eusthenopteron*, but less so than *Acanthostega*. Not only did it have long limb bones, but it had small bones corresponding to those in our wrists (**Figure 4.2**). It also had a neck—something that lobe-fins lack—but it did not have toes, as *Acanthostega* did. The discovery of *Tiktaalik* shows how paleontologists can make predictions from hypotheses, even if they're studying events that took place hundreds of millions of years ago.

Figure 4.3 shows the relationship of *Tiktaalik* to other lobe-fins and early tetrapods. By looking at this evolutionary tree, we can see some of the stages by which the tetrapod body plan arose.

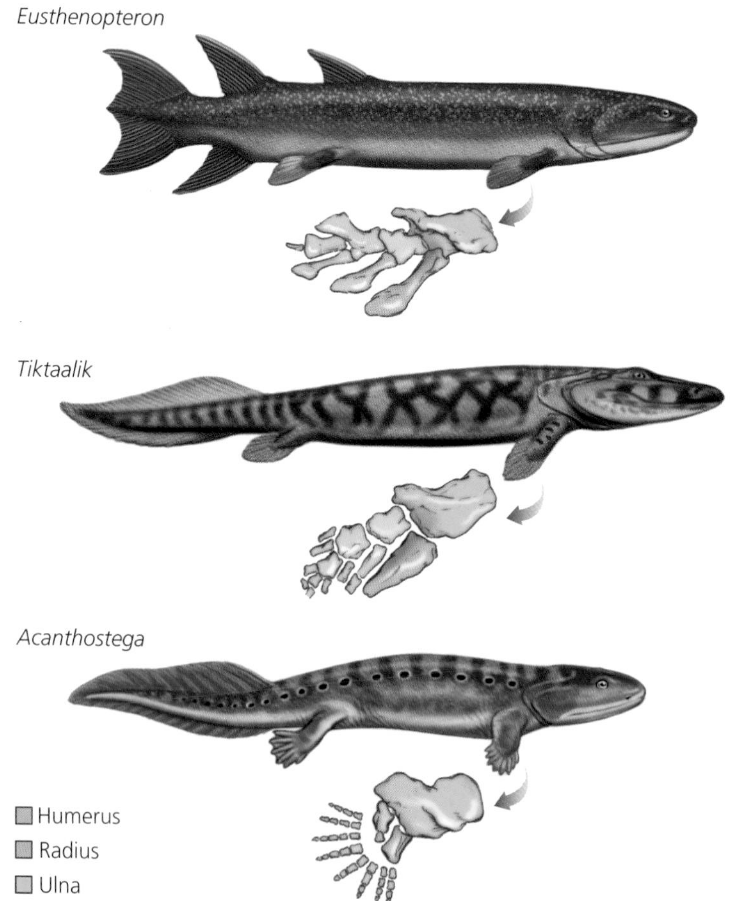

Figure 4.2 Fossils of lobe-fins and early tetrapods reveal the homologies in their limb bones. *Eusthenopteron* had bones that were homologous to the long bones (the humerus, radius, and ulna) of our arms. *Tiktaalik* shared more homologies, including wrist bones. *Acanthostega*, an early tetrapod, had distinct digits at the ends of its limbs. While all tetrapods today have only five or fewer digits, *Acanthostega* had eight. (Adapted from Friedman and Coates, 2008)

Figure 4.3 Facing page: This tree shows the relationship of lobe-fins to tetrapods, and how new tetrapod traits evolved over time. The tetrapod "body plan" evolved gradually, over perhaps 20 million years. The earliest tetrapods probably still lived mainly underwater. This tree includes only a few representative species; paleontologists have discovered many others that provide even more detail about this transition from sea to land.

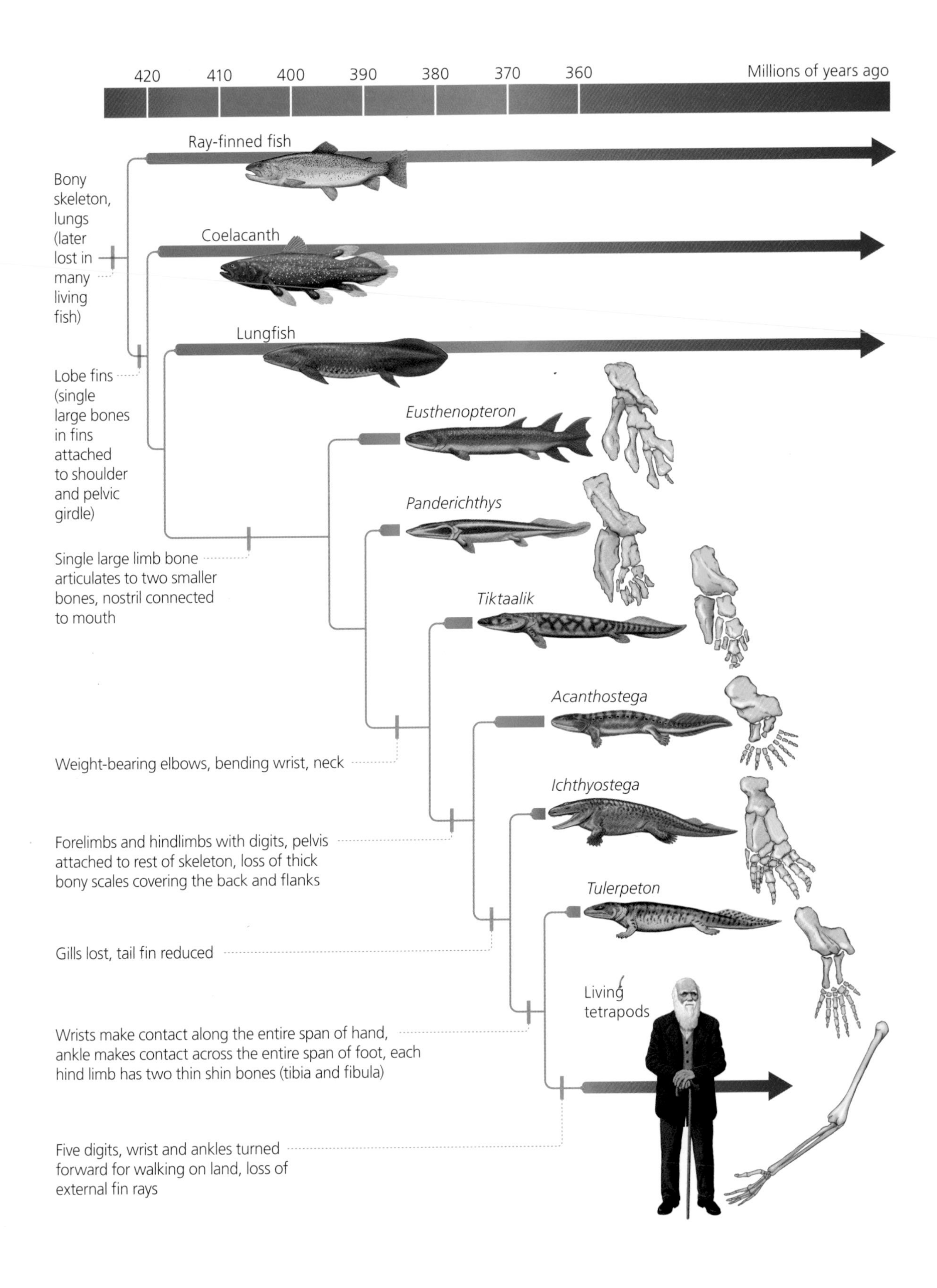

420
410
400
390
380
370
360
Millions of years ago
Ray-finned fish
Coelacanth
Lungfish
Eusthenopteron
Panderichthys
Tiktaalik
Acanthostega
Ichthyostega
Tulerpeton
Living tetrapods
Bony skeleton, lungs (later lost in many living fish)
Lobe fins (single large bones in fins attached to shoulder and pelvic girdle)
Single large limb bone articulates to two smaller bones, nostril connected to mouth
Weight-bearing elbows, bending wrist, neck
Forelimbs and hindlimbs with digits, pelvis attached to rest of skeleton, loss of thick bony scales covering the back and flanks
Gills lost, tail fin reduced
Wrists make contact along the entire span of hand, ankle makes contact across the entire span of foot, each hind limb has two thin shin bones (tibia and fibula)
Five digits, wrist and ankles turned forward for walking on land, loss of external fin rays

When you look at this illustration, bear in mind that these species do not form a continuous line of ancestors and descendants. *Tiktaalik* has a number of odd features that are not seen in other lobe-fins and that tetrapods do not share. It probably evolved these peculiar traits after its ancestors branched off from other lobe-fins. Nevertheless, this tree shows us things about the evolution of tetrapods that we'd never know if not for fossils.

For example, living lobe-fins aren't very good guides to what our fish ancestors looked like and what they did with their fins. The common ancestor of tetrapods and their closest relatives evolved stout, paddle-shaped fins. *Tiktaalik* could probably have done a push-up, judging from its bones and the attachments for muscles. Species that branched off after *Tiktaalik* show the further evolution of lobe-fins. By the time *Acanthostega* evolved, for example, this lineage had evolved not just the long bones of the legs and wrist bones, but also full-fledged digits. *Acanthostega* had eight; other species had six or seven. All living land vertebrates have five or less. Only fossils can show us that this five-finger rule took millions of years to emerge.

This evolutionary tree also provides clues to the environment in which legs and feet first evolved. *Tiktaalik*, like other lobe-fins of its time, lived in shallow coastal waters. At the time, the first large forests were forming and creating a rich supply of organic matter in these waters—the first wetlands, in other words. *Tiktaalik* may have used its fins to move along the bottom of the wetlands. It may also have used them to lift up its mouth to the surface to breathe. *Tiktaalik*, like other lobe-fins at the time, had gills, but it probably also had lungs. Even after tetrapods like *Acanthostega* had evolved fully formed tetrapod limbs, they still probably lived in the water. *Acanthostega* had bones for supporting gills that it may have used to get oxygen. Its shoulder and pelvic bones were so slender that they probably couldn't have supported its weight on land. *Acanthostega's* tail remained lined with delicate fin rays that would have been damaged if *Acanthostega* had dragged it along the ground. Scientists have hypothesized that *Acanthostega* used its digits to move around underwater, perhaps holding onto underwater vegetation or clambering over submerged rocks.

If we could examine only living tetrapods, we might assume that full-blown limbs with digits had evolved as an adaptation to moving about on land. Thanks to the fossil record, however, we can now see that much of this anatomy had already evolved before the transition from water. It is an adaptation that evolved under one set of conditions and that later served another function.

Evolution as Tinkering

Cats and humans have more traits in common than hair and milk. Their ears, for example, contain a delicate chain of bones that vibrate when they are struck by sound waves and that transmit those vibrations to nerve cells. Birds and iguanas, the closest relatives on the tree in Figure 4.1, don't have this arrangement of bones in their ears (**Figure 4.4**). Here again, we meet a complicated adaptation that seems at first to have arisen from nowhere—and, here again, fossils can reveal its origins. The bones we use to hear once helped our ancestors bite.

The oldest fossils in the mammal lineage wouldn't have looked much like today's cats, humans, or any of the other hairy creatures we're familiar with. Instead, these 320-million-year-old tetrapods were sprawling beasts. However, they had certain features in their skeletons—particularly the way in which the bones of the skull fit together—that are found today only among mammals. Together, mammals and these relatives of mammals are known as synapsids.

For their first 100 million years, the synapsids evolved into many different forms. *Dimetrodon* had a strange sail-shaped back. Others looked like turtles with fangs. Others had peg-shaped teeth for grinding up plants. One lineage of synapsids, known as the cynodonts, evolved a more upright stance and other mammal-like traits not found in other synapsids. By about 200 million years ago, the basic mammal body plan seen today had evolved, as seen in fossils of such mammals as *Morganucodon*.

There are three living branches of mammals. Monotremes, which branched off first, include the duck-billed platypus and the echidna. They produce milk, but only through a loose network of glands rather than through a nipple; and, like reptiles and birds, they lay eggs. Living mammals that are not monotremes are known as eutherians. The eutherian lineage has two living branches. One branch, the marsupials, includes the opossum, kangaroos, and koalas. Marsupial young are born live, but are still quite tiny; after they are born, they crawl

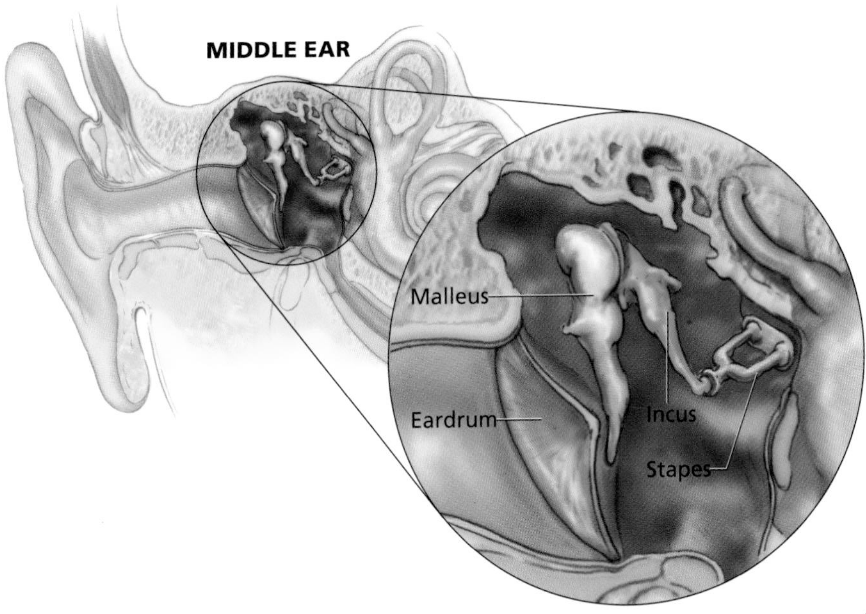

Figure 4.4 Living mammals have three small bones in the middle ear that transmit sound from the eardrum to the inner ear. These bones develop in mammal embryos as part of the lower jaw. As Figure 4.5 shows, middle ear bones evolved from jaw bones.

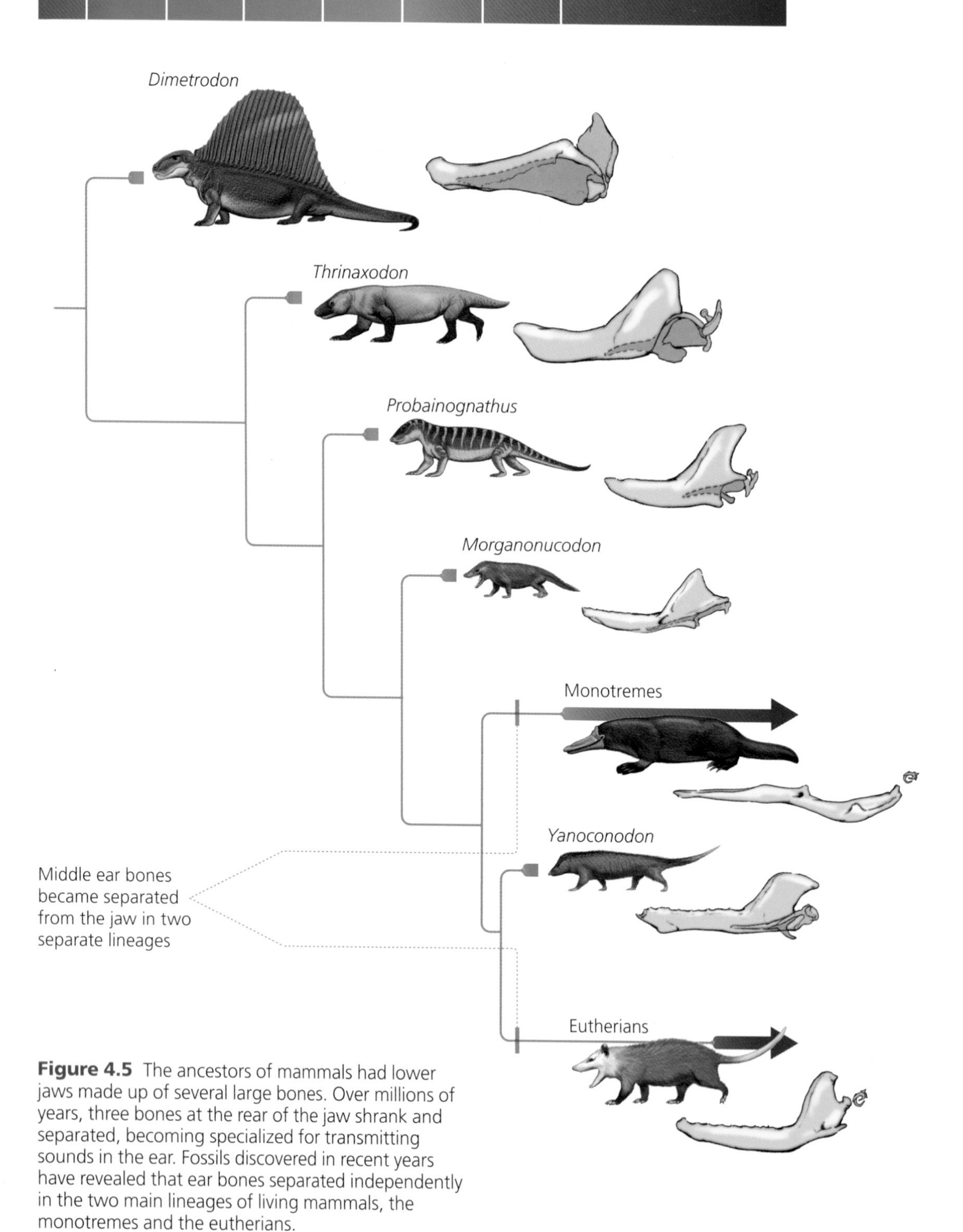

Figure 4.5 The ancestors of mammals had lower jaws made up of several large bones. Over millions of years, three bones at the rear of the jaw shrank and separated, becoming specialized for transmitting sounds in the ear. Fossils discovered in recent years have revealed that ear bones separated independently in the two main lineages of living mammals, the monotremes and the eutherians.

into a pouch on the mother's belly, where they can be carried until they're big enough to survive on their own. Placental mammals make up the other branch of living eutherians. They include us and all other mammals that develop a placenta to feed embryos in the uterus. Fossils indicate that the common ancestor of all living mammals must have lived before 180 million years ago.

In other words, all of the traits found in living mammals did not emerge together. Some traits evolved before others, and they all changed gradually. The ancestors of today's mammals didn't simply switch from a sprawling gait to an upright one. Their fossils reveal other changes. Their noses were connected to cavities that became increasingly big and complex, a sign that their body temperature was rising, allowing them to warm the air as they breathed it in.

The fossil record also documents the emergence of the mammalian ear. **Figure 4.5** shows the ear and the surrounding region in mammals and some of their relatives. It reveals that the bones of the middle ear started out as part of the lower jaw. In early synapsids, the jaw was a collection of interlocking bones. The front-most bone, the dentary, held many teeth, while the bones in the rear formed a hinge against the back of the skull. Like many reptiles today, early synapsids had simple ears. They may have picked up vibrations through their jaws that were then relayed back to the middle ear.

As the ancestors of mammals evolved, the dentary became larger. Its growing size may have been an adaptation to chewing, because a single large bone could provide more strength than a group of smaller bones. As the dentary expanded, the bones in the back of the jaw shrank. At first, they still helped to anchor the lower jaw to the skull, but eventually the dentary took over this job completely.

In time, some of the bones at the rear of the lower jaw disappeared. Others took on a new role. They became part of the system of bones in the ear. These old jaw bones made the mammal ear a better listening organ. They evolved into a series of levers that could amplify faint high-frequency sounds. At first, the fossils show, this chain of bones remained tethered to the lower jaw. But, in the ancestors of the living mammals, it eventually broke free.

Paleontologists have documented many of the steps in this transition with fossils of extinct relatives of today's mammals. But the evidence for this transformation also comes from living mammals. When the bones of the ear first develop in mammal embryos, they are anchored to the lower jaw, as they were in the ancestors of mammals and other tetrapods. Only later do they break free.

Feathered Dinosaurs Take Flight

In 1860, just a year after *The Origin of Species* was published, German quarry workers discovered the fossil of a bird like nothing alive today. It was clearly a bird, with a bird's skeleton and a bird's feathers. (The impressions of the feathers

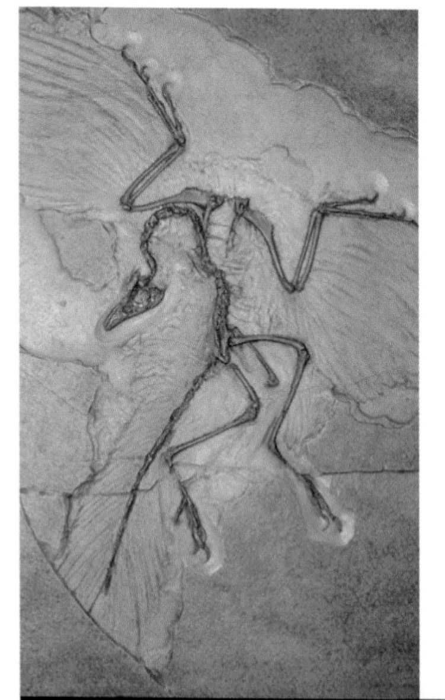

In 1860, just after Darwin published *The Origin of Species*, German quarry workers discovered a fossil of a bird with reptile traits such as teeth and claws on its hands. Known as *Archaeopteryx*, it lived 145 million years ago.

were preserved, thanks to the still, oxygen-free swamp into which the bird fell when it died.) But the bird, which would turn out to be 145 million years old, also had teeth in its beak, claws on its wings, and a long, reptilian tail. Scientists dubbed it *Archaeopteryx* ("ancient wing").

Before the discovery of *Archaeopteryx*, birds seemed profoundly different from all other living things. They share many unique traits, such as feathers and fused arm bones, not found in other tetrapods. *Archaeopteryx* offered clues to how birds had evolved from reptile ancestors. But *Archaeopteryx* alone left many questions unanswered. Did feathers evolve first, before flight? Which reptile ancestor did birds evolve from? For a century, those questions remained very much open. But over the past 40 years, thanks to new fossil discoveries and careful comparisons of fossils and living birds, a consensus has emerged. Birds, paleontologists now agree, are dinosaurs.

Figure 4.6 shows how birds are related to dinosaurs and other reptiles. In the 1970s and 1980s, paleontologists observed that the skeletons of birds share many traits with those of one group of dinosaurs in particular, a group known as the theropods. Theropods were bipedal meat-eating dinosaurs whose ranks included *Tyrannosaurus rex* and *Velociraptor*.

But, in the late 1990s, a new trait emerged to link theropods and birds: feathers. New fossils, mostly found in China, revealed ground-running theropods

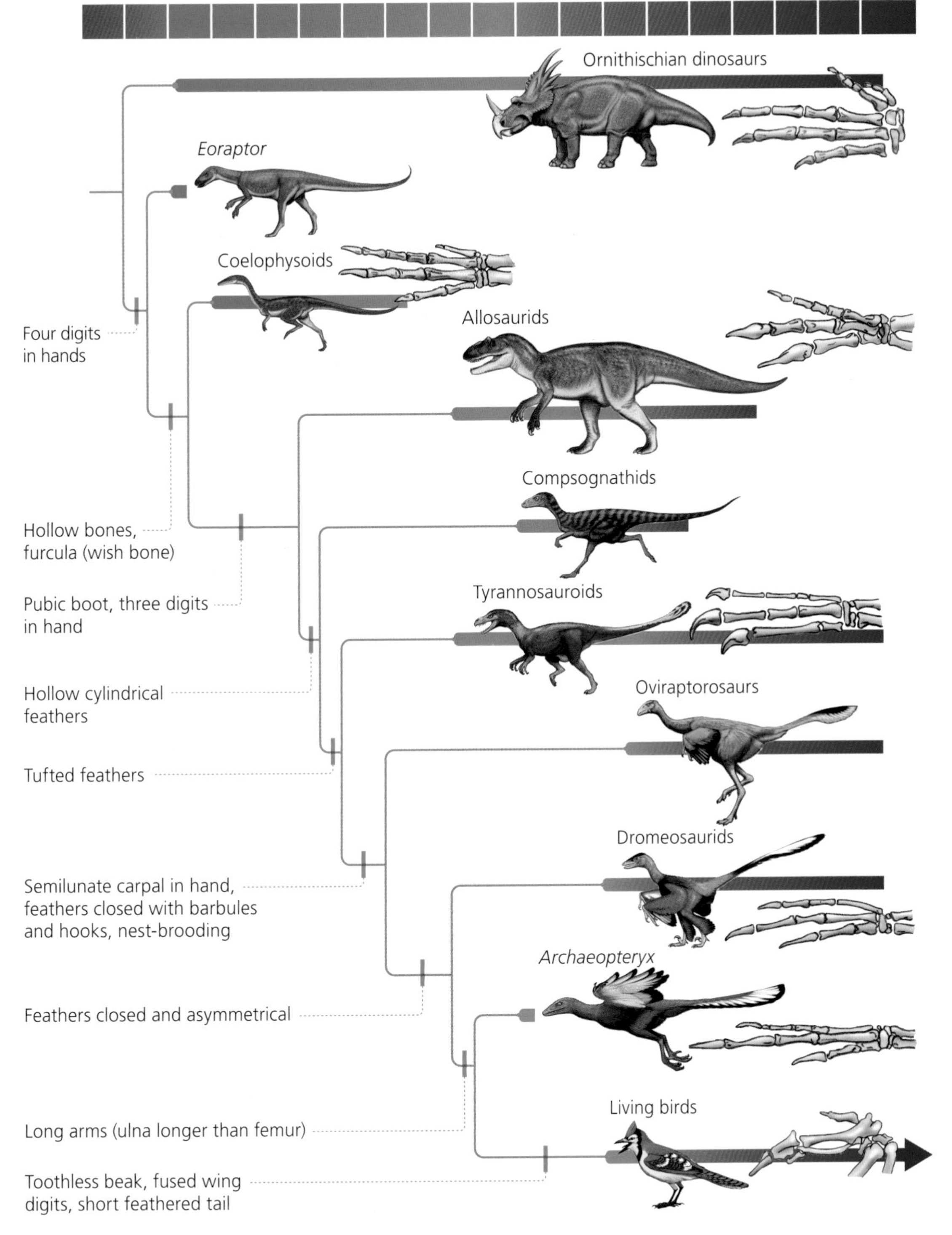

Figure 4.6 Birds evolved from feathered dinosaurs. Long before birds could fly, increasingly complex feathers evolved, as well as other traits found today only in birds.

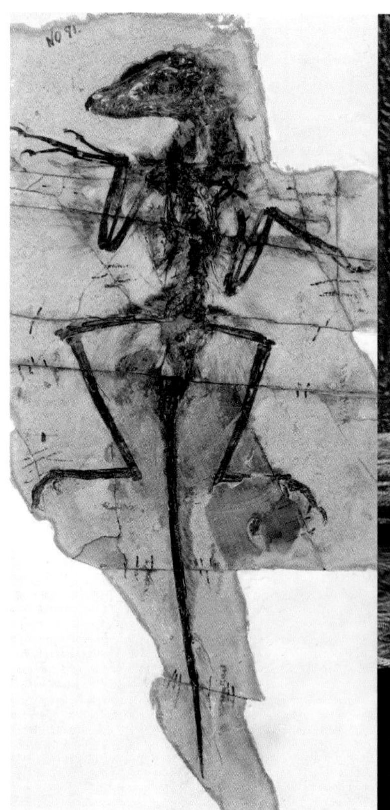

Paleontologists have discovered a number of fossils of dinosaurs with feathers, such as this 130-million-year-old fossil of *Sinornithosaurus* (left). This reconstruction (right) shows what paleontologists think *Sinornithosaurus* looked like in life. It could not fly; instead, it may have used feathers to stay warm or attract mates.

that were covered with strange outgrowths. Close analysis revealed features on them that are found today only on feathers, such as vanes, barbs, and a central stalk. Not only did the feathers have the same structures as feathers on birds, but one fossil of a dinosaur even had raised bumps, just like the quill nodes on bird bones where feathers are anchored. Closer relatives of birds had many types of feathers, while more distantly related theropods only had simple tufts.

The evolution of the traits of birds began long before birds existed, in other words. Early theropods obviously couldn't have used feathers for flight, because their arms were too short and their feathers couldn't lift them off the ground. But birds also use feathers for other things, such as insulation and attracting mates, hiding from predators, and even insulating eggs. This last possibility is particularly striking when you consider a fossil discovered in 1993. A theropod known as an oviraptor lay over its nest of eggs, its arms spread out in much the same way modern birds spread their wings.

Knowing that these feathered dinosaurs are the closest known relatives of birds, scientists can study them to make hypotheses about how flight evolved. Dinosaurs such as *Deinonychus* had slender, clawed arms that could swing out and down, probably in order to catch prey. But studies on living birds hint that this kind of movement may have helped dinosaurs to run faster. By flapping primitive feathered arms, these dinosaurs could generate a small amount of thrust. Small

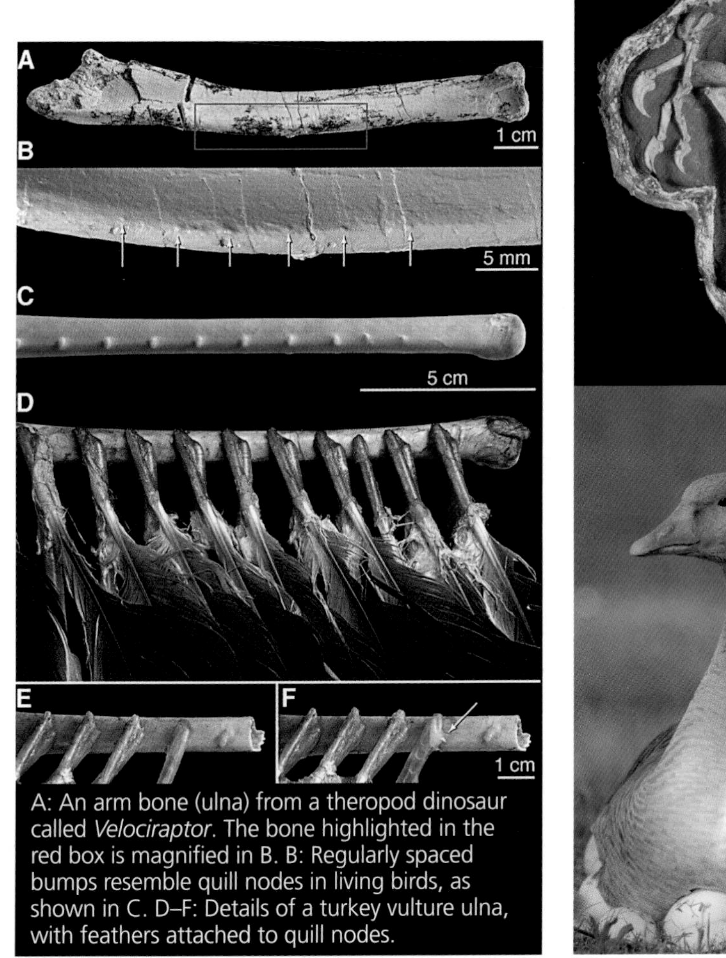

A: An arm bone (ulna) from a theropod dinosaur called *Velociraptor*. The bone highlighted in the red box is magnified in B. B: Regularly spaced bumps resemble quill nodes in living birds, as shown in C. D–F: Details of a turkey vulture ulna, with feathers attached to quill nodes.

Birds sit on their eggs to keep them warm as they develop. In Mongolia, paleontologists have discovered nesting fossils of dinosaurs in a similar posture. This discovery suggests the nesting behavior seen today in living birds evolved over 150 million years ago in non-flying feathered dinosaurs.

feathered dinosaurs may have been able to have run up nearly vertical surfaces to escape predators, just as many baby birds can today. In one lineage of small, feathered dinosaurs, this speed-boosting flapping evolved into true flight.

A New Ape

When Charles Darwin wrote *The Origin of Species*, he wrote next to nothing about the evolution of humans. "Light will be thrown on the origin of man and his history" was about all he had to say. Privately, Darwin wrote to a friend that he did not want to delve too deeply into human evolution because it would prejudice readers against his theory. It was one thing to say that the finches of the

Galápagos Islands had evolved into new species from a common ancestor. It was quite another to say that humans were related to apes.

But Darwin had much to say on the topic of human evolution, and in 1871 he felt the time was right to say it, in a book entitled *The Descent of Man*. There, Darwin observed that human anatomy contained much evidence of its evolution. We had vestigial features on our bodies, for example, such as a stump of a tail (**Figure 4.7**). Some babies were even born with the stump of a tail emerging from their backs. The closest living relatives to humans, Darwin argued, were the African apes. (Darwin knew of gorillas and chimpanzees, but today scientists recognize a third African species, the bonobos, which diverged from chimpanzees about three million years ago.) Darwin noted many similarities between humans and apes, from the details of our skeletons to our similar facial expressions.

At some point after our ancestors split from those of other apes, Darwin argued, they must have evolved to walk upright. But Darwin had no evidence on which to test this hypothesis. Only one fossil hominid had been discovered at the time—what came to be known as the Neanderthal. And it wasn't even clear if Neanderthals were a separate species or just a population of humans with some unusual features, such as thick brow ridges. (We'll return to the mysterious Neanderthals later—see pages 205 and 325.) So Darwin's hypothesis made a prediction: In the future, paleontologists might find fossils of species that had evolved some of the traits found today only in humans while also retaining some of the traits shared by other apes but that have been lost in humans.

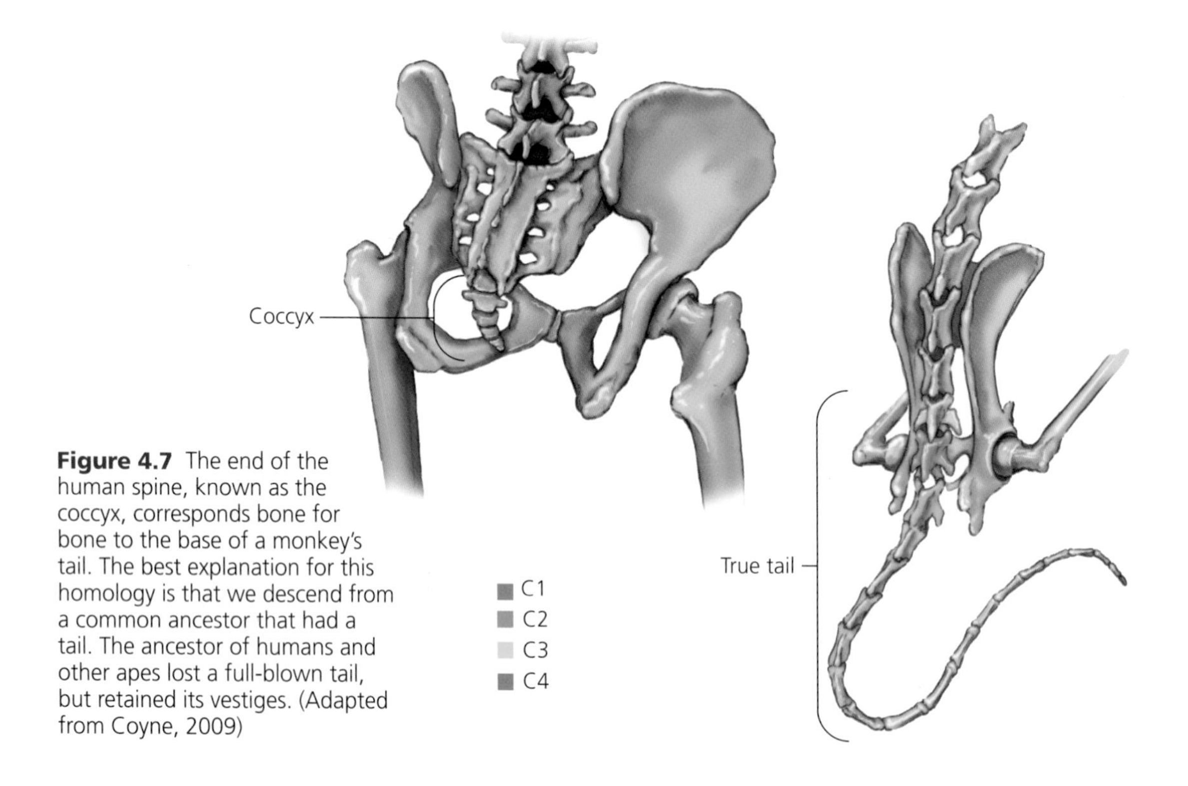

Figure 4.7 The end of the human spine, known as the coccyx, corresponds bone for bone to the base of a monkey's tail. The best explanation for this homology is that we descend from a common ancestor that had a tail. The ancestor of humans and other apes lost a full-blown tail, but retained its vestiges. (Adapted from Coyne, 2009)

By the early 1900s, paleontologists were finding fossils that met that prediction. In 1924, a South African physician named Raymond Dart identified the skull of a child with forward-facing eyes and a small jaw, much like humans have. But the skull also had many traits that linked it with apes, such as a very small braincase. Since then, scientists have identified the fossils of some 20 different kinds of hominids. In many of those cases, paleontologists have found numerous bones from a number of individuals, all of them belonging to the same species. Paleontologists now know enough about these fossils to reconstruct a preliminary phylogeny. The tree in **Figure 4.8** represents their current understanding of the evolution of hominids.

The fact that chimpanzees, bonobos, and gorillas are the closest living relatives of humans offers some evidence about what the first hominids were like. All of these apes have small brains, compared with those of humans, so it's likely the first hominids did too. Since these living apes all walk mainly on both their feet and their hands, it's likely the earliest hominids inherited this way of walking as well.

The earliest fossils of hominids date back to a time between 6 and 7.5 million years ago. By that time hominids had already evolved a few traits that set them apart from other apes. For example, a hominid called *Sahelanthropus tchadensis* had smaller canine teeth than other apes, and its cheek teeth were thicker. Another difference lies in the hole where the spinal cord exits the back of the skull, known as the foramen magnum. In chimpanzees and other apes, this hole is oriented backwards. In *Sahelanthropus* the hole was oriented downwards, much like the human foramen magnum. That suggests that the head of *Sahelanthropus* sat atop its neck like ours, rather than extending forward like a chimpanzee's.

This kind of stance would be consistent with standing on two legs rather than on four. Unfortunately, when paleontologists discovered *Sahelanthropus* in Chad in 2001, they only found parts of its skull and jaw. They have yet to find bones from the rest of its body. But other researchers have found leg bones in Kenya from a six-million-year-old hominid called *Orrorin tugenensis*. And still other researchers have found the remains of a hominid called *Ardipithecus* in Ethiopia dating back 4.5 to 4.8 million years. These hominids also share some key traits with humans, such as small canines and bigger cheek teeth. And their leg bones hint that they walked upright. *Orrorin's* thigh bone (the femur) had a ball at its top that was oriented much like the ball on the femur of a modern human. It's possible that this was an adaptation for bearing the weight of an upright upper body.

To find more clues to the anatomy of early hominids, paleontologists turn to more complete fossils of hominids that lived between four million and two million years ago. Dart's small-brained child belonged to one of these species, known as *Australopithecus africanus*. (*Australopithecus* means "ape of the south.") *A. africanus* and other hominids from the same period still had some chimplike traits, such as a long snout and a small brain. Weighing between 25 and 50 kilograms, they were about the weight of a chimpanzee, too. The males were much taller than the females, which is a common pattern in mammals in which males fight each other for the opportunity to mate with females. Hominids such as *A.*

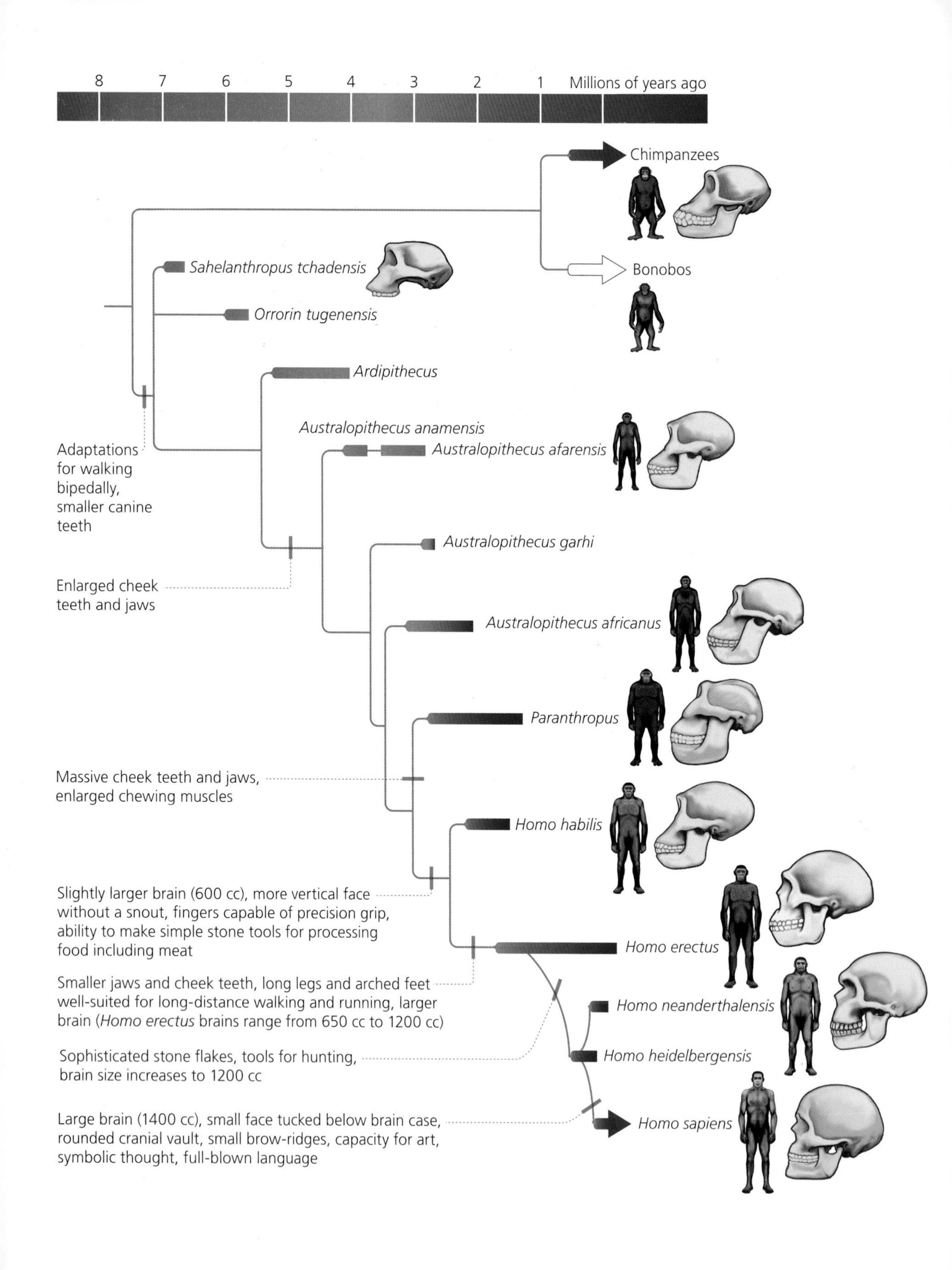

8
7
6
5
4
3
2
1
Millions of years ago
Chimpanzees
Bonobos
Sahelanthropus tchadensis
Orrorin tugenensis
Ardipithecus
Australopithecus anamensis
Australopithecus afarensis
Adaptations for walking bipedally, smaller canine teeth
Australopithecus garhi
Enlarged cheek teeth and jaws
Australopithecus africanus
Paranthropus
Massive cheek teeth and jaws, enlarged chewing muscles
Homo habilis
Slightly larger brain (600 cc), more vertical face without a snout, fingers capable of precision grip, ability to make simple stone tools for processing food including meat
Homo erectus
Smaller jaws and cheek teeth, long legs and arched feet well-suited for long-distance walking and running, larger brain (Homo erectus brains range from 650 cc to 1200 cc)
Homo neanderthalensis
Homo heidelbergensis
Sophisticated stone flakes, tools for hunting, brain size increases to 1200 cc
Homo sapiens
Large brain (1400 cc), small face tucked below brain case, rounded cranial vault, small brow-ridges, capacity for art, symbolic thought, full-blown language

Orrorin tugenensis was a hominid that lived in Kenya 6 million years ago. Its femur, shown here, appears to have been adapted for supporting an upright torso. This suggests that *Orrorin* and other early hominids were walking at least partially upright.

africanus also had traits that a number of paleontologists have interpreted as adaptations for climbing trees. For example, they had long, curved toes and fingers that would have helped them to grasp branches. Their arms were relatively long, and their ankles could rotate more freely than ours.

On the other hand, hominids from this period had many adaptations for bipedal walking that chimpanzees and gorillas lack. Their spines curved so that the upper body sat above the hips rather than extending forward. Their knees were located close to the midline of the body. Their feet bore many traits that are important for walking upright, such as a stout heel and the beginnings of an arch instead of a flat sole. Paleoanthropologists have even found hominid footprints from this period. A bed of volcanic ash that formed in Tanzania 3.6 million years ago preserved the marks of bipedal walkers—probably a species known as *Australopithecus afarensis* that lived in the area at the time.

Why did these hominids evolve to walk upright? A lot of evidence suggests that hominids were shifting to a new environment. Chimpanzees live mainly in forests, where they can pluck fruit from trees. By looking at the plant and animal fossils found alongside the early hominid fossils, paleontologists have determined that they were living in more open woodlands. In this new environment, hominids may have had to have travel farther to find food. Upright walking is more efficient than walking on knuckles, and so bipedal hominids could have saved energy on long walks.

Between three million and two million years ago, a new wave of hominids emerged. Some, such as *Paranthropus*, belonged to a separate branch of the

Figure 4.8 Facing page: This tree shows how humans are related to some species of living apes and extinct hominids. The earliest fossils of hominids suggest that our ancestors were partially bipedal seven million years ago, but this hypothesis is still based on few fossils. More evidence for bipedality emerges in hominids that lived between four and three million years ago. Later hominids evolved large brains and sophisticated tools. As this tree shows, at many times in hominid history, several species of hominid co-existed. (Phylogeny based on Strait, Grine, and Fleagle, 2007)

hominid tree from our own. But others had traits that reveal their close kinship to us. *Homo erectus*, which emerged 1.9 million years ago, had a flatter face than earlier hominids and slender arms and legs. The curved toes and other adaptations for tree climbing were replaced in *Homo erectus* with flat feet and other adaptations for walking—and perhaps running—bipedally. The hominid brain was also evolving: *Homo erectus* had a brain ranging between 600 and 900 cubic centimeters, about twice as big as a chimpanzee's brain. (Our own brain is around 1,400 cubic centimeters.)

This change in anatomy comes along with a new kind of fossil record: stone tools. The oldest stone tools known, which are about 2.5 million years old, were little more than small rocks with chipped edges. Over time, the tools become more sophisticated, including stones that hominids fashioned for chopping or crushing. Paleontologists have found some of these tools near fossilized bones of wildebeest and other large grazing mammals with cut marks on them. Like forensic scientists at a crime scene, paleoanthropologists have used this evidence to generate a hypothesis: the hominids were butchering the animals for their meat.

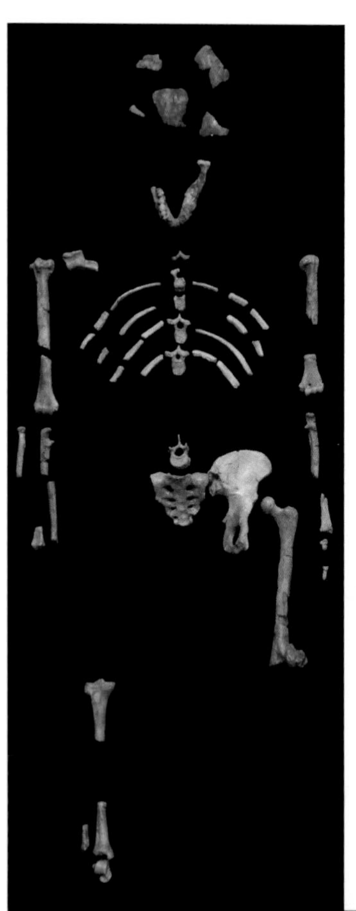

Australopithecus afarensis lived in East Africa between about 4 and 3 million years ago. It was short and small-brained, but could walk upright.

Hominids probably started out as scavengers, but at some point they began to hunt. Harvard paleoanthropologist Daniel Lieberman and University of Utah biologist Dennis Bramble have argued that early species of the genus *Homo* were already hunting, using their long legs and other new adaptations to run long distances to exhaust their prey. It's also possible that this new anatomy was the reason *Homo erectus* was able to become the first hominid to spread out of Africa to Asia and Europe. *Homo erectus* was still alive in Indonesia as recently as 50,000 years ago. And on the Indonesian island of Flores, paleoanthropologists have discovered the bones of tiny, small-brained hominids that lived as recently as 19,000 years ago. Some researchers argue they are a species that evolved from *Homo erectus*, and have dubbed them *Homo floresiensis*. (Others believe they were humans.) But these hominids were our distant cousins, not our ancestors.

A volcanic eruption 3.6 million years ago in Tanzania preserved the footprints of bipedal hominids—probably *Australopithecus afarensis*. In 2009, scientists discovered a 1.5 million-year-old trackway in Kenya that preserved footprints nearly identical to those of living humans.

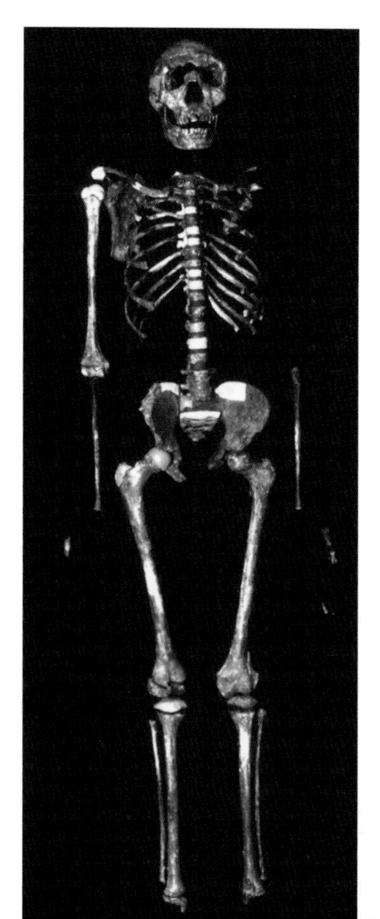

About 2 million years ago, hominids evolved that were taller and more bipedal than their ancestors. Known as *Homo erectus*, this species rapidly spread out of Africa and across Asia.

Clues to our own ancestry come from more humanlike hominids that evolved in Africa and Asia several hundred thousand years ago. *Homo heidelbergensis,* for example, had a higher skull than earlier hominids and a jaw that no longer projected as far forward. *Homo heidelbergensis* may have given rise to the Neanderthals in Europe and Asia about 300,000 years ago. Neanderthals were stocky and still had thick brow ridges, along with a number of other traits that set them off from humans, including bowed arm bones and a slight depression on the back of the skull. But as different as Neanderthals might or might not have been from us, they still left behind an impressive array of stone tools. They also made spears they used to take down big game.

In Africa, meanwhile, *Homo heidelbergensis* may also have given rise to our own species, *Homo sapiens.* The oldest fossils that share enough key traits with living humans to be considered part of our own species come from a site called Omo in southwestern Ethiopia, dating back to about 200,000 years ago. Along with increasingly human-looking fossils, paleontologists are finding increasingly sophisticated tools from this period. By 270,000 years ago African hominids were making finely crafted obsidian blades. But it was only after the emergence of *Homo sapiens* that our lineage began to leave behind symbols, such as elaborate cave paintings.

Fossils, stone tools, and the anatomy of living apes can tell us a lot about human evolution. But humans are different from other animals, and not simply because we can stand upright. We also use language, for example. Words don't fossilize, but there are other ways to study the evolution of language and other aspects of human nature. We'll return to human evolution from time to time in this book to flesh out our history in increasing detail.

These four examples—tetrapods, mammals, birds, and humans—show how scientists use evolutionary trees to get more information out of the fossil record. We can see how complex body plans emerge gradually from older ones. They offer further support for Darwin's argument that homology, in all its guises of adaptation, is the result of common ancestry. But like all insights in science, they also raise new questions of their own. What are the genetic changes that produced the tetrapod body plan, for example, or gave rise to new structures, such as feathers? The answers, as we'll see in the next chapter, lie in the molecules that make heredity possible.

TO SUM UP...

- **A theory is an overarching set of mechanisms or principles that explain a major aspect of the natural world.**
- **A phylogeny is the evolutionary relationships among groups of organisms.**
- **A cladogram is a branching diagram that charts the phylogenetic relationships of groups of organisms.**
- **Scientists reconstruct evolutionary trees to develop and test hypotheses about how major evolutionary transformations took place.**
- **Tetrapods belong to a clade called lobe-fins, which includes coelacanths and lungfishes.**
- **Some tetrapod traits, such as toes and legs, evolved while the ancestors of tetrapods still lived in water.**
- **The bones of the mammal ear evolved from bones of the lower jaw.**
- **Birds evolved from ground-running dinosaurs.**
- **Feathers evolved before flight. Dinosaurs used these early feathers for other functions, perhaps for insulation, courtship display, and nest-brooding.**
- **Human anatomy and behavior have evolved gradually over the past seven million years. Hominids were bipedal millions of years before they evolved large brains.**

Evolution's Molecules

5

At the heart of Charles Darwin's theory of evolution was variation. Because every generation contained variation, some individuals could be selected over others. And because some of that variation could be inherited, species could change over time. Recognizing the importance of variation, Darwin sought to understand it as well as anyone could in the mid-1800s. He

Left: Bao Xishun, shown here with his wife, Xia Shujian, is the tallest human alive today. He stands 2.36 meters (7 feet 9 inches) tall. Above: Joel Hirschhorn of Harvard University and his colleagues study DNA from tens of thousands of individuals to figure out why some people are tall and some are short.

would skin rabbits and line up their bones on his billiard table to measure their lengths. In his study, he pored over a microscope, observing the anatomy of barnacles, noting the variations in size and shape within a single species.

Today, biologists carry on Darwin's tradition, but they now have powerful tools that did not exist in Darwin's day. They can study variation in exquisite detail, and they can also probe the molecules that produce those patterns. At Harvard University, for example, Joel Hirschhorn studies why some people are tall and some are short.

As a species, we humans have a remarkable range of heights. In central Africa, Pygmies never grow taller than 1.5 meters (about 4.5 feet). In Denmark, men grow to an average height of 1.8 meters (5 feet, 11 inches). And the tallest person ever recorded, Robert Wadlow, stood 2.7 meters tall (8 feet, 11 inches) before his death in 1940.

Compared to other traits, height is easy to measure—after all, you just need a single number. But that doesn't mean that the biology underlying height is simple. How tall you are depends on many things. Your experience can add centimeters or take them away—the quality of the food you eat, and even the chemistry of your mother's womb. On the other hand, much of the variation from person to person is hereditary. That is, short parents tend to have short children, and tall children tend to be born to tall parents.

Children tend to look like their parents because they inherit certain molecular instructions for building bodies, known as genes. Hirschhorn and an international network of colleagues are pinpointing some of the genes that influence height. They are sifting through the genes of tens of thousands of individuals, searching for genetic variations that are associated with being taller or shorter than average.

In 2007, Joel Hirschhorn and his colleagues made headlines for a dramatic discovery. For the first time in the history of genetics, they had found a version of a gene present in many people that has a strong effect on height. One version of the gene, named *HMGA2*, can increase a person's height by a centimeter. The following year, Hirschhorn and his colleagues expanded the study by adding genetic information from 2,189 people who were either very tall or very short. They found 11 more genes clearly linked to height. But none of the scientists would claim to have found *the* "height genes." In fact, they now recognize that their search for the genetic influence on height has only begun.

In this chapter we'll investigate genes and their role in building bodies. In the next chapter, we'll see how their role changes over time, making evolution possible.

Proteins, DNA, and RNA

The human body is made up of about a trillion cells, each of which is made up of millions of molecules. Three kinds of molecules are especially important: proteins, DNA, and RNA. Here we'll take a look at each kind.

Proteins

Proteins help to give the body its structure and to carry out many of the chemical reactions that make life possible. There are 100,000 different kinds of proteins in the human body, ranging from keratin, which makes your fingernails hard, to hemoglobin, which carries oxygen in your blood. Some proteins cut apart molecules in the food we eat. Other proteins act as signals, either within a cell or from one cell to another. If you've ever felt a rush on a roller coaster, you were experiencing a flood of adrenaline. It was produced by the adrenal gland cells atop your kidneys, and it spread quickly around your entire body in your bloodstream.

Although proteins come in a staggering number of forms, all of them are made of the same building blocks. These building blocks, known as amino acids, can be joined together end-to-end to form long chains (**Figure 5.1**). There are 20 different amino acids that all living things use to build their proteins, and the particular sequence of amino acids in a protein determines the protein's function. The charges carried by an amino acid attract it to some of the other amino acids in a protein, and repels it from others. As a result of these forces, a protein chain will fold spontaneously into a complex three-dimensional structure. It may fold into a sheet, a cylinder, or some other shape that allows it to carry out its function.

There are an unimaginably huge number of different proteins that can be assembled from 20 different amino acids. So why is it that each of your cells makes only the few proteins it actually needs? Why, for example, don't the cells in your skin make proteins for moose antlers? The answer lies in another kind of molecule in our cells, known as deoxyribonucleic acid, or DNA for short.

DNA

DNA serves as a kind of cookbook for the cell. It stores recipes for each of the cell's proteins as well as many other molecules. Like proteins, DNA is made up of building blocks. But, instead of amino acids, DNA is made up of compounds called nucleotides. One end of a nucleotide links it to other nucleotides to form a backbone for DNA. The other end of a nucleotide, known as a base, helps to store the information necessary to build proteins and other molecules. There are four different bases in DNA: adenine, cytosine, guanine, and thymine (A, C, G,

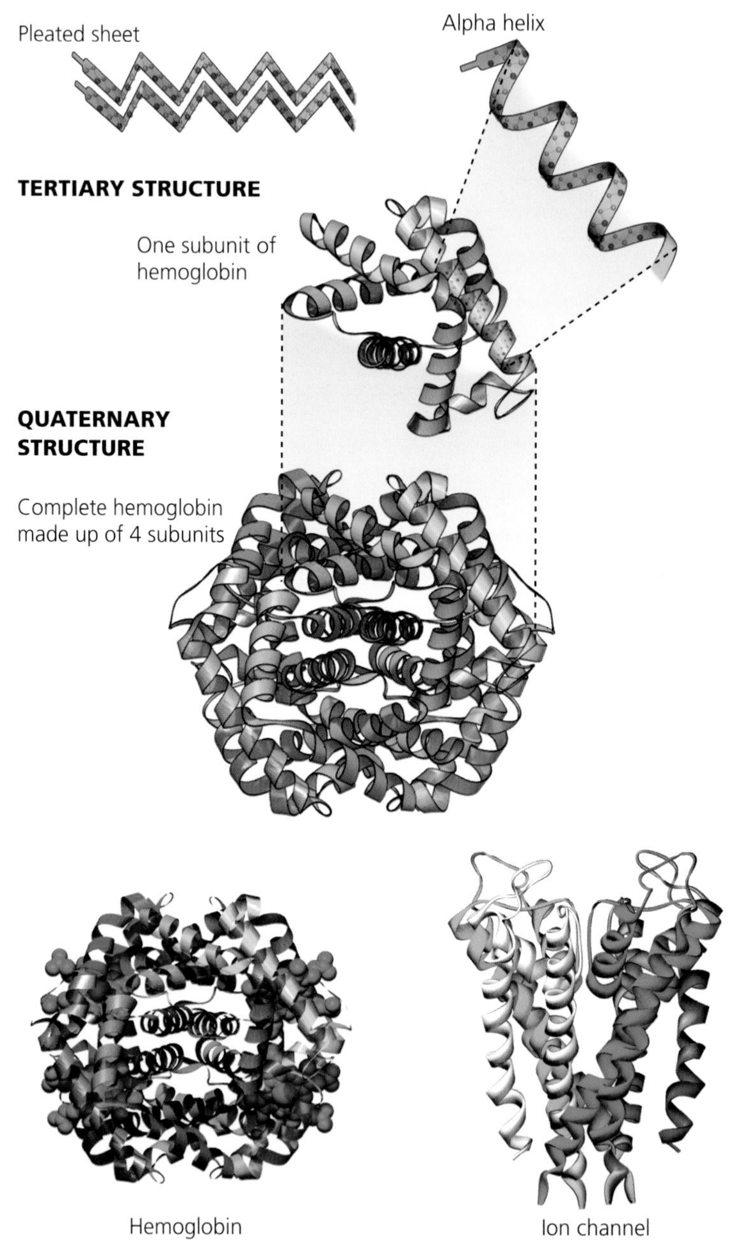

Figure 5.1 Proteins are chains of building blocks known as amino acids. Each protein folds in on itself to produce a new structure, such as the pleated sheet and helix shown here. These proteins can then bend further into more complex shapes or join with other proteins into even larger structures. On the lower left is hemoglobin, a protein that carries oxygen in red blood cells. On the lower right is a tunnel-like protein through which atoms can move in or out of a cell.

and T for short). You can think of these bases as letters that spell out different genetic recipes.

A molecule of DNA consists of two strings of nucleotides that form a double helix. The strings are held together by a weak attraction between the bases. Human DNA contains about 3.5 billion base pairs. If you could stretch out the DNA from a single cell, it would measure two meters long. But our cells keep their DNA in tightly bundled clumps called chromosomes. Almost all human chromosomes belong to nearly identical pairs. In women, all 23 chromosomes form matching pairs. In men, however, 22 pairs match, but the remaining two chromosomes do not. These chromosomes are known as X and Y, for their shapes. (Women, on the other hand, carry two X chromosomes.)

From DNA to Protein

When a cell produces a new protein, it must first untwist certain stretches of its DNA. Segments of DNA that encode proteins are known as genes. An average human gene contains around 20,000 bases, although some genes have as many as 2.3 million bases. To read the gene, a number of proteins land on the DNA at a particular spot near the beginning of the gene's sequence (**Figure 5.2**). The proteins then creep along the gene, assembling a string of nucleotides whose sequence matches that of the template DNA. This process is known as transcription, and the single-stranded molecule that is produced is called messenger ribonucleic acid, or mRNA. In humans and other eukaryotes, stretches of a gene are edited out of the RNA molecule, leaving behind a shorter transcript.

Once the RNA molecule has formed, the cell uses it as a template for building a protein, in a process known as translation. A cluster of proteins and RNA molecules, called a ribosome, grabs onto the newly formed RNA molecule and essentially reads its sequence of bases. The ribosome reads three bases at a time and uses that information to select a particular amino acid to add to the end of a new protein. These trios of bases are known as codons, and all living things use almost precisely the same genetic code to translate codons into amino acids. This common code is evidence that all living things share a common ancestry.

Beyond Protein-Coding Genes

All of the DNA in a human nucleus (the genome) is about 3.5 billion bases long. But of all that DNA, the protein-coding segments make up only 1.2%. The rest of the DNA is a hodge-podge of segments, some of which are essential for our survival and many of which have no known function (**Figure 5.3**).

Some of these segments are transcribed into RNA molecules. But these RNAs are never translated into proteins. Instead, they have their own functions in the cell. For example, some of these RNAs form the core of the ribosome, where they stick new amino acids to the end of a protein. Other RNA molecules can silence genes by binding to RNA molecules that would otherwise be translated into proteins. RNAs can act like switches, turning genes on and off in response to

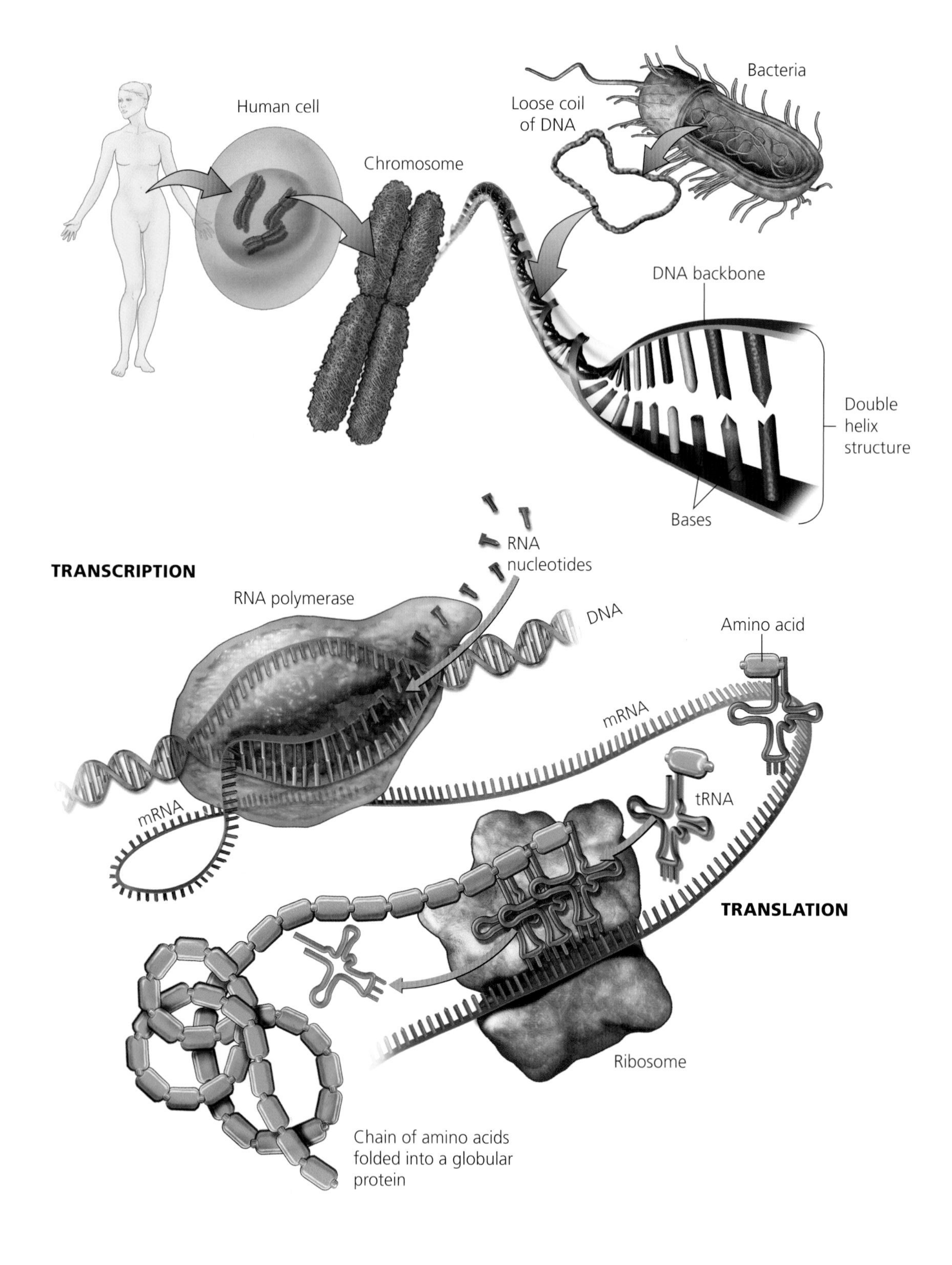
Human cell
Chromosome
Loose coil
of DNA
Bacteria
DNA backbone
Double
helix
structure
Bases
TRANSCRIPTION
RNA
nucleotides
RNA polymerase
DNA
Amino acid
mRNA
mRNA
tRNA
TRANSLATION
Ribosome
Chain of amino acids
folded into a globular
protein

Figure 5.2 Facing page: Top: DNA stores genetic material in all cellular organisms (some viruses use RNA). A pair of twisted backbones are held together by paired bases (A with T and G with C; see page 87). In bacteria, the DNA is typically arranged in a circle. In a human cell, it is wound into chromosomes, which are stored in a sac called the nucleus. Bottom: Each protein is encoded in a stretch of DNA called a gene. A new protein is produced when proteins called RNA polymerases assemble a single-stranded version of the gene, called messenger RNA (mRNA). An mRNA molecule acts as a template for building a protein. Transfer RNA (tRNA) molecules bind amino acids, the building blocks of proteins. A complex called a ribosome joins tRNA molecules to the mRNA according to a genetic code. In the process, amino acids are attached to a growing chain of amino acids that will fold into a protein.

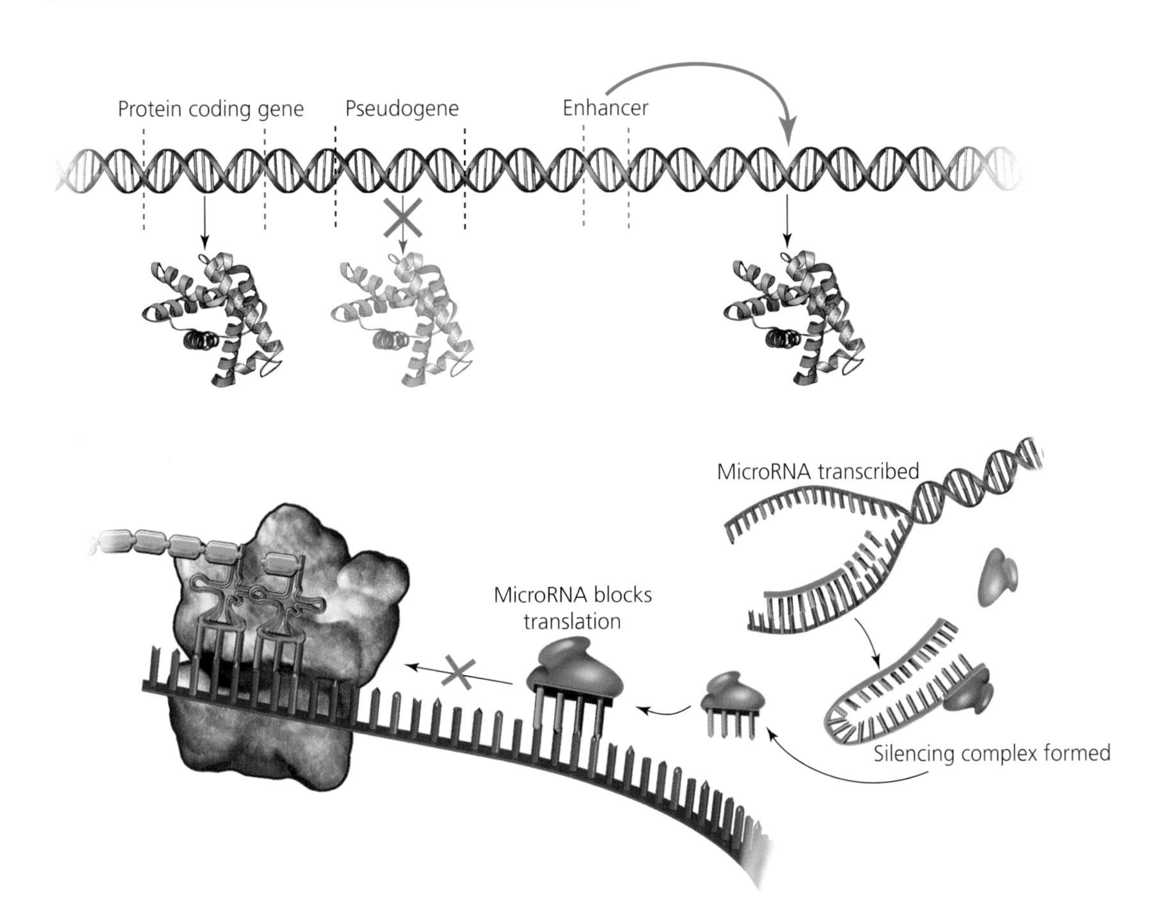

Figure 5.3 Protein-coding genes (upper left) only make up 1.2% of all our DNA. This figure shows a few of the other kinds of elements in the other 98.8% of the genome. Pseudogenes are "dead genes." Millions of years ago they could encode proteins, but genetic changes have since disabled them. Enhancers (top right) are short segments of DNA to which certain proteins must attach before a neighboring gene can be transcribed. They are known as regulatory elements, because they regulate gene expression. Other regulatory elements can shut genes off. Lower right: Some segments of DNA encode RNA that are not translated into proteins. Instead, these RNA molecules carry out protein-like functions. Here, a short segment of micro-RNA is transcribed and joins proteins. Together, this complex attaches to DNA and blocks the expression of another gene.

changes in the environment. Some of them help to coordinate the development of embryos. In order for a human embryo to develop different tissues and organs, for example, certain genes must make proteins in certain cells, while other genes must be blocked.

Many proteins, known as transcription factors, also regulate genes. In order to do so, they clamp onto short segments of DNA near those genes. Some of these transcription factors bend the surrounding DNA so that other proteins can begin to read a nearby gene. Other transcription factors shut genes down by clamping to neighboring DNA and preventing the gene-reading proteins from reaching them.

As we'll see in the next chapter, DNA can change over the course of generations. In many cases, the sequence of a protein-coding gene will change to a sequence that can no longer encode a working protein. It may instead make a functional RNA molecule. Or it may simply stop making any functional molecule whatsoever. Such formerly functional genes are called pseudogenes ("fake genes"), and they're very common. In a recent survey, scientists estimated that there are 11,800 pseudogenes in the human genome.

More than half of the human genome evolved from pieces of DNA that were able to make new copies of themselves that were then inserted back into the genome. Many of these so-called mobile elements got their start as viruses infecting our ancestors. They became part of their hosts' genomes, but they could still replicate themselves. In Chapter 11, we'll explore how mobile elements have shaped human evolution.

Heredity

Heredity is made possible by the unusual chemistry of DNA. The bases on one strand of DNA form weak bonds with the bases on the other strand, but each base is able to form a bond with only one other kind of base. A and T can bond, as can C and G. But neither A nor T can form a bond with C or G. When a cell divides, it first pulls apart the two strands of its DNA and builds new, complementary strands on each one. Thanks to the chemistry of base pairs, each new strand can be an identical match to the original one. When the cell divides, the two new cells can thus inherit an identical genome. (Sometimes cells make copying errors, which are crucial for evolution. We'll turn to those mistakes in the next chapter.)

The simplest kind of heredity is found in bacteria (**Figure 5.4**). Once a bacterium has grown large enough, proteins pull the twin strands of its DNA apart. It then assembles new strands using the separated strands as templates. Once the two new strands are assembled, the microbe now has an identical pair of complete DNA molecules. The new DNA molecules move to opposite ends of the cell, which then divides in half. Now there are two bacteria where there was once one, each with its own identical copy of the original DNA.

These two new bacteria can make the same kinds of proteins as their ancestor. That ancestor might, for example, carry a gene that encoded a protein that made it resistant to a bacteria-killing drug. Its descendants would be resistant as well, because they carry the same gene.

It's also possible for bacteria to acquire genes without inheriting them. Some bacteria, for example, can insert tubes into other bacteria, through which they can pump some of their DNA. Some microbes can slurp up naked DNA molecules that have been released by dead organisms. And viruses, which insert their own DNA into the DNA of their hosts, can also move some DNA from one host to another. These ways of acquiring genes are known as horizontal gene transfer.

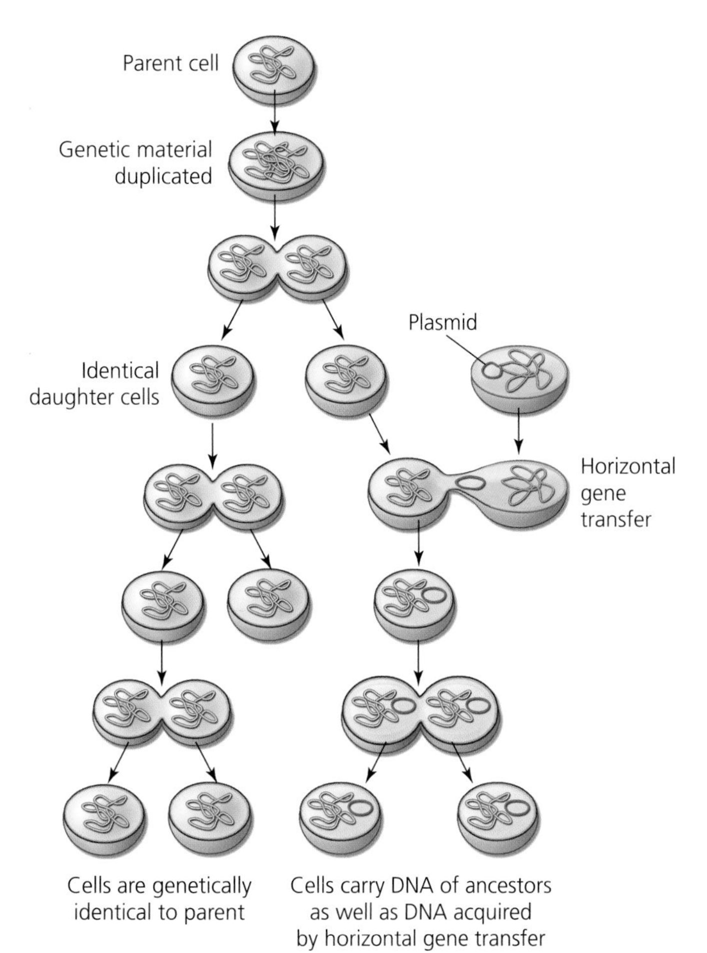

Figure 5.4 Bacteria reproduce by dividing in two. As they prepare for the division, they separate the two strands of DNA and add a new strand to each one, creating two new DNA molecules, which are typically identical to the original one. On rare occasion, however, genes can be passed from one bacterium to another, through a process known as horizontal gene transfer.

When a cell inside our bodies divides, it also produces a new copy of DNA, much as bacteria do. But the lineages of cells that multiply in our skin or our heart will not survive our own death. The only cells that can pass on their DNA to the next generation of humans are sperm cells and egg cells (known collectively as gametes). This extra step in sexual reproduction adds a lot of complexity to heredity.

Gametes are produced through a distinctive kind of cell division known as meiosis (**Figure 5.5**). During meiosis, each pair of chromosomes cross over and exchange segments of DNA in a process known as recombination. The progenitor cells divide again to produce gametes, but each gamete only inherits one chromosome from each pair. When a sperm cell fertilizes an egg, their chromosomes combine to produce a new set of 23 pairs.

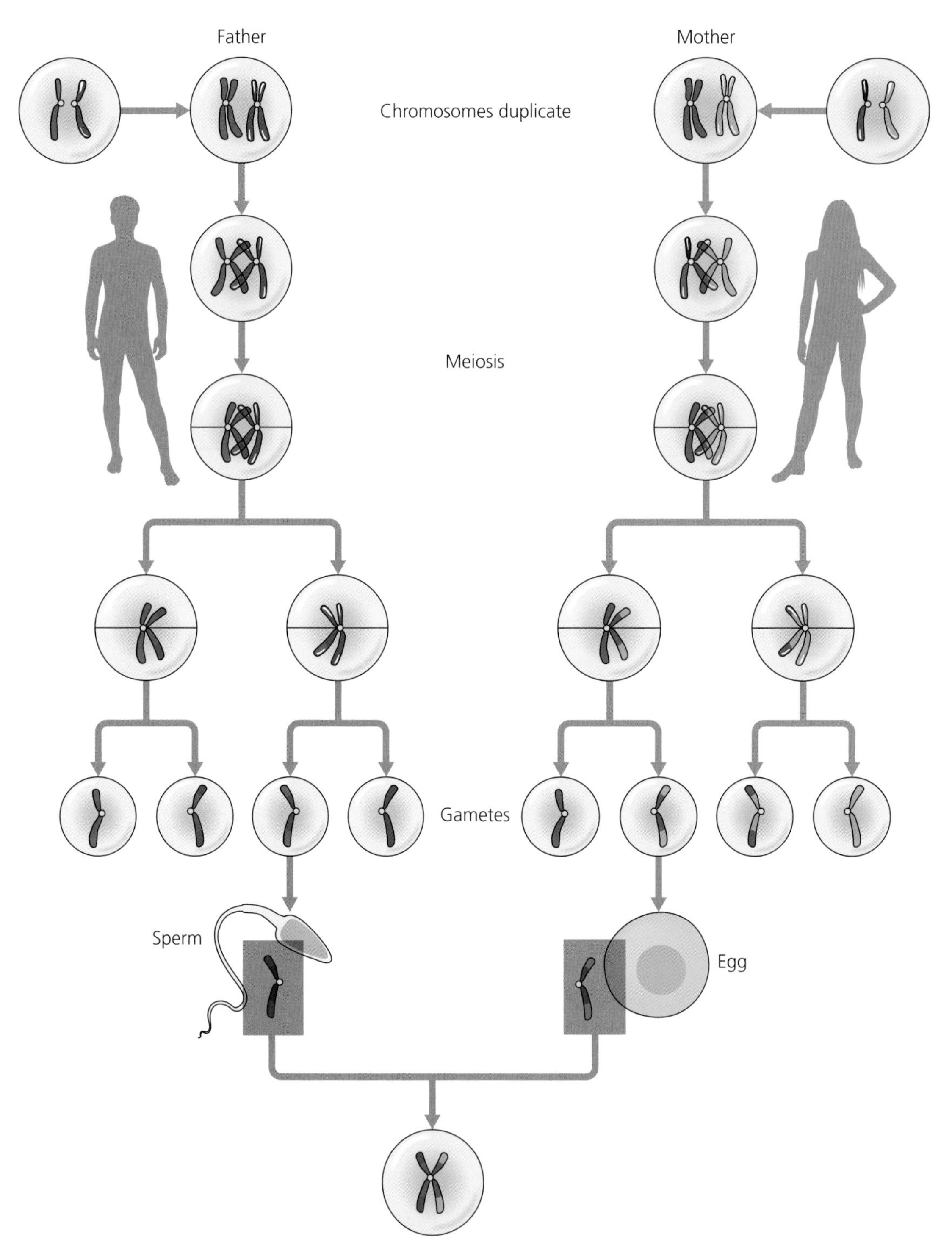

Figure 5.5 Among sexually reproducing organisms, like humans, males and females combine their gametes to reproduce. During the production of gametes, each pair of chromosomes cross over and exchange segments of DNA. Each gamete only receives one copy from each pair of chromosomes. As a result, each child carries a unique combination of the DNA of his or her parents.

When bacteria reproduce, they usually produce genetically identical offspring. Sexual reproduction, on the other hand, guarantees that children will not be clones of their parents. It also means that siblings, who develop from different sperm and egg cells, will not be identical to one another. (An exception to that rule, of course, is identical twins, who develop from a single fertilized egg.) The reason that parents do not produce families of identical children is that their own chromosome pairs are not identical. One chromosome may have one version of a gene (an allele) while the other chromosome has an allele with a slightly different DNA sequence.

The best estimate of the differences between our chromosomes comes from Craig Venter, a genome-sequencing pioneer. In 2007, he and his colleagues published the complete sequence of his own genome. They compared each pair of chromosomes, tallying up the differences. The researchers identified 3.2 million places where a single nucleotide in one chromosome did not match the corresponding nucleotide in its partner. The scientists also found about a million segments of DNA on one chromosome that were missing from its partner, or that were duplicated.

When people have children, they only pass down one copy of each chromosome. Which one they pass down is pretty much a matter of chance. Sexual reproduction also produces genetic variation in another way. As progenitor cells divide to form sperm or egg cells, the pairs of chromosomes cross over one another, and they swap chunks of DNA. The chromosomes in the egg cells and sperm cells end up with unique combinations of DNA as a result.

Genetics in the Garden

Scientists draw a distinction between the genetic material in an organism and the traits that the genetic material encodes. The genetic makeup of an organism is known as its genotype, and the manifestation of the genotype is known as the phenotype. Organisms do not inherit a phenotype; they inherit genes, which together constitute a genotype, which gives rise to a phenotype.

Understanding how phenotypes emerge from genotypes is no easy task. A trait does not come with a label on it, detailing all the genes that helped to build it and what specific role each of the genes played. Scientists rely instead on indirect observations. Some of the best methods for discovering the link from genotype to phenotype require scientists to study large populations, rather than a single organism.

Gregor Mendel (1822–1884) first recognized that inherited traits were made possible by factors (now called genes) passed down from parents to offspring.

It was in a population of pea plants that the father of genetics, Gregor Mendel, got some of the first clues about how genes work. Mendel (1822–1884) lived most of his adult life as a monk in a monastery in what is now the Czech Republic. Before entering the monastery, he had attended the University of Vienna and had become fascinated by heredity. After many years of reflection, he concluded that heredity was not a blending of traits, as many naturalists then believed. Instead, he believed that it came about by the combination of discrete factors from each parent.

To test his idea, Mendel planned out an experiment to cross different varieties of pea plants and to keep track of the color, size, and shape of the new generations of pea plants that they produced. For two years, he collected varieties and tested them to see if they would breed true. Mendel settled on 22 different varieties and chose seven different traits to track. His peas were either round or wrinkled and either yellow or green. Their pods were yellow or green as well, and they were also either smooth or ridged. The plants themselves might be tall or short, and their flowers, which could be violet or white, might blossom at their tips or along their stems.

Delicately placing the pollen of one plant on another, Mendel created thousands of hybrids, which he then interbred. After crossing smooth and wrinkled peas, for example, he shucked the pods a few months later and found that the hybrid peas were all smooth. The wrinkled trait had utterly disappeared from sight. Mendel then bred these smooth hybrids together and grew a second generation. While most of the peas were smooth, some were wrinkled—just as deeply wrinkled as their wrinkled grandparents. The wrinkled trait had not been destroyed during the round generation, in other words: it had gone into hiding in the hybrids and then reappeared when the hybrids were interbred.

The number of peas that ended up wrinkled would vary on each plant, but, as Mendel counted up more and more of them, he ended up with a ratio of one wrinkled seed for every three smooth ones. He crossed varieties to follow the fate of other traits, and the same pattern emerged: one green seed for every three yellow ones, and one white flower for every three violet ones. Again and again, the peas produced a three-to-one ratio of the traits.

Mendel realized that he had found an underlying regularity to the confusion of heredity, but his contemporaries ignored his work. He died at his monastery in 1884 with a reputation as having been little more than a charming putterer. But he was actually a pioneer in genetics, a field that didn't even come into formal existence until 16 years after his death.

After a hundred years of research, it is now clear why Mendel's peas grew the way they did (**Figure 5.6**). The difference between a smooth pea and a wrinkled pea is determined by a single gene that encodes a protein that helps to break down sugar. This protein is called starch-branching enzyme (SBEI). Mendel's peas had different combinations of two alleles of this gene, which are now called *R* and *r*. Mendel's wrinkly peas carried two copies of *r*, while the smooth peas either had *RR*, or an *R* from one parent and an *r* from the other.

Scientists refer to wrinkles on peas as a recessive trait, because two copies of the same allele are required to produce it. Smoothness is a dominant trait, because only a single *R* allele is enough to produce it. An organism that carries two copies of the same allele (*rr* or *RR* in this case) is called a homozygote. An organism with two different alleles of a gene is called a heterozygote.

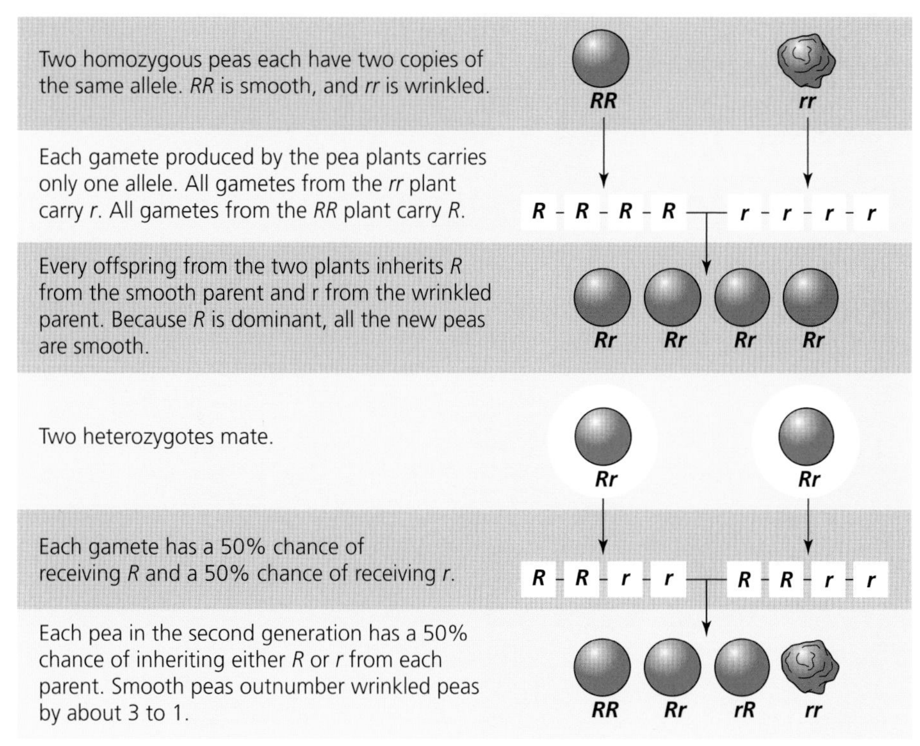

Figure 5.6 Gregor Mendel noticed that heterozygote peas produced offspring with a three-to-one ratio of certain traits. This figure shows how this ratio emerges as alleles for wrinkly or smooth peas are passed down through three generations.

The DNA sequence of the *R* and *r* alleles differs in one crucial way: the *r* allele contains an extra 800 base pairs of repetitive DNA. This extra DNA prevents a pea cell from producing SBEI. Without SBEI, a pea can't break down sugar effectively, and so sugar levels go up. A sugary seed absorbs extra water as it develops, so that it swells to a larger size. Later, when the pea begins to dry out, it shrinks and its surface folds in on itself, forming wrinkles.

Things go differently in *RR* peas. They make the SBEI enzyme, so they can break down sugar and don't swell as much. When the *RR* peas dry, their smaller surface does not wrinkle, leaving them smooth. In heterozygotes, the *R* allele makes its normal supply of SBEI, while the *r* allele makes none. Apparently, the SBEI produced from a single *R* allele is enough to keep peas from becoming wrinkled.

Now we can understand why Mendel discovered his three-to-one ratio of smooth and wrinkled peas produced from hybrids. Each *Rr* hybrid has a 50% chance of passing down either allele to its offspring. So a quarter of those offspring will be *rr*—wrinkly. The rest will be *RR*, *Rr*, or *rR*—in all cases, smooth. Looking at a single pea plant cannot tell you what probability it has of inheriting a particular allele, nor can it tell you the different effects of its *R* and *r* alleles. Only in a population—such as the population of peas in Mendel's garden—can these patterns emerge.

The Complex Path from Genotype to Phenotype

When headlines declare that scientists have discovered "a gene for" some particular trait, chances are that the headline writer needs a brush-up class on genetics. There is rarely a simple one-to-one link between a gene and a particular trait.

To begin with, a single gene often has several functions. It may be active in several organs at once, playing different roles in each one. If people are born with a defective version of a gene, the change may manifest itself in several ways. For example, a mutant form of the gene *HoxA13* causes a condition called hand-foot-genital syndrome. Women with this condition have tiny big toes and a second uterus. This combination of deformities shows that *HoxA13* has important roles in building both structures. This genetic multitasking is known as pleiotropy.

A complex trait like a toe or a uterus is not built by a single gene like *HoxA13*. Many genes influence its development. The variations in complex traits from one individual to another are also influenced by the environment in which they develop. These complexities are what make it so hard to pin down the factors that affect traits that seem to be very simple, such as height.

In 1885, Francis Galton (Charles Darwin's first cousin) plotted the height of 1,329 men on a graph and produced a relatively smooth curve (**Figure 5.7**). Most of the men were of average or near-average height. Taller and shorter men were equally rare. These results are in striking contrast with Mendel's wrinkled and smooth peas. That trait was in either one state or the other, with nothing in between—in other words, it is a discontinuous trait. Human height, on the other hand, is continuous.

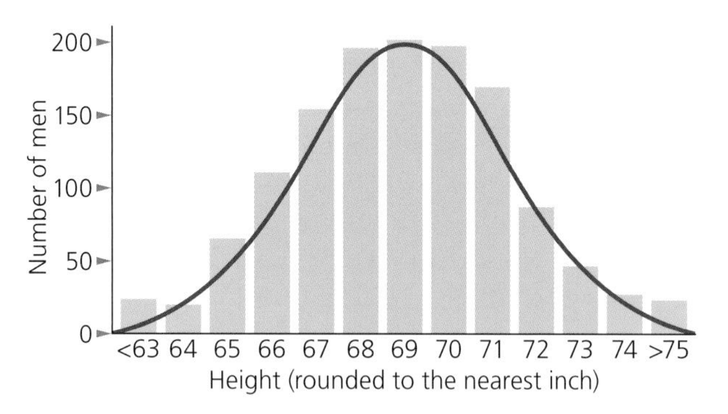

Figure 5.7 Francis Galton graphed the heights of 1,329 men and found that most were around average height, with the same proportions of taller and shorter men. Many traits have a similar distribution through a population. (Adapted from Hartl, 2003)

The total variation of a phenotypic trait such as height is the sum of the variation caused by genetic factors and variations caused by environmental factors. Height is strongly influenced by the environment. People who grow up in more affluent families, for example, tend to be taller than those who grow up in poorer ones. Mothers who suffer from malnutrition while pregnant tend to give birth to babies who will grow up to be shorter than people with well-fed mothers.

One of the clearest illustrations of the power of the environment comes from the research of Barry Bogin, an anthropologist who teaches at Loughborough University in England. In the 1970s, Bogin began to study the short stature of the Mayan people of Guatemala. Some scholars called them the Pygmies of Central America, because the men averaged only 1.6 meters (five feet, two inches) tall, and the women 1.4 meters (four feet, eight inches). The other major ethnic group in Guatemala is the Ladinos, who are of Spanish descent. Ladinos are of average height.

The biggest factor in the difference between the Ladinos and the Mayas was not genetic. It was poverty. The Mayas had less food and less access to modern medicine, which caused them to be shorter. During the Guatemalan civil war, a million refugees came to the United States. By 2000, Bogin found, American Mayans were 10 centimeters (four inches) taller than Guatemalan Mayans, making them the same height as Guatemalan Ladinos. The so-called Pygmies only needed access to a decent diet to grow much taller.

Yet scientists have also long known that genes also help to make some people tall and other people short. In 1903, for example, the British statistician Karl Pearson published data on 1,100 families and showed that tall fathers tended to have tall children. Recently, David Duffy of the Queensland Institute of Medical Research and his colleagues surveyed twins (**Figure 5.8**). They compared identical twins (who share the same set of genes) to fraternal twins (who develop from separate eggs and thus share only some of the same alleles). Identical twins grow to be much closer in height than fraternal twins. Both sorts of twins grow up in the same environment, so the main source of such a difference must be genes.

It has only been recently that scientists such as Joel Hirschhorn, whom we met at the beginning of this chapter, have begun zeroing in on the genes that influence height. Part of their recent success comes from advances in technology. To compare all 3.5 billion base pairs in every subject's genome would take centuries. Instead, scientists scan the human genome with the help of so-called genetic maps. They have identified millions of genetic markers scattered along the human genome. Each marker is a distinctive stretch of DNA. Just like all other human DNA, these genetic markers exist as many different variants in the human population. Scientists have discovered that variations in a few genetic markers tend to be found in people who are taller or shorter than average. This association suggests that there's a gene in the neighborhood of these markers that has an effect on height.

The gene *HMGA2* turns out to have a very strong association with differences in height. Carrying two copies of a particular allele can add a centimeter

Figure 5.8 Top: This graph shows the heights of identical twins (one twin's height is marked along the x axis, and his or her twin sibling's height is marked on the y axis). Identical twins tend to be of the same height. Bottom: Fraternal twins do not show as strong a tendency to be of similar height. The difference between these two results is due to the strong influence of genes on height. Because identical twins inherit identical sets of genes, they are more likely to be of similar height than fraternal twins, which develop from separate eggs. (Courtesy of David Duffy)

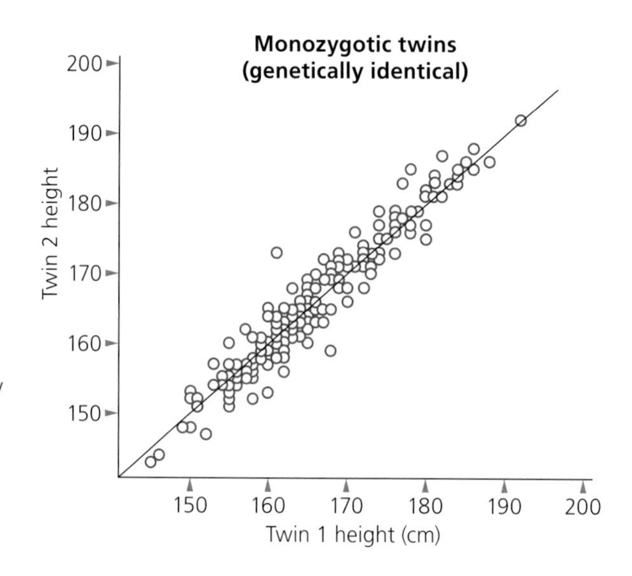

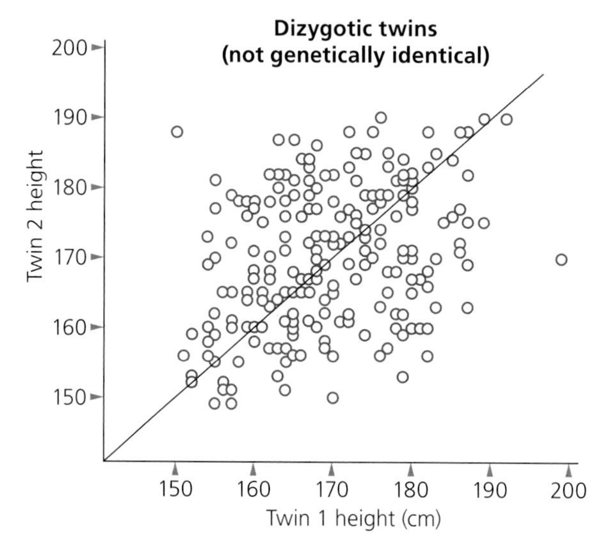

to a person's stature. Scientists can't say for sure how *HMGA2* influences phenotypes; one suggestive hint is the fact that it plays some kind of role in cell division. It would be a mistake to call *HMGA2* "*the* tall gene." More than 96% of the variation in height was *not* associated with different *HMGA2*. Many short people lack the "short" allele of *HMGA2*, and many tall people lack the "tall" variant.

Hirschhorn's study, and similar ones by other teams, all hint that there are hundreds of genes associated with height. *HMGA2* alleles add an extra centimeter *on average* in the entire set of people the scientists studied. There are a vast number of combinations of different height-associated alleles in that population. As a result, many people with the height-boosting *HMGA2* allele are *shorter* than people without it. The research Hirschhorn and his colleagues are carrying out on height show how even a seemingly simple trait can reveal the complex link between genotypes and phenotypes.

TO SUM UP...

- **Genetic information is stored in DNA.**
- **DNA is copied into RNA, which plays many functions in the cell.**
- **Some RNA molecules serve as templates for proteins.**
- **The phenotype is the manifestation of the genotype.**
- **Organisms reproduce sexually or asexually.**
- **Complex traits are influenced by many genes.**
- **Many individual genes influence several traits.**
- **The environment can produce large variations in the expression of a trait.**
- **There is no "height gene." Any particular height-linked gene probably accounts for only a small amount of variation in the height of a population. This is true for other complex traits as well.**

The Ways of Change

6

Mutation, Drift, and Selection

In the Adriatic Sea, off the coast of Croatia, lies a scrub-covered, seven-acre patch of rock known as Pod Mrčaru. There are no people on this tiny island, but it is home to a healthy population of six-inch-long lizards. They appear to be ordinary Italian wall lizards (*Podarcis sicula*). But these lizards have a remarkable story to tell. They're part of a striking experiment that's documenting evolution in our own lifetime.

Left: Scientists have documented rapid natural selection in lizards that live on a tiny island in the Adriatic Sea. Above: Biologist Duncan Irschick catches the lizards with a fishing pole to study them.

Before 1971, *Podarcis sicula* did not exist on the island. In that year, Eviatar Nevo, an Israeli biologist, and his colleagues picked up five pairs of the lizards from a nearby island called Pod Kopište and brought them to Pod Mrčaru. On Pod Kopište, the diet of the lizards is made up mostly of insects. The ecology of the island makes that diet possible—with few plants to provide cover, the insects are easy targets on the barren rocks. On Pod Mrčaru, however, Nevo's lizards faced a very different ecosystem. There, shrubs are abundant and insects are harder to find. Nevo wanted to see how the lizards would fare in this new environment.

Nevo didn't get the chance to find out, because Yugoslavia descended into decades of political upheavals and civil wars. It was not until 2004 that a team of scientists from the University of Massachusetts and the University of Antwerp in the Netherlands returned to Pod Mrčaru to take stock of the lizards. A genetic test confirmed that the lizards on the island were the descendants of Nevo's 10 founders. But the scientists discovered that the lizards had become significantly different from their cousins on Pod Kopište.

For one thing, the lizards on Pod Mrčaru were eating a lot more plant material. As much as 61% of their food was made up of leaves and stems, compared with 7% or less on Pod Kopište. The tough, fibrous tissues of plants are much harder to eat and to digest than crunchy insects. The lizards on Pod Mrčaru had bodies that were adapted to their new diet. Their heads were bigger, allowing them to generate stronger bites. Adapting to eating plants had made them less efficient at hunting insects, however: their legs had become shorter and they had become slower.

When the scientists dissected the lizards, they discovered other changes inside. The digestive tract of each lizard had developed new muscular rings that pinched off a section of the gut, forming a new chamber. In this chamber, plant-feeding worms and bacteria had taken up residence, breaking down tough leaves and stems that lizards could not digest on their own.

The transformation of the anatomy of the lizards altered their ecology as well. The lizards of Pod Mrčaru could get more energy from the plants than the ones on Pod Kopište get from insects. That extra energy could support a bigger population of lizards per acre of island. On Pod Kopište, the lizards defended patches of the island to protect their supply of insects. With a more abundant supply of food on Pod Mrčaru, however, the lizards stopped fighting with each other over territory.

The lizards of Pod Mrčaru have done what Darwin predicted: they have adapted to their environment through natural selection. But Darwin would have been surprised by the lizards of Pod Mrčaru. After all, it took them no more than 33 years to achieve their changes. Just as water carves a canyon grain by grain, Darwin argued, life must evolve too slowly for us to perceive. It turns out, however, that scientists can observe natural selection in their own lifetimes.

Mutations: Creating Variation

Evolution is possible because sometimes DNA spontaneously changes as it is being replicated. These changes, known as mutations, have many causes. Radioactive particles pass through our bodies every day. If one of these particles strikes a molecule of DNA, it can damage the molecule's structure. When a cell copies the DNA, it may misread the garbled sequence and add the wrong nucleotide. Our cells have proofreading proteins that can fix some of these errors, but they sometimes let mistakes slip by.

Another source of mutations comes from within our own genomes. There are millions of segments of DNA in the human genome that can make copies of themselves, which are then inserted back somewhere else in the genome. These pieces of "jumping DNA," known as mobile elements, can sometimes insert themselves into the middle of a gene. Mobile elements are responsible, for example, for the *r* allele that caused some of Gregor Mendel's peas to develop wrinkly surfaces (page 96). At some point in the distant past, a mobile element measuring 800 base pairs long was inserted into the gene that makes the SBEI protein, preventing peas from making SBEI.

Mutations alter DNA in several different ways (**Figure 6.1**):

- A point mutation changes a single base to another.
- A segment of DNA may be inserted into the middle of an existing sequence. The insertion may be as short as a single base or as long as thousands of bases (including entire genes).
- A segment of DNA may be accidentally deleted. A deletion can be as big or as small as an insertion. A small portion of a gene may disappear, or an entire set of genes may vanish.
- A segment of DNA may be accidentally duplicated. This duplication may give rise to two copies of the same gene. Sometimes an entire genome can even be duplicated.
- A segment of DNA may be inverted (flipped around and inserted backwards into its original position).
- Chromosomes can be fused together.
- Genes can be passed from one organism to another through horizontal gene transfer (page 93). This occurs much more often in bacteria than animals or plants. Horizontal gene transfer can even pass DNA between species.

Each time a cell divides, there's a tiny chance that any particular piece of its DNA will mutate. Scientists can measure the rate of mutation with several methods. In 2008, for example, Michael Lynch of Indiana University and his colleagues reared colonies of yeast, a single-celled fungus that bakers use to make bread dough rise. From a single ancestor, Lynch and his colleagues reared hundreds of genetically identical populations of yeast. They then allowed these lines

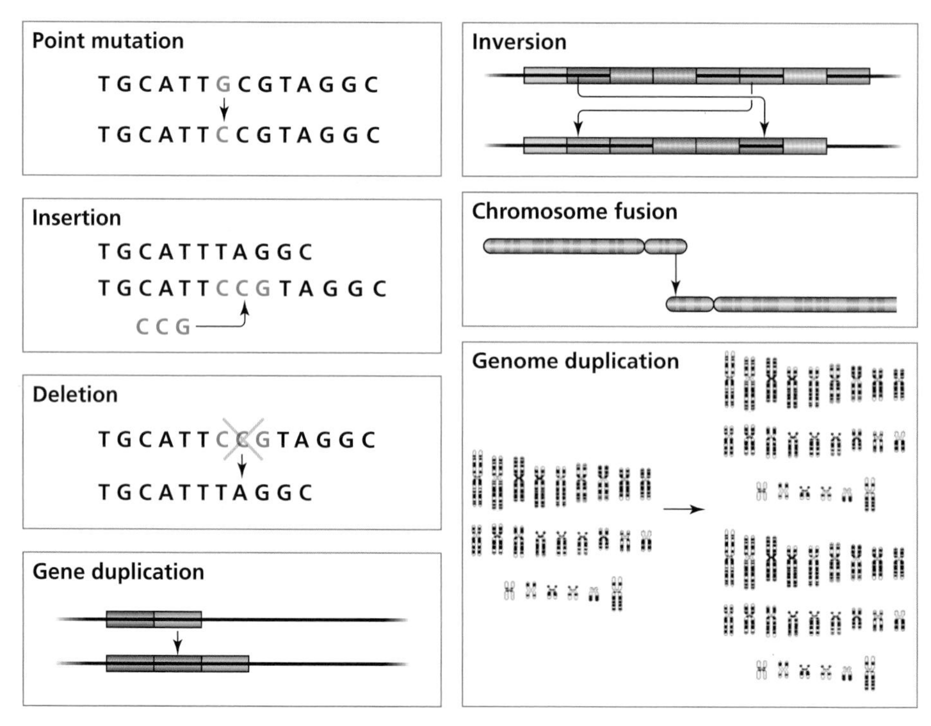

Figure 6.1 DNA can experience several different kinds of mutations, some of which are shown in this figure.

to reproduce for 4,800 generations. The scientists then sequenced the 12 million base pairs of DNA in the genomes of some of their descendants.

The scientists found that, each time a yeast cell divides, each site in its DNA has a 0.0000003% chance of experiencing a point mutation. Since a yeast genome has only 12 million base pairs, a typical yeast cell may not experience any point mutations at all. But in a population of millions, each new generation is likely to have thousands of yeasts with point mutations.

Different kinds of mutations have different mutation rates, as Lynch's research demonstrates. Lynch's team found that many gene duplications and deletions arose in the yeast. They estimate that each gene has about a one-in-a-million chance of being lost or being duplicated each time a cell divides. Duplications and deletions are still rare, in other words, but they're also about a thousand times more likely than point mutations.

Estimating mutation rates in multicellular organisms like us gets more complicated, because cells in our bodies can mutate in different ways. Any individual cell in our body has a chance of mutating as it divides. If it's a skin cell, the skin cells that descend from it will continue to carry that mutation. But this lineage of cells will end when we die. Such mutations are known as somatic mutations, because they occur in the "soma," or body. If, on the other hand, a mutation arises in the line of cells that gives rise to sperm or egg cells, it may be passed on to offspring. And those offspring, in turn, may pass the mutation down to their own offspring. These mutations are known as germ-line mutations. When scientists estimate the rate of mutations in humans or other multicellular organisms, they typically measure germ-line mutations. By one estimate, each new human acquires about 60 new germ-line mutations in his or her genome.

From Harmful to Helpful

A mutation is a change to an organism's genotype. In some cases, a mutation can also change an organism's phenotype. A single point mutation in a gene may change a codon so that it codes a new amino acid. That new amino acid may change the way a protein works. A mobile element may insert itself in the middle of a gene, garbling its sequence and making it impossible for the cell to produce a protein from it. Insertions and deletions can also add or remove segments of proteins. If a gene is duplicated, a cell may start to make twice as much of the corresponding protein.

Mutations to noncoding DNA may also alter an organism's phenotype. Transcription factors switch genes on and off by binding to regions nearby. If those regions mutate, a transcription factor may no longer be able to grab hold. A mutation can cause a micro-RNA to stop silencing one gene and start silencing another.

The best-studied mutations are the ones that make people sick, because they are so medically important. Cystic fibrosis, for example, is the result of a mutation to a gene that encodes a channel on the surfaces of cells, causing cells in the lungs to fill with fluid. Other mutations disable proteins for breaking down food or synthesizing proteins. Some mutations prevent embryos from developing normally. Other mutations can have harmful effects that are less obvious. They can leave people more vulnerable to diseases triggered by environmental causes, such as cancer.

But most mutations do not cause death or disease. Some alter noncoding regions of DNA that serve no vital function. Many mutations alter protein-coding genes without changing their function. In Lynch's study on yeast, the vast majority of mutations he and his colleagues detected were neutral: in other words, they had no effect on survival. Studies on human DNA show that most mutations we acquire are neutral as well. Because people safely carry many mutations, they can pass some of those mutations down to their children. These new mutations add to the genetic variation of our species. If you select two people at random, their genomes will differ at three million single nucleotides. Humans also vary in the number of copies of many genes they carry.

In addition to harmful mutations and neutral ones, some mutations can have beneficial effects. A beneficial mutation may provide resistance to diseases, for example, or allow an organism to get more energy from its food. Beneficial mutations do not arise to give an organism a trait it "needs," however. The causes of mutations—incorrect copying of DNA, imperfect repair, radiation, and so on—are distinct from the effects of mutations. Exactly what sort of mutation strikes an individual is, in an important sense, a matter of chance.

Spreading Mutations

Once a mutation has altered the DNA in a single-celled asexual organism its descendants will pass it down from one generation to the next. The fate of mutations in a multicellular species like our own is more complicated. Mutations arise every day in our bodies, and each time our cells divide, they pass those mutations down to their descendants. But once we die, our mutant cells die with us. The only way a mutation can survive our death is if it arises during the development of sperm or eggs. When a mutant gamete gives rise to a new human being, all of the cells in that person will carry the mutation—including the new human being's own gametes.

Exactly which combination of mutations humans pass down to their children also depends on how their DNA is recombined during meiosis (page 94). Recombination does not create new genetic sequences the way a point mutation or a gene duplication does, but it does bring together alleles of different genes into new combinations. And because genes work together, these new combinations can have significant effects on phenotypes.

Once a mutated gene begins to be passed down from generation to generation in a sexually reproducing species, things get even more complicated. If a new mutation arises in a father, for example, his child will carry his mutated version of the gene and another allele inherited from the mother. If the mutant allele is recessive, it will not have any effect if the mother's allele is dominant. On the other hand, if the mutant allele is dominant, it will always alter the phenotype of a child that inherits it. But the precise effects of an allele also depend on the other alleles a child inherits. That's why siblings can look so different from one another. The final complication of sexual reproduction comes when that child then has children of his or her own. There is only a 50% chance that any given grandchild will inherit the father's original allele.

Turning Biology into Equations

Mutated genes are a bit like newfangled words. Every year a flood of new English words come into vogue, but only a few of them prove successful, sweeping across the country and, ultimately, ending up in the dictionary. They may endure for centuries, long after the people who invented them have died. But most new words fail to take. Some of them may linger on in small circles, whose members use them to describe obscure things. Many new words survive only for a few years before sinking into oblivion.

Mutations suffer similar fates as organisms pass them down through the generations. One way to study those fates is to build elegant mathematical models of populations and their genes, and then to test those models with experiments and observations. This branch of biology is known as population genetics. One of

How Do Scientists Study Evolution?

Scientists use the same basic methods to study evolution as they would to study any other aspect of nature. They use theories to guide their research, testing hypotheses with experiments and observations. But the different branches of science also have their own quirks and demands, and evolutionary biology is no exception.

Evolution plays out on many different scales all at once. The molecular biology of DNA is crucial to the evolutionary process, but, in order to become part of long-term transformations to life, any changes to DNA must have a significant effect on how cells work and how organisms function. To understand those effects is no simple matter, not even when scientists are studying single-celled microbes. In the case of multicellular organisms like us, studying those effects calls for studying the complicated ways in which an egg made of a single cell develops into an adult made of trillions of cells. Natural selection may change how a species fits into its ecosystem, taking evolution up to an even higher level of complexity.

Meanwhile, the very planet on which evolution takes place is changing as well. The climate changes regionally and globally, over the course of decades, millennia, and millions of years. Continents split and collide. Some researchers argue that, just as individual organisms are pitted against each other through natural selection, entire species or groups of species are pitted against each other over vast stretches of time.

Evolutionary biologists have to integrate these scales into their research. They also have to grapple with many different processes acting at once. The reproductive success of an individual shark, for example, may be affected by the diseases it suffers, the changing supply of fishes it can hunt, and the success it has in finding a mate. The thousands of sharks in a single population all have their own story to tell; it's only in their combination that evolution works.

...evolutionary biologists must draw on many different kinds of evidence, drawn from different branches of science

To grapple with this level of complexity, evolutionary biologists must draw on many different kinds of evidence, drawn from different branches of science. Mathematical biologists create theoretical models of evolution; geochemists reconstruct ancient environments; virologists track the ongoing evolution of a new disease. As new technology emerges, whether it's ground-penetrating radar for finding fossils or robots that can sequence DNA, evolutionary biologists make it a part of their tool kit. As a result, evolutionary biology is a vibrant area of research, with many new papers coming out every week on a vast range of subjects.

the most important lessons of population genetics is that the frequency of an allele can change in unexpected ways. Mendel found, for example, that two hybrid pea plants would produce three offspring with smooth peas for every one with wrinkled peas. But that doesn't mean that an entire population of pea plants would have the same three-to-one ratio. A simple mathematical model reveals why (**Figure 6.2**).

The Hardy–Weinberg Model

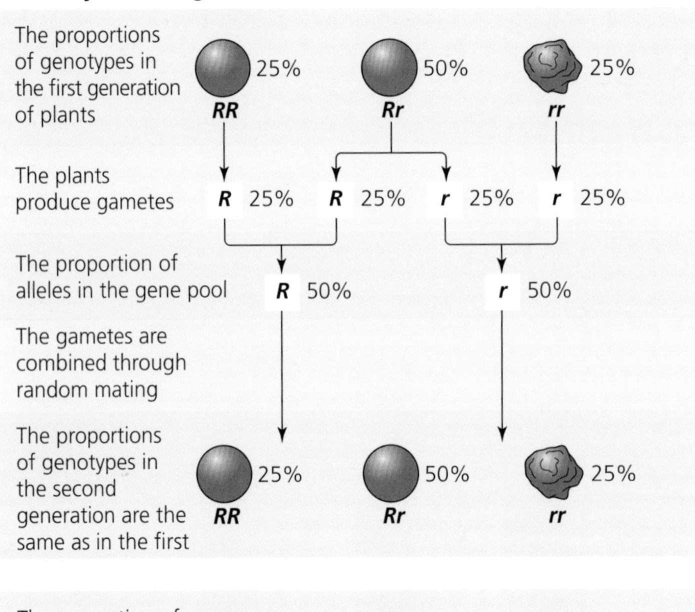

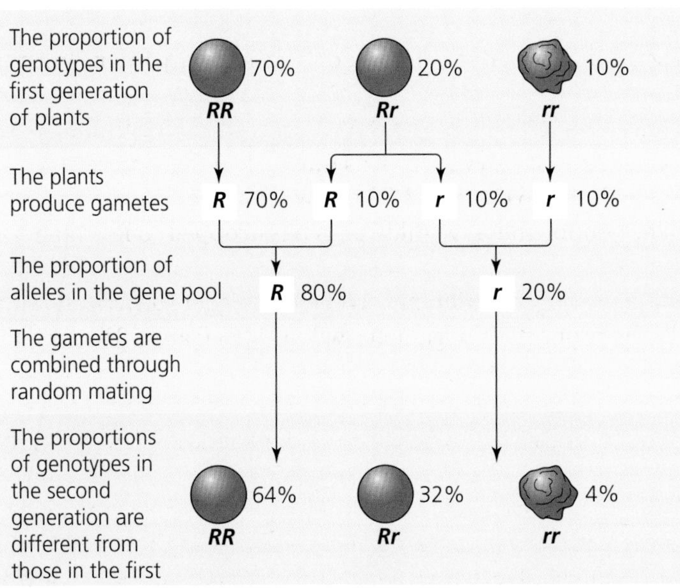

Figure 6.2 The Hardy–Weinberg model shows how the frequency of alleles can change from one generation to the next. Top: In the first generation, 25% of the peas are wrinkled (*rr* homozygotes). Half of the peas are heterzygotes (*Rr*), and 25% are *RR* homozygotes. Each pea produces gametes. For the sake of simplicity, the model assumes that the alleles in a pea's gametes are proportional to the alleles in the pea. (For example, half of the gametes produced by *Rr* heterozygotes are *R* and the other half are *r*.) As the third row shows, half of all the gametes produced by all the peas are *R*, and the other half are *r*. If we assume that these gametes all combine randomly with other gametes, half of the peas will inherit *R* and half will inherit *r*. As a result, the second generation will have exactly the same frequency of genotypes as the first. Bottom: In this case, the proportions of genotypes are different, with many more *RR* homozygotes, and fewer *Rr* heterozygotes and *rr* homozygotes. If the Hardy–Weinberg model begins with these peas, we end up with a different frequency of genotypes, with fewer homozygotes and more heterozygotes. If this model is extended to the third generation, the proportions will be the same as in the second. Biologists call this unchanging condition Hardy–Weinberg equilibrium.

Imagine that you have a huge garden full of peas. Some of them are *rr* homozygotes, some *RR* homozygotes, and the rest heterozygotes. Let's say that 25% of the peas are *rr*, 50% *Rr*, and 25% *RR*. Now imagine that all of them mate randomly. What will the next generation of peas look like?

First the peas must produce gametes. Each gamete receives only one copy of the gene, with a 50% chance of it being either copy carried by the plant. Obviously, all of the gametes produced by *RR* plants will have the *R* allele, and *rr* plants will only produce gametes with *r*. As for the hybrids, let's assume that half of the gametes from all the hybrids carry *R* and the other half have *r*. A little arithmetic, illustrated in Figure 6.2, tells us that 50% of all the gametes produced by the entire population will carry *R*, and the other half will carry *r*.

Now let's look at what happens when these gametes combine at random, producing fertilized eggs that will grow up into a new generation of pea plants. Each new plant ends up with two alleles, randomly selected from the copies in the previous step. Since half of the gametes carry *R* and the other half carry *r*, we'll assume that each gamete has a 50% chance of carrying either allele. That leaves us with 25% *rr*, 25% *RR*, and 50% *Rr*. In other words, each generation passes on the same proportion of alleles to the next. Since *r* is recessive, 75% of the peas will be smooth and 25% will be wrinkly.

But what if you start off with a different mix of peas in your garden? Imagine that only 10% of the peas carry two copies of *r*. Only 20% are heterozygotes (*Rr*), and 70% carry two copies of *R*. Only 20% of the gametes produced by this population will carry *r*, and 80% will carry *R*. It turns out that, in the next generation, only 4% are *rr*, 32% are *Rr*, and 64% are *RR*. The new generation is different from the previous one. The proportion of wrinkly peas has dropped to less than half its original level.

What happens in the next generation may come as a bigger surprise: nothing new happens at all. The proportion of gametes will be 80% *R* and 20% *r*, and the plants they give rise to will be 4% *rr*, 32% *Rr*, and 64% *RR*—exactly the same proportions as in the previous generation. The same will hold true of the next generation and all the ones that follow, as long as the conditions of the model remain the same. The population has reached an equilibrium.

Population geneticists refer to this balance as Hardy–Weinberg equilibrium, after the British mathematician G. H. Hardy and the German physiologist Wilhelm Weinberg, who independently created similar versions of this model in 1908. Hardy and Weinberg demonstrated that, even though pairs of parents may follow Mendel's rules of genetics, a population taken as a whole may not. Their model helped to show how genetic variation—a crucial ingredient for evolution—can endure thanks to simple Mendelian genetics.

The Hardy–Weinberg principle is also a useful tool for figuring out why a particular allele is rare or common in a population. In the 1970s, for example, Luigi Luca Cavalli-Sforza of Stanford University and his colleagues studied a gene that encodes part of hemoglobin, the molecule that transports oxygen through the bloodstream. They surveyed 12,387 people in Nigeria and found that they had two

Genotype	Observed	Expected by HW
SS	29 (0.2%)	187.4 (1.5%)
AS	2,993 (24.2%)	2,672.4 (21.6%)
AA	9,365 (75.6%)	9,527.2 (76.9%)
Total	12,387	12,387

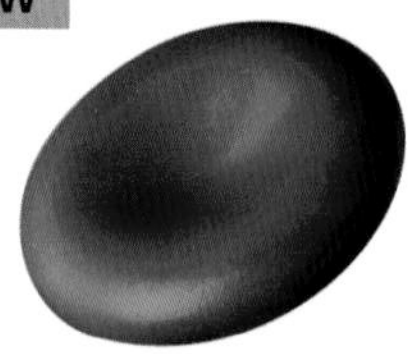

Figure 6.3 Geneticists measured the frequencies of two alleles of a gene that encodes part of hemoglobin, known as A and S. They then determined the frequencies of homozygotes and heterozygotes that would be expected if the alleles were in Hardy–Weinberg equilibrium. As this chart shows, the frequencies are significantly different from the expected values. As we'll see later, natural selection is responsible for the difference between these expectations and the actual frequencies. (Adapted from Cavalli-Sforza, 1975)

different alleles, which are known as *S* and *A*. **Figure 6.3** shows the numbers of people with two copies of *S*, two copies of *A*, and one copy of each of those two alleles. We can put those numbers into the Hardy–Weinberg model. If they're in equilibrium, they should stay the same when we move to the next generation. But that's not what happens. In the next generation, there are fewer people with *SS* or *AA* than you'd expect from the Hardy–Weinberg equilibrium, and more people with *AS*.

A few processes can push a population out of Hardy–Weinberg equilibrium. Here we'll take a look at two of the most important: genetic drift and natural selection.

Genetic Drift

In the 1950s, Peter Buri, a biologist at Iowa State University, bred thousands of flies of the species *Drosophila melanogaster*. His flies carried two different alleles, called *bw* and *bw75*, for a gene that influences their color. A fly with two copies of *bw* is white. A *bw/bw75* heterozygote is light orange. A fly with two copies of *bw75* is bright red-orange.

Buri established 107 populations of light-orange flies, each with one *bw* and one *bw75* allele. He started each population with eight males and eight females, and he let them reproduce. From the next generation, he randomly selected eight new males and eight new females to breed. Buri bred the flies for 19 generations, and each time he tallied up the number of *bw* and *bw75* alleles in the 16 flies he chose.

If Buri's flies were in Hardy–Weinberg equilibrium, the frequency of the alleles should not have changed at all. Half of the alleles in each generation should have been *bw* and half should be *bw75*. But that's not what happened. Instead, as **Figure 6.4** shows, *bw* became rarer and rarer in some populations until it disappeared completely, leaving only bright red-orange *bw75/bw75* flies. In other populations, *bw75* disappeared, leaving only white *bw*/*bw* flies. The rest of the populations spanned the range between these two extremes.

How did all of this change take place in Buri's experiment? A simple game can illustrate what's going on. You'll need a large bowl of jellybeans that are of the same size and shape, but that are equally divided between two colors (**Figure 6.5**).

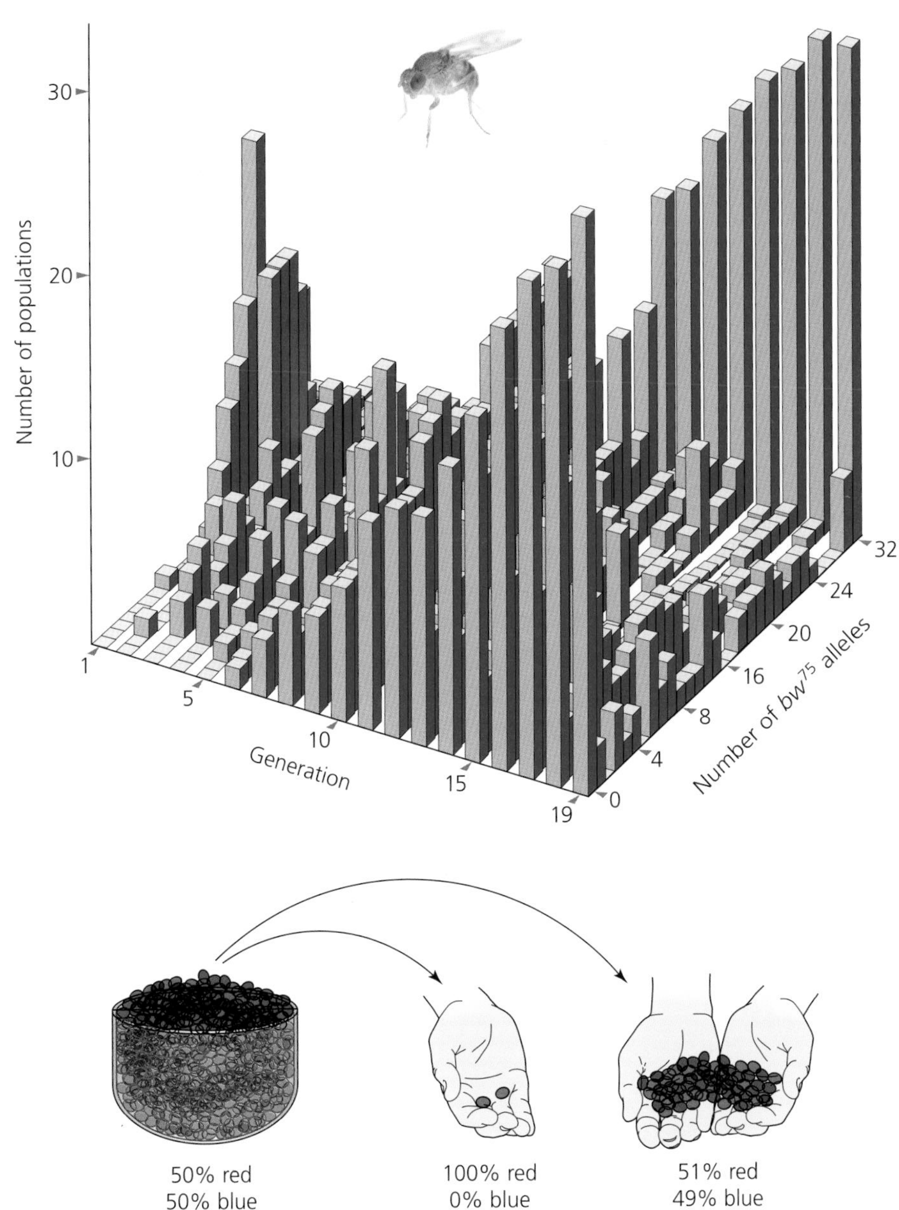

Figure 6.4 This graph charts a series of experiments in which 107 populations of 16 fruit flies apiece were allowed to reproduce for 19 generations. At the end of the experiments, the *bw* allele had either become fixed or disappeared from many populations. (Adapted from Hartl, 2007)

Figure 6.5 To understand genetic drift, imagine alleles in a population as jellybeans in a bowl. If you only select a few jellybeans, you have a good chance of ending up with a different proportion of colors in your hand than in the bowl. If you grab a large handful, you're more likely to end up with similar proportions. In small populations, alleles may become more or less common from one generation to the next through a similar random process.

Now grab some jellybeans from the bowl. If you have a bowl made up of 50% red jellybeans and 50% blue jellybeans, there's a 50% chance that any jellybean you pick out at random will be red. But, if you pick only two jellybeans, you might very well pick two blue ones instead. Even if you were to pick out four jellybeans without looking, it wouldn't be a shock to find you had not drawn two red ones and two blue ones. But the more jellybeans you grab, the higher the odds become that you'll end up with something close to a 50-50 split between the two colors.

In Buri's experiment, the *bw* alleles in the flies were a lot like jellybeans in a bowl. The flies could make a vast number of eggs and sperm, and half of them carried *bw*, while the other half carried *bw75*. If these gametes were mixed together randomly to produce a large number of new flies, it would be like taking a large handful of jellybeans. In other words, chances would be very good that the next generation would have about the same proportions of *bw* and *bw75* alleles as the previous one.

But Buri did not breed a lot of flies. Instead, he selected only 32 gametes in each generation. He was taking a few jellybeans from the genetic bowl, in other words, and as a result the odds were much greater that he would end up with something other than a 50-50 split between the alleles.

These strokes of luck meant that in some populations, *bw* became rarer. And in some populations, *bw* became so rare that only a single fly still carried a single copy. And then, in the next generation, that single fly passed down its copy of *bw75*, and *bw* was gone for good. The same flukes left other populations without any copies of *bw75*. In each case, biologists would say that one of the alleles had become fixed in the population—in other words, all its members now carried it.

The rise and fall of alleles by chance is known as genetic drift. Genetic drift is an important element of the evolutionary process. Many of the fixed segments of DNA in the human genome owe their success to it.

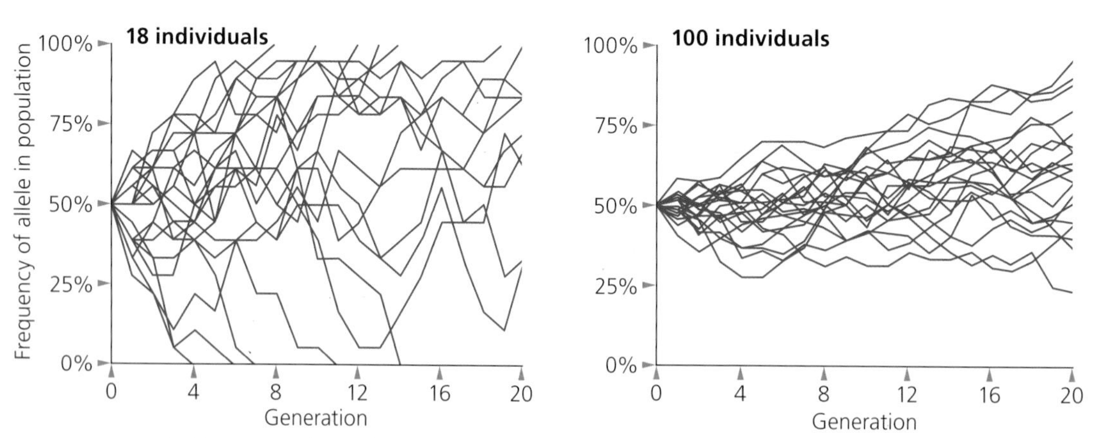

Figure 6.6 Left: This graph shows the results of several computer simulations of a population of 18 individuals. They tend to experience strong genetic drift, as an allele either becomes fixed or disappears. Right: In simulations with 100 individuals, genetic drift is weaker. Each new generation is much more likely to have the same frequency of the allele as the previous one. (Adapted from Hartl, 2007)

While genetic drift is random to some extent, it does follow certain statistical rules. The time it takes for an allele to become fixed is shorter in a small population than in a large one. **Figure 6.6** illustrates this rule with computer simulations of a sexual population. There are two alleles of a gene in this population, and, at the start of each simulation, half of the individuals carry one allele and the other half carry the other. In the first set of simulations, each generation is made up of 18 individuals, and each individual produces only one offspring. The simulations show how one of the alleles becomes fixed. The second set of simulations is run with a population of 100 individuals. In many of these runs, one allele becomes more common than the other, but in none of them does one of the alleles become fixed. An intuitive way to think about genetic drift and population size is to picture a population like a bowling lane. On a narrow lane, a bowling ball may end up in the gutter very quickly. On a wider one, there is more room for the ball to veer back and forth and yet remain on the lane.

Selection

Charles Darwin argued that natural selection came about because some individuals were more successful in their environment and survived to reproduce greater numbers of offspring than less successful individuals. Their descendants then inherited their traits, including the ones that made them more successful.

Population geneticists have developed mathematically precise ways to measure natural selection. The rate at which a genotype increases in a population is known as its fitness. Some genotypes increase faster than others, and the difference between them is known as their relative fitness. Scientists usually give the genotype with the highest fitness in a population a value of 1. The relative fitness of every other genotype is some fraction of 1. If a strain of bacteria grows 80% as fast as the fastest strain under certain conditions, for example, its relative fitness is 0.8.

Bacteria are a good place to begin a discussion of fitness, because their asexual reproduction makes it simpler to understand. If a mutation arises in a microbe it will pass on its new allele to both of its descendants. If the allele has no effect on the reproduction of the bacteria, the mutant lineage will grow at the same rate as other lineages without the mutation. In many cases, however, mutations raise a microbe's fitness or lower it. Selection occurs when two or more genotypes have different levels of fitness.

Negative selection occurs when an allele lowers the relative fitness of a genotype (**Figure 6.7**). In its most extreme form, a mutation may completely disable an essential gene, causing a microbe to die instantly. The growth rate of its lineage instantly drops to zero. But mutations don't have to be lethal to lead to negative fitness. They only have to lower the bacteria's rate of growth. These mutations may cause a

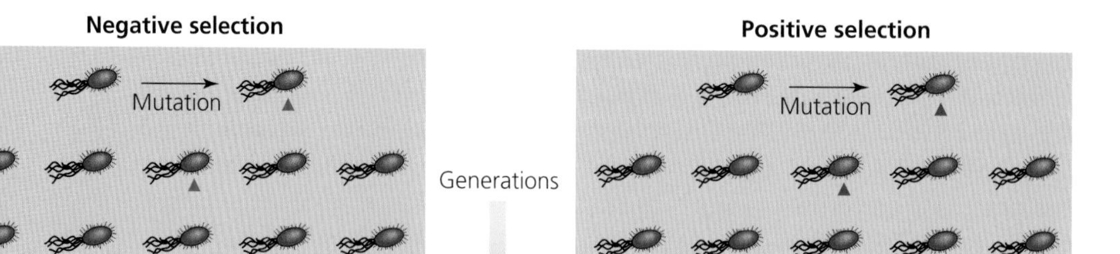

Figure 6.7 A mutation that raises the reproductive fitness of an individual will tend to spread through a population over time. A mutation that lowers fitness will tend to become less common and disappear.

microbe to grow more slowly, for example, or to become more vulnerable to a virus. Over the course of many generations, microbes with these mutations will have fewer offspring and become less and less common. Eventually, they will be replaced completely by other strains of microbes.

Positive selection, on the other hand, takes place when an allele increases the reproductive success of an individual. An individual with the allele has, on average, more offspring than individuals that don't. Positive selection can occur if a mutation allows bacteria to metabolize their food more efficiently. It can also occur on old mutations if the environment changes. When we take antibiotics to treat bacterial infections, for example, some bacteria will be less vulnerable to the antibiotic, thanks to the alleles they carry. The positive selection they experience will make them more common in our bodies. But once the antibiotics are gone, resistance alleles may no longer create positive selection. They may be neutral. In some cases, resistance genes are even harmful in the absence of antibiotics, because they cause bacteria to grow more slowly.

Another form of selection, known as stabilizing selection, occurs when extreme versions of a trait are selected against. Very small newborn babies, for example, are more vulnerable to many health problems and are therefore more likely to die than larger babies. If babies are too big when they are born, however, they are more likely to have trouble passing through the birth canal. Their mothers are more likely to die during childbirth, raising the risk that the babies will die as well. Thus, natural selection stabilizes the size of babies at an intermediate average.

Small Differences, Big Results

Selection is sometimes dramatic, and other times it is subtle. The most dramatic cases of selection come about when a single mutation creates a lethal genetic disorder during childhood. Geneticists have identified many such disorders, but each one typically only affects a tiny fraction of the population. That's because children who die of these disorders cannot pass on the mutation to their offspring. Negative selection immediately removes these alleles from the population.

But even when genotypes are separated by a tiny difference in their fitness, selection can have big long-term effects. That's because populations grow like investments earning interest. Let's say you invest $100 in a fund that earns 5% interest each year. In the first year, the fund will increase by $5. In the second, it will increase by $5.25. In every subsequent year, the fund will increase by a larger and larger amount. In 50 years, you'll have more than $1,146. Because of this accelerating growth, even a small change in the interest rate can have a big effect over time. If the interest rate on your fund is 7% instead of 5%, you'll only make an extra $2 in the first year. But, in 50 years, the fund will be more than $2,945—close to triple as much as a 5% interest rate would yield. Slight differences in fitness get magnified in a similar way. Over time, a genotype with a slightly higher relative fitness can come to dominate a population.

The power of natural selection is stronger in larger populations than in smaller ones. In small populations, random changes can obliterate beneficial mutations. In large populations, genetic drift has a weaker effect. **Figure 6.8** shows a computer simulation that illustrates this effect in which an allele with a selective advantage of 5% is added to populations of different sizes. In the big population of 10,000 individuals, it becomes more common in all the simulations. In a population of 10 individuals, however, it disappears from half of the simulations. Fitness, in other words, is not a guarantee of survival.

While computer simulations and mathematical models can help reveal the workings of natural selection, experiments are essential as well. It's possible to run these experiments on animals, but, as the lizards of Pod Mrčaru demonstrate, it can be many years before the results of such experiments emerge. As a result, a number of scientists have turned instead to microbes. They reproduce very quickly—the gut microbe *Escherichia coli* can divide as quickly as three times in an hour. And, because bacteria are very small, a scientist can rear hundreds of millions of them in a single laboratory flask.

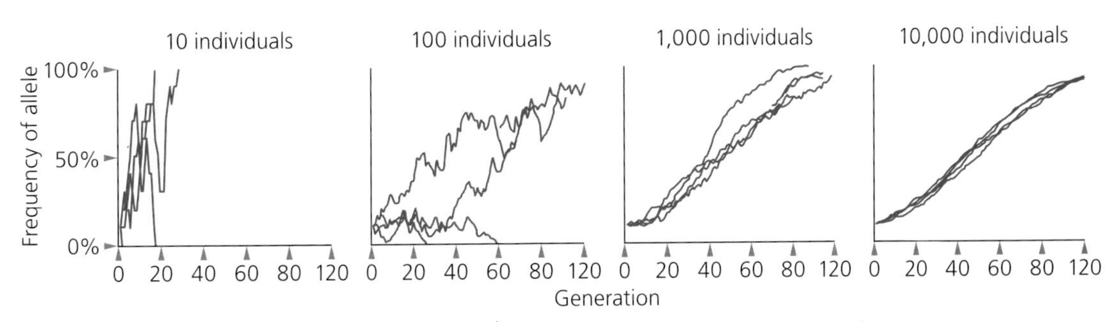

Figure 6.8 Natural selection is weak in small populations and strong in large ones. These graphs show the results of computer simulations in which an allele that raises fitness by 5% is added to populations of different sizes. In the smallest populations, it disappears from half the simulations. But in large populations the allele becomes more common in all of them. (Adapted from Bell, 2008)

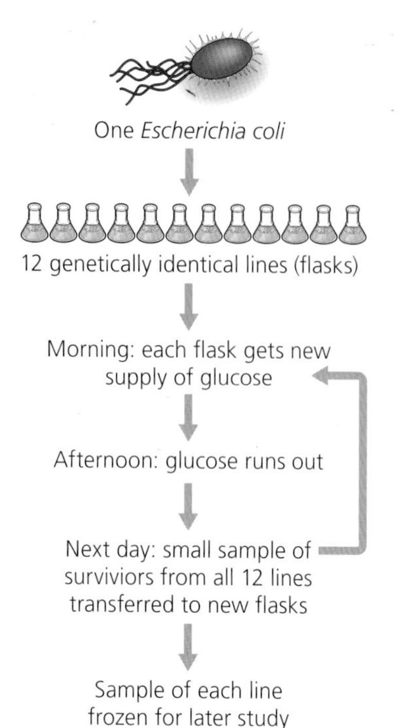

Figure 6.9 Richard Lenski and his colleagues have bred bacteria for 20 years using this method.

The longest-running of these microbial experiments is taking place in the laboratory of Richard Lenski at Michigan State University (**Figure 6.9**). In 1988, he used a single *E. coli* bacterium to start 12 genetically identical lines of bacteria. He reared each of those lines in a flask, giving them a meager diet of glucose. The bacteria would run out of sugar by the afternoon, and, the following morning, Lenski's students would transfer 1% of the broth to a new flask with a fresh supply of glucose. The original strain Lenski used had been reared for decades on an abundant supply of sugar. As a result, the bacteria could grow only very slowly at first on their restricted diet. Any mutations that would speed up their growth or boost their survival rate might be favored by natural selection.

Every 500 generations, Lenski stored away some of the bacteria from each of the 12 lines in a freezer. They became a frozen fossil record. By thawing them out later, Lenski could directly compare them with younger bacteria, comparing how quickly they grew under the same conditions—a direct measurement of their relative fitness. (This shows another big advantage that bacteria have over lizards!)

The experiment has now progressed for 44,000 generations. (It would have taken about a million years if Lenski were using humans as experimental organisms instead of bacteria.) **Figure 6.10** shows the evolution of Lenski's *E. coli* over the first 20,000 generations. The bacteria became more fit in the new environment than their ancestors in all 12 lines. (The vertical bars show the range of relative fitness in the lines.) In other words, all 12 of the lines have experienced natural selection: they have acquired mutations that make them more efficient at growing under the conditions that Lenski set up.

Because Lenski and his colleagues have reared trillions of bacteria, the microbes have experienced a vast number of mutations. Most of the mutations have been harmful or neutral. Only a few have been beneficial enough to spread by natural selection. But spread they have, and Lenski and other researchers are beginning to pinpoint the individual mutations.

In 2007, for example, Bernard Palsson, a biologist at the University of California at San Diego, fed *E. coli* glycerol, an ingredient in soap on which the bacteria normally grow very slowly. Like Lenski, he reared genetically identical lines of *E. coli* from a common ancestor on their tough new diet. Within a few hundred generations, all of his lines of bacteria were growing two to three times faster on glycerol than their common ancestor.

Palsson sequenced the genome of the ancestor and compared it to some of the genomes of its descendants. He discovered a few mutations had become fixed in each line. A mutation would arise in a single microbe, and, if that mutation was favored by natural selection, it allowed the microbe's descendants to

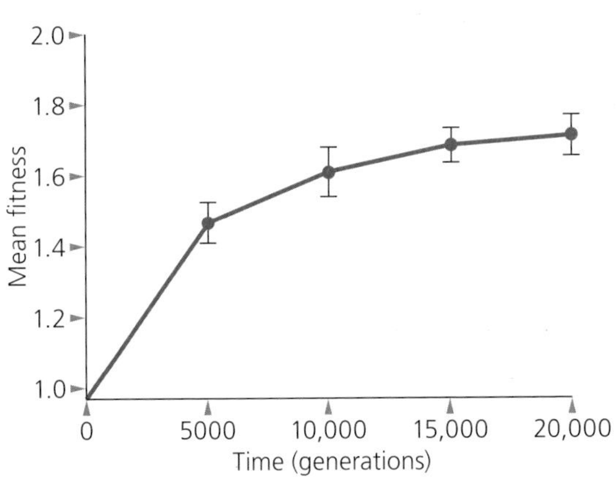

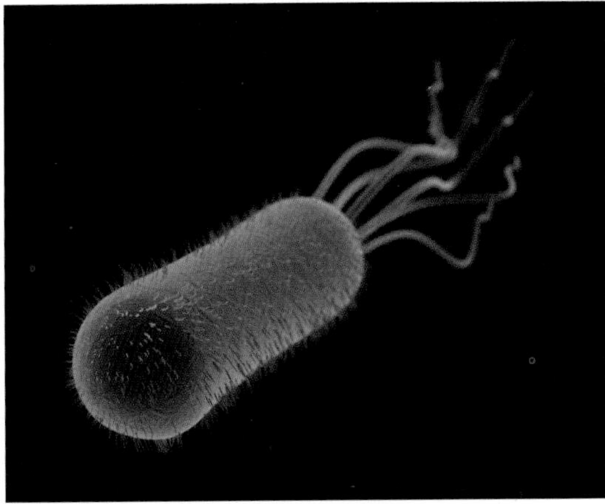

Figure 6.10 The bacteria in Lenski's experiment have experienced natural selection. New mutations have caused the descendants to reproduce faster under the conditions of the experiment than their ancestors. (Adapted from Ostrowski et al., 2008)

take over the entire population. A new mutation in one of its descendants would lead to a similar takeover.

Instead of thawing out the ancestors of these evolved bacteria, Palsson used a high-tech method. He made copies of the mutated genes and then inserted them, one by one, into the ancestral *E. coli* genome. Palsson found that the "reevolved" ancestor grew faster, thanks to the mutations, than it had with its original DNA. Experiments like those by Lenski and Palsson are turning evolutionary biology into a precise experimental science, in which individual mutations favored by natural selection can be identified and verified.

Selection in the Balance

Lenski and Palsson study natural selection in asexual bacteria. Sexual reproduction introduces some new wrinkles to evolution. It can bring together beneficial mutations from two parents in a single offspring, which will be more fit than individuals that carry deleterious mutations. As the fit individuals come to dominate the population, deleterious mutations are gradually purged. The same is not true for asexual organisms. Bacteria that only inherit DNA from their ancestors can't combine beneficial mutations.

Sexual reproduction adds another wrinkle to natural selection by giving every new offspring two copies of each gene. Sometimes one copy of a particular allele raises an organism's fitness, but two copies lower it. That's the case with the *S* and *A* alleles for hemoglobin. The reason that there are so few *SS* carriers in Nigeria is that the *S* allele gives rise to a deformed hemoglobin molecule. The red blood cells that carry them are deformed as well, taking on a long, curved shape like the blade of a sickle. This deformity leads to a dangerous condition, known as sickle-cell anemia, in which many red blood cells die and others clump together, damaging blood vessels, organs, and joints. Without medical treatment, sickle-cell anemia is often fatal. Sickle-cell anemia is the reason why *SS* carriers are out of Hardy–Weinberg equilibrium. Few people with sickle-cell anemia live long enough to have children. As a result, the *S* allele experiences strong negative selection.

But there are also fewer people with *AA* alleles than expected, and there are more people with *AS*. That's because the *S* allele does more than just cause red blood cells to sickle. It also protects people from malaria, a disease that kills 881,000 people a year and infects an estimated 247 million. Malaria is caused by a single-celled protozoan called *Plasmodium,* which is carried by mosquitoes. When a mosquito bites a victim, *Plasmodium* slips into the bloodstream. It then invades red blood cells and replicates inside them. The infected red blood cells become deformed and tend to clog small blood vessels, sometimes leading to fatal bleeding. For reasons that aren't yet clear, the *S* allele prevents infected cells from becoming deformed, reducing the risk of dying from malaria.

People with one copy of each allele enjoy the protection against malaria that comes from the *S* allele, but they don't suffer the potentially lethal sickle-cell anemia that can come with having two copies (**Figure 6.11**). As a result, the *AS* genotype is more common in Nigeria (a country with high rates of malaria) than expected, and the *AA* genotype is less common.

Instead of favoring only one allele, natural selection in this case promotes genetic variation in a population. Biologists call this special form of selection balancing selection. In the case of *S* and *A* alleles, balancing selection makes the population more resistant overall to malaria. Unfortunately, it also leaves millions of people suffering from sickle-cell anemia. If the *S* allele had no malaria-protecting qualities, it would rapidly become very rare, because people with two copies of it would have fewer children. But heterozygotes have so much reproductive success that they raise the number of *S* alleles circulating in the population, raising the odds of some people being born with two copies of the allele.

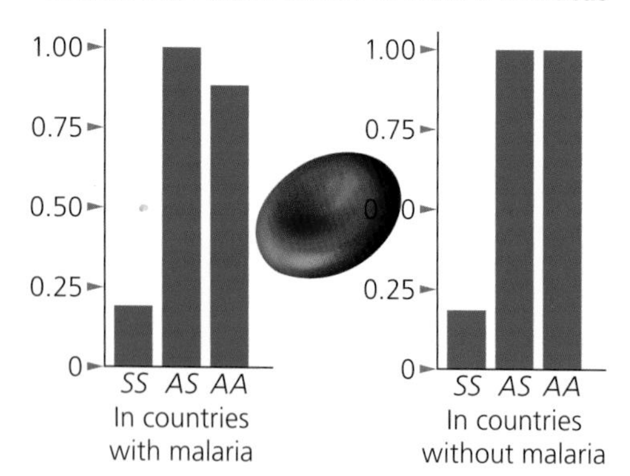

Figure 6.11 People with one copy of the *S* allele are more likely to survive malaria than people who are *AA* homozygotes. People who are *SS* homozygotes suffer sickle-cell anemia and have much lower fitness. The higher fitness of *AS* heterozygotes has the unfortunate effect of raising the frequency of *SS* homozygotes.

Sickle-cell anemia drives home a sobering truth about the nature of fitness: fitness is not an inherent quality. It emerges from the relationship of organisms to their environment. If malaria were eliminated tomorrow, the *AS* genotype would immediately lose its relative fitness, and the *S* allele would begin to disappear.

The Speed of Evolution

When a population experiences selection, it may evolve quickly or slowly in response. The speed of the response depends on the strength of selection, the amount of variation in the population, and how much of that variation is inherited.

Selection is strong when a small fraction of a population reproduces much more than the rest. Imagine that only the big fish in a pond can survive a drought and reproduce. **Figure 6.12** shows a graph of the sizes of fish in a hypothetical population. When selection is strong, only the biggest fish are able to reproduce. In milder conditions, smaller fish might be able to survive and reproduce. In that case, selection is weaker.

The effect that selection has on a population depends on the population's variation. As we saw in the last chapter, the total variation in a trait has two sources: variation from genetic factors and variation from environmental factors. The fraction of the variation due to genetic factors is known as the heritability of a

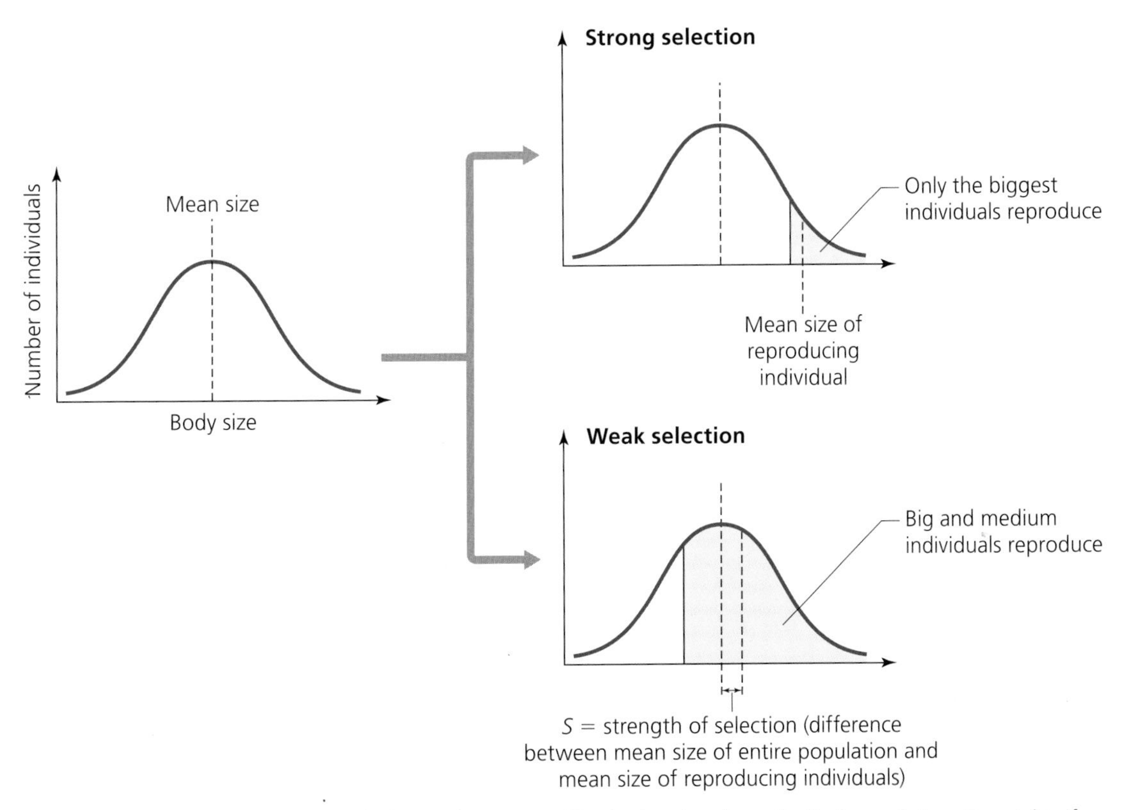

Figure 6.12 The graph on the left shows the range of body sizes in a hypothetical population. Top right: If the population experiences strong selection for large body size, only the very biggest individuals reproduce. The mean size of the reproducing individuals is much bigger than the mean of the entire population. Bottom right: If selection is weak, big and medium individuals reproduce, and the mean size of the reproducing individuals is much closer to the mean size of the entire population.

trait. If body size is only the result of environmental factors—the temperature of the water, for example, or how much food a fish larva finds—then the heritability of body size is zero (**Figure 6.13**). All of the big fish that get to reproduce will give rise to a generation of new fish that has the same variation. At the other extreme, when the heritability equals one, all of the variation is due to genetic factors. In such a case, the average size of the fish will increase.

The interplay between these factors can be summed up in an elegantly brief equation: The response of a population is the product of the heritability of a trait times the strength of selection. If selection is strong, a population can respond with a rapid change, even if a trait is only weakly heritable. And even weak selection can lead to significant evolutionary change if a trait's heritability is high.

A trait such as body size is not the only thing that evolves when a population experiences natural selection. The individuals with the greatest reproductive success are the ones with the alleles conferring on them the greatest fitness. With each generation, those alleles become more common in the population. As a result, the average fitness of the population rises as well. As those alleles become more common, less fit ones become rare. In other words, there's less genetic variation in the population. Less genetic variation means the heritability of the trait declines. And less heritability, in turn, means the evolutionary change with each passing generation becomes smaller.

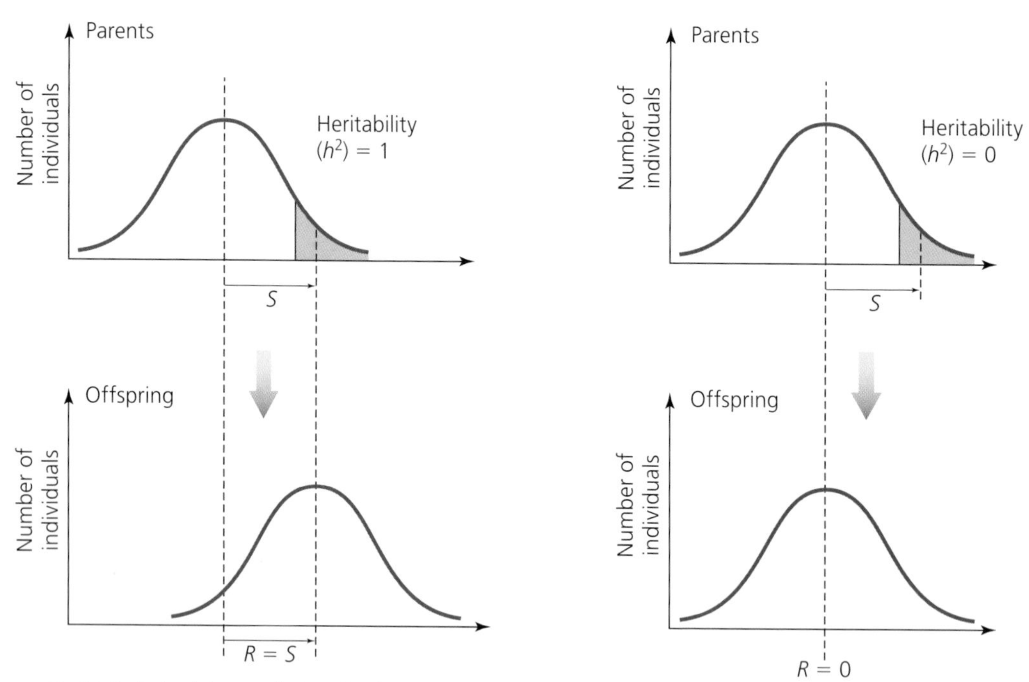

Figure 6.13 Left: In this population, body size is a completely heritable trait. If the population experiences strong selection for body size, the next generation's mean body size increases dramatically. The response (*R*) to selection is equal to the selection. Right: In this population, body size has no heritability. In other words, the size of parents is not correlated with the size of their offspring. Even if the population experiences strong selection for body size, the mean size does not change in the next generation. The response is zero.

One way to think about the relationship between genetic variation and natural selection is to imagine that a population is moving across a landscape of mountains and valleys. The landscape is actually a graph, with the *x* and *y* axes representing two traits (see **Figure 6.14**). The height of the landscape (the *z* axis) is the fitness of a particular combination of those traits. By increasing the average fitness of a population, natural selection pushes it uphill. As the genetic variation in the population shrinks, its fitness rises more slowly. Eventually, it stops at a peak.

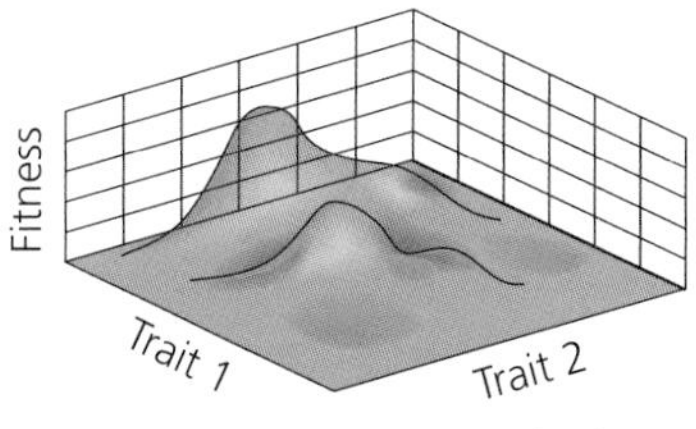

Real populations—like the *E. coli* cultures in Richard Lenski's lab—climb these adaptive hills. But ultimately, this fitness landscape is only a useful metaphor. Like all metaphors, it has limits. Genetic variation can be replenished by new mutations or interbreeding with other populations, for example. If the environment changes, the fitness of different traits changes as well. Instead of a fixed terrain, the fitness landscape is more like a surging ocean.

Figure 6.14 One way to think about fitness is as an evolutionary landscape. Certain combinations of traits represent peaks of fitness, surrounded by valleys of lower fitness. Natural selection can push a population up to one of these peaks.

Natural Selection All Around

Carrying out an experiment on selection in a laboratory brings many advantages. Scientists can carefully control the conditions under which selection occurs. In Richard Lenski's experiment, for example, all 12 lines of bacteria grew on exactly the same diet of sugar. Every single microbe in the study descended from a common ancestor. Lenski's frozen fossil record lets him step back in time.

Field biologists who study populations in the wild do not enjoy these luxuries, but, with other methods, they can measure natural selection as well. They determine how heritable a trait is, and then they figure out how variations in that trait raise or lower an organism's fitness.

It can take many years of hard labor to discover these things. For three decades, for example, Peter and Rosemary Grant of Princeton University have been visiting the Galápagos Islands in the eastern Pacific. It was on those islands that Darwin discovered the finches that made such an impression on him (page 27). Darwin's finches, as they're now called, have been assigned to 14 species. On a given island, a few hundred birds may be born in a given year, and most spend their entire lives there. The Grants have made some of the most precise measurements of natural selection on these birds.

The Grants have focused much of their attention on the medium ground finch (*Geospiza fortis*), which lives on a small island called Daphne Major. They survey every bird, measuring its body mass, the width of its beak, and many other vital statistics. They can trace families, determining how many offspring each bird has. From year to year, the Grants can also compare individual finches to their off-

For over 30 years, Peter and Rosemary Grant have studied natural selection in Darwin's finches on the island of Daphne Major.

spring to determine how strongly inherited each kind of variation is.

For a medium ground finch, the Grants found, the size of its beak can make the difference between life and death. Medium ground finches use their heavy beaks to crack seeds. On Daphne Major, the birds have a choice—small seeds from a plant called *Chamaesyce amplexicaulis* (spurge) and hard, woody seeds from a plant known as *Tribulus cistoides*, commonly called caltrop. Finches with big beaks (11 millimeters deep) can crack open the caltrop seeds in 10 seconds. Finches with beaks 10.5 millimeters deep need 15 seconds. If a bird's beak is 8 millimeters deep or less, it takes so long to crack caltrop seeds that the bird gives up on them altogether. Instead, it eats only small spurge seeds.

The Grants found that genes help to produce the range of big and small beaks. Big-beaked birds tend to produce chicks with big beaks, and small-beaked birds tend to produce chicks with small beaks. That means that the average beak size on Daphne Major can be altered by natural selection (**Figure 6.15**).

In 1977, Daphne Major was hit by a major drought. Most of the *C. amplexicaulis* plants died, leaving the medium ground finches without any small seeds to eat. Many of the birds died because they couldn't crack open caltrop seeds. The Grants discovered that within a few years the population of finches had recovered. But now their beaks were, on average, 4% deeper (about a fifth of a millimeter). Finches with bigger beaks had a better chance of surviving the drought and could therefore produce a bigger fraction of the next generation. In other words, natural selection caused the average size of the beaks of medium ground finches to increase.

Five years later, the Grants were able to see natural selection at work again. At the end of 1982, heavy rains came to the islands. *C. amplexicaulis* bloomed, producing lots of small seeds. Now small-beaked birds had the advantage. They could eat small seeds more efficiently than the big-beaked ones, allowing them to grow faster and have more energy for producing offspring. Now natural selection favored them. The average size of beaks decreased by 2.5% (about a tenth of a millimeter).

The work of the Grants represents one of the first large-scale studies of natural selection in the wild. Since they began publishing their first results, many other scientists have followed suit. Today there are hundreds of studies documenting selection in wild populations.

In southern France, for example, scientists have shown how natural selection is adapting a plant to city life. A small flower called *Crepis sancta* has colonized the city of Marseilles, growing in the patches of ground around trees planted along the streets. The plants can make two different kinds of seeds—one that can drift off in the wind, and another that simply drops to the ground.

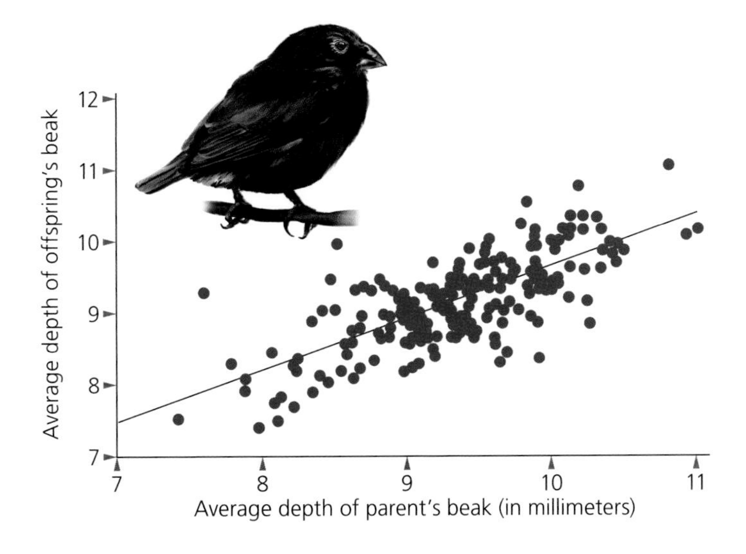

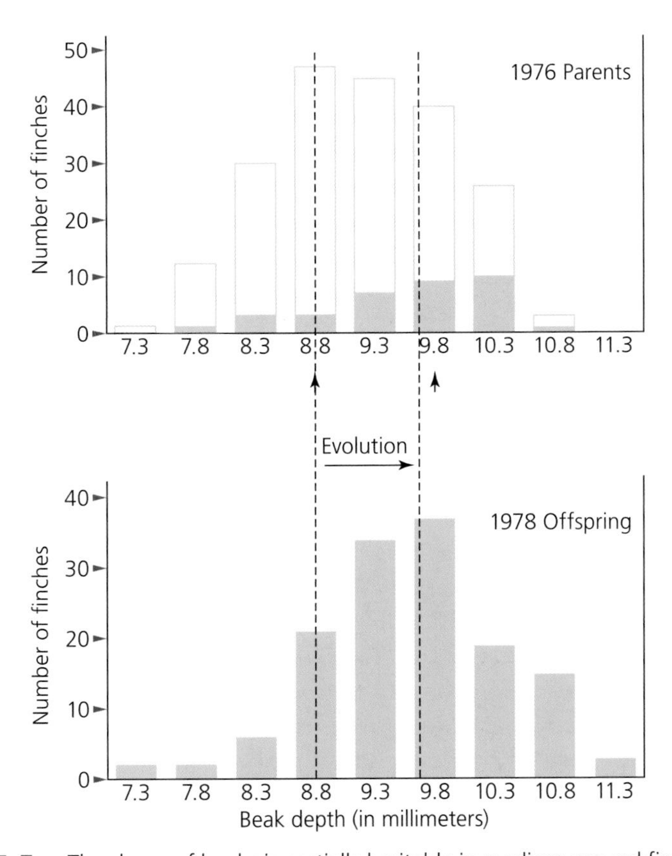

Figure 6.15 Top: The shape of beaks is partially heritable in medium ground finches. Middle: During a drought in 1976, birds with deeper beaks had more chicks than birds with smaller beaks. (The white bars show the total number of medium ground finches on Daphne Major. The blue bars show the number of birds that reproduced.) Bottom: The average beak sized increased in the offspring born to birds during the drought. (Adapted from Grant, 2007)

The scientists hypothesized that wind-carried seeds would be a burden to plants living in the city because they would end up stranded on concrete. Dropped seeds would have a better chance of surviving, because they'd fall onto the patch of ground where the parent plants grew. The scientists raised *C. sancta* from Marseilles in a greenhouse alongside *C. sancta* from the countryside. Under the same conditions, the scientists found that the city plants were making 4.5% more nondispersing seeds than the ones in the countryside. The scientists estimated that about 25% of the variation in the ratio of the two types of seeds is controlled by genetic differences. With these levels of heritability and selection, it should have taken about 12 generations to produce the observed change of 4.5% in the seed ratio. As predicted, about 12 generations of plants have lived in Marseilles since the sidewalks were built.

Without intending it, humans created a new environment for these plants, which are now adapting to it. As more time passes, the city plants may continue to make more dropping seeds and fewer wind-borne ones.

Drinking Milk: A Fingerprint of Natural Selection

If your ancestors hail from western Europe, chances are you can digest milk. If you're Chinese, chances are you can't. It turns out that the difference is the result of natural selection on humans over the past few thousand years.

Humans are mammals, and one of the hallmarks of living mammals is the production of milk. Young mammals drink milk from their mothers and digest it by producing a protein called lactase. Lactase acts like a pair of scissors, snipping apart molecules of a sugar in milk called lactose. The fragments of the lactose can then be absorbed into the bloodstream. As mothers wean their young, however, mammals typically stop producing lactase. They don't waste energy making a protein they no longer need.

About 70% of humans go through this same process. As a result, they can digest milk when they're young, but they have a difficult time with it when they're adults. When they drink milk, the lactose builds up in their guts, where bacteria can feed upon it. The waste released by the bacteria causes indigestion and gas.

In about 30% of people, however, the gut continues to make lactase into adulthood. They can consume milk and other dairy products into adulthood without any discomfort.

Scientists have searched for many years for the alleles that make people lactose-tolerant or lactose-intolerant. They've found several alleles of a segment of DNA called *LCT*, which contain the genes humans use to produce lactase. The alleles show a strong association with the ability to digest milk into adulthood.

How did 30% of people end up with alleles for lactose tolerance? As we've seen, two possible explanations are genetic drift and natural selection. To distinguish between these two possibilities, scientists at Finland's National Public Health Institute took advantage of the way recombination scrambles genes. Every time parents pass down genes to their children, their chromosomes swap chunks of DNA. If a new allele arises through a mutation, it's passed down with some surrounding DNA. But, over time, as chromosomes are cut and swapped again and again, the allele remains linked to fewer and fewer of its original neighbors.

Natural selection spreads alleles much faster than genetic drift, giving recombination less time to separate an allele from neighboring regions of DNA. An allele that undergoes rapid natural selection will be very common, and it will also be linked with the same neighboring alleles in many people (**Figure 6.16**). That's exactly the pattern that the Finnish scientists found in people with *LCT* alleles that confer lactose tolerance. Large pieces of DNA surrounding the *LCT* gene were shared by all of the people who were found to be tolerant of lactose.

Another important clue comes from the cultures of the people in which these alleles are found—people who live in places like northern Europe, East Africa, and Saudi Arabia. These places are home to cultures that took up cattle herding over the past 10,000 years and made cows or camels crucial to their diet.

The best-supported hypothesis about lactose tolerance that scientists have come up with is that mutations arose in these herders that prevented the *LCT* from shutting off as it had in the past. In their herding cultures, the ability to digest milk brought huge benefits to the herders, simply because milk was so plentiful. People who could get protein and other nutrients from milk were more

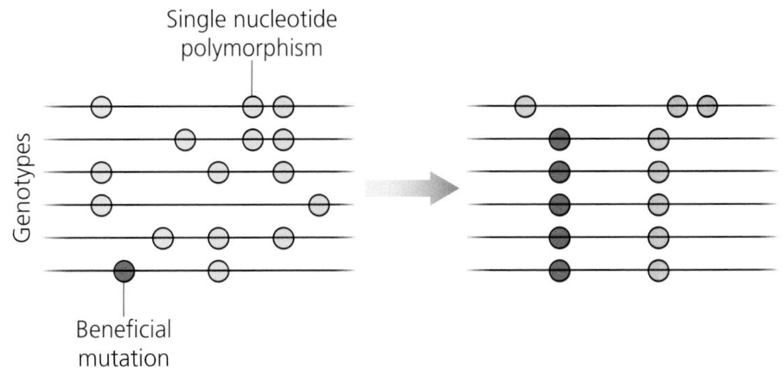

Figure 6.16 During recombination, segments of chromosomes get swapped. As a result, long blocks of DNA gradually get broken up and shuffled into new combinations in a population. But natural selection can interrupt this shuffling. Left: Each line represents a block of DNA in an individual. The blue dots represent genetic markers found in only some members of the population. A new mutation (red) arises in one individual and raises its fitness. Right: Individuals who inherit a block of DNA with the new mutation also have higher fitness, and so the mutation spreads rapidly through the population, along with its neighboring DNA. As a result, an unusually high number of individuals will carry the same block of DNA containing the mutation. This unusual pattern is a good clue that strong natural selection has taken place.

likely to survive and to pass on their mutant copy of *LCT* to their offspring. In other cultures, similar mutations arose from time to time, but they offered no advantage, because milk was not a common part of their diet.

The Geography of Fitness

As an allele spreads through a population, it also spreads through space. If a giraffe carrying a beneficial allele strolls across the Serengeti and joins a new herd, the allele may then become common in the new herd as well. Another giraffe can then carry it on to an even more distant herd. The rate at which alleles move between populations (a process known as gene flow) is controlled by many variables. The amount of gene flow depends on how far individual organisms move, for example, and how far their gametes move. A tree obviously will not pull up its roots and go for a stroll, but its pollen can drift far and wide. Seeds can get stuck to the feet of birds and cross entire oceans.

Many of the genes that flow between populations are neutral. They don't raise or lower the fitness of organisms, no matter where those organisms live. But gene flow also carries beneficial and deleterious alleles between populations. Once in a new population, an allele that previously raised fitness may actually lower it. In this way, new copies of alleles may arise in one population, only to disappear in another—like water coming out of a faucet and going down a drain.

This complex movement of genes accounts for many patterns in nature. Take, for example, the scarlet kingsnake (*Lampropeltis triangulum elapsoides*), which lives in the eastern United States. In the southeastern part of their range, in such states as Florida and Georgia, scarlet kingsnakes have a colorful pattern of red, yellow, and black rings. But in the northern part of their range (in Tennessee, Kentucky, and Virginia), scarlet kingsnakes are much more red.

In 2008, George Harper and David Pfennig, two biologists at the University of North Carolina, discovered why the same species of snake looks so different in different places (**Figure 6.17**). In the southeastern portion of their range, scarlet kingsnakes live alongside eastern coral snakes (*Micrurus fulvius*). Eastern coral snakes have a potentially fatal bite, and, like many species of venomous animals, they have evolved a bright pattern. Predators, such as carnivorous mammals, recognize the pattern on the coral snakes and avoid them. Scarlet kingsnakes are not venomous, but the ones that live alongside coral snakes have evolved a similar pattern. As a result, predators avoid the scarlet kingsnakes even though they're harmless.

The range of scarlet kingsnake is much larger than coral snakes. In the northern part of their range, where scarlet kingsnakes don't live alongside coral snakes, their pattern is different from that of coral snakes. The variation between the northern and souther populations of kingsnakes is not a result of their being genetically isolated from each other. Collecting DNA from scarlet kingsnakes throughout their range. Harper and Pfennig documented gene flow across the entire range of the snakes. Scarlet kingsnakes from the south migrate to other populations, bringing their alleles for mimicking coral snakes. So why haven't all the scarlet kingsnakes evolved into mimics?

Because, it turns out, mimicking a coral snake only provides protection from predators that live in the range of the coral snake. Predators outside the coral snake's range are more willing to attack snakes with a coral-snake–like pattern. What serves as an effective warning in one place becomes a way to draw the

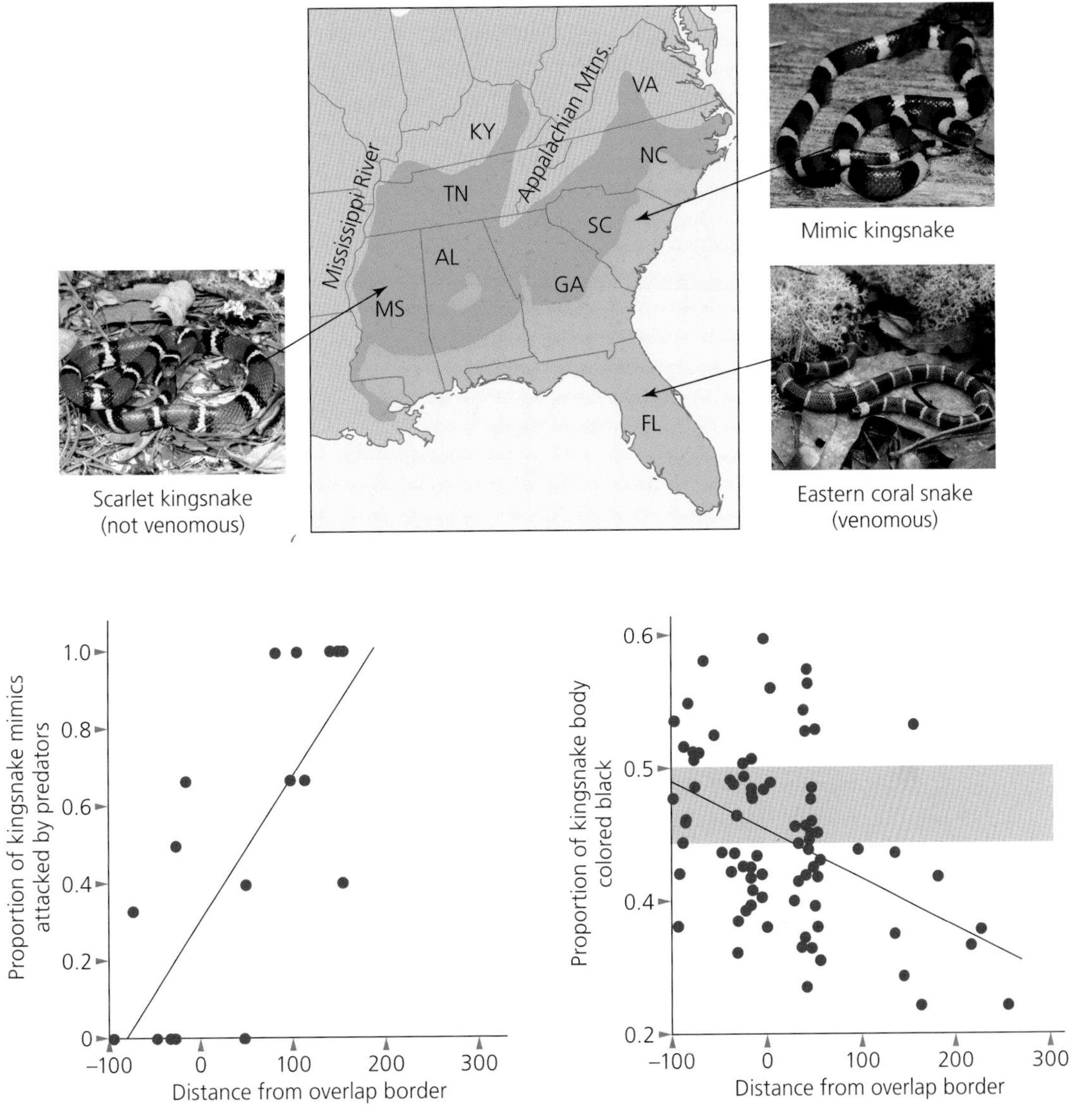

Figure 6.17 Scarlet kingsnakes have evolved to mimic the venomous coral snakes in regions where they overlap. But farther away from these regions, the kingsnakes look less like the coral snakes. Natural selection favors genes for mimicry in the regions of overlap, but as these genes flow to other regions, they get eliminated. Predators that live in coral-snake territory learn to avoid their bright color pattern. But in other regions, they are more likely to attack kingsnakes with this pattern, because it's easy to spot. (Adapted from Harper and Pfennig, 2008)

attention of predators in another. The farther away scarlet kingsnakes are from the overlap zone, Harper and Pfennig found, the more strongly natural selection works against coral-snake mimics. As a result, the farther you move from the overlap zone, the less like coral snakes the scarlet kingsnakes look.

The Limits of Selection

Natural selection does not bestow every adaptation a species may need to survive some particular set of conditions. It does not achieve perfection as it increases fitness. Balancing selection actually fosters the persistence of sickle-cell anemia in human populations. Thanks to pleiotropy, genes can have several effects, some beneficial and some harmful.

Understanding the constraints on natural selection is important not just for understanding how nature works but also for predicting how biotechnological inventions will work. Over the past decade, many farmers have begun to plant genetically modified crops that can fight off insects. The plants carry a gene from a bacterium that makes a toxin that's lethal to insects. The gene is known as Bt, which comes from the name of the species to which it belongs, *Bacillus thuringiensis*.

When Bt was applied to cotton and to other crops, some biologists warned that insects might evolve with a resistance to the toxin. It's a lesson that scientists have learned time and again over the past century, as many insects have evolved resistance to conventional pesticides. But an understanding of evolution also let them come up with a strategy to keep the development of resistance down.

In a field planted with Bt-treated crops, insects that could resist Bt were able to flourish. But the mutations that gave them their resistance also imposed a cost. In a field planted with ordinary crops, the resistant insects would be outcompeted by other insects that didn't invest so much in detoxifying Bt. In the two fields, evolution moves in opposite directions. And if the two fields are next to each other, the insects can mate and mix their genes. The resistant insects mate with vulnerable neighbors and produce less resistant offspring.

Scientists advised farmers planting crops that were to be treated with Bt to also plant some ordinary crops. The fields of ordinary crops would serve as "refuges," where Bt-susceptible insects could outcompete resistant ones. The susceptible insects would then spread their genes into the surrounding fields and slow the rise of resistance to Bt.

Several years after Bt crops were introduced, Bt-resistant insects began to evolve in significant numbers. In 2008, scientists surveyed the rise of resistance. In states with large areas of refuge, resistance evolved much more slowly than in states with small areas of refuge. The farmers had carried out a giant experiment in evolution, and it had turned out as the evolutionary biologists had predicted.

TO SUM UP...

- Mutations are spontaneous changes in the genetic information of an organism. Offspring inherit mutations from their parents.
- Mutations are rare, but they arise at a roughly steady rate.
- Mutations can have harmful or beneficial effects, but the majority of mutations are neutral.
- Mutations can become more or less common in a population over generations.
- The Hardy–Weinberg model shows how genetic variation can be sustained in a population. But it is based on the assumption that there is no migration into a population, no genetic drift, and no selection.
- Genetic drift is the random rise or fall of alleles in a population. It is stronger in smaller populations than in larger ones.
- Relative fitness is the ratio of population growth rates between two genotypes.
- Natural selection can rapidly fix an allele in a population or cause it to disappear.
- A slightly higher degree of fitness can allow one genotype to outcompete another.
- Balancing selection can help to maintain genetic variation in a population.
- The rate of evolution depends both on the heritability of traits and on the strength of selection.
- Examples of natural selection include the changing size of beaks on Darwin's finches and the evolution of lactose tolerance in humans.
- The fitness of alleles can vary across the range of a population.

The History in Our Genes

7

Sarah Tishkoff has been traveling through Africa from end to end for more than a decade. She took her first trip there as a graduate student in genetics at Yale University, and she still goes back, now that she's a professor at the University of Pennsylvania. She has bounced along cratered roads in Tanzania, and she has traveled aboard hand-cranked ferries in the jungles of Cameroon. On her journeys, Tishkoff takes

Left: Scientists are using DNA to determine how the peoples of the world are related to one another. Above: Sarah Tishkoff of the University of Pennsylvania criss-crosses Africa to gather genetic samples to study human diversity.

syringes, vials, and centrifuges. Her goal is to create a genetic portrait of the 922 million people who live in Africa. She and her colleagues have gone a long way toward that goal, having collected DNA from more than 7,000 people from more than 100 ethnic groups.

There are many things that Tishkoff hopes to learn from this portrait. She and her colleagues are beginning to identify alleles that make some Africans more vulnerable to certain diseases and resistant to others. But she also has come to Africa to understand history—not just the history of Africans, but the history of us all. Tishkoff and her colleagues have created a detailed genealogy of the human race. Our species got its start in Africa about 200,000 years ago, and it diversified there for tens of thousands of years before a few Africans migrated into Asia, Europe, and the New World.

In Chapter 4, we saw how scientists use morphological traits to construct evolutionary trees. Until the 1990s, that sort of information was the only kind available to evolutionary biologists. But powerful computers and gene-sequencing technology have brought a revolution to the study of the tree of life. A remarkable amount of history is encoded in the DNA of living species, and scientists are using it to get answers to a range of questions about biology—from the origin of diseases to the functions of mysterious genes.

The Genetic Archive

Homologies such as hands and wings are evidence of the common ancestry of people and birds. But we also have homologies in the DNA that encodes our bodies—similar genetic sequences we inherited from common ancestors. Over the past 20 years, evolutionary biologists have discovered how to use DNA to draw evolutionary trees, and today molecular phylogenies are published by the thousands every year.

In 1992, David Hillis and his colleagues at the University of Texas showed how to read history in genes by running an elegant experiment. They tracked the evolution of a virus called T7, which uses *E. coli* as its host. The scientists established a lineage of T7 from a single ancestor, rearing billions of descendants. Then they moved two of the viruses to new petri dishes, where they started new lineages. Once again, the scientists allowed the viruses to replicate (and to mutate). Later, they used these two petri dishes to start four new populations. And then they used those four to start eight petri dishes (**Figure 7.1**).

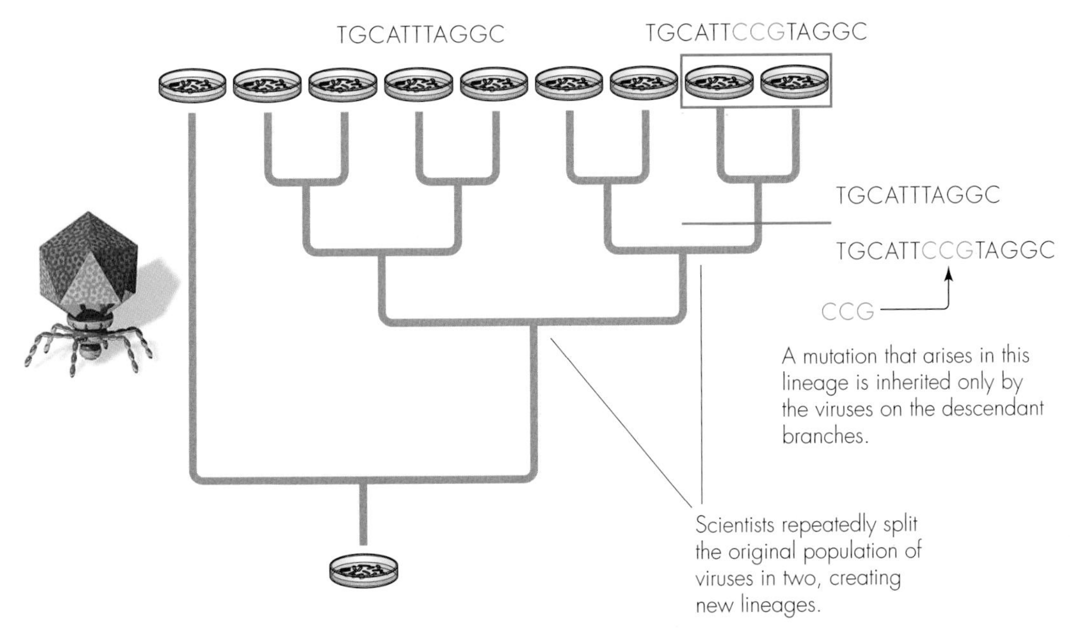

Figure 7.1 Researchers produced nine separate lineages of viruses from a common ancestor. By comparing the DNA from all the lineages, they were able to reconstruct the evolutionary tree of the viruses. (Adapted from Hillis, 1992)

You can think of the nine populations of viruses as nine tips of an evolutionary tree, with the original virus at its root. In each lineage, new mutations arose and were spread through natural selection or genetic drift. If a mutation arose in the common ancestor of two lineages, it would be inherited by both of them. The researchers set out to reconstruct the evolutionary tree of the viruses only by comparing the differences in their DNA.

It was very unlikely they'd be able to get the right answer by chance. With every new branch included in an evolutionary tree, there are more possible arrangements. In fact, the number of possibilities explodes as the number of branches in a tree increases. A tree with nine branches, like the one in Hillis's experiment, can be arranged in 135,135 different ways.

Today, scientists can easily sequence the entire genome of a virus and search it for mutations. Back in 1992, however, the technology for reading genomes did not yet exist. Instead, the scientists took advantage of a kind of natural DNA-recognizing technology. Bacteria make proteins known as restriction enzymes that can destroy invading viruses. They do so by latching onto specific short sequences of DNA in a virus and slicing it into fragments.

Hillis and his colleagues exposed viruses from each lineage to 34 different restriction enzymes. Each enzyme recognized a different sequence of DNA, and thus produced fragments of different sizes from a given virus. DNA from two genetically identical viruses would be cut into the same set of fragments. But mutations would prevent some restriction enzymes from cutting DNA at certain sites, while allowing them to cut it at new sites.

By analyzing the fragments of DNA, Hillis and his colleagues determined which viruses shared the same sites where the same restriction enzymes made their cuts. They then determined what sort of evolutionary history would be most likely to account for these results. They used five different methods to reconstruct the phylogeny of each of the viruses. In one method, they calculated which tree required the minimum number of steps to produce the restriction-enzyme sites in the different viruses. Imagine, for example, that two viruses have their DNA cut in precisely the same spot by the same restriction enzyme, and that two other viruses do not. It's possible that the first two independently evolved precisely the same DNA sequence. That would require two steps, one in each virus's lineage. It's also possible that the ancestor of all four strains acquired the mutation and then two of the strains lost it—three steps. And it's also possible that the common ancestor of the two strains with the mutation acquired it—one step. The third scenario offers the fewest number of steps. It also implies that the two lineages of viruses with DNA cut by the same enzyme at the same site are more closely related to each other than they are to any other virus lineages.

Using this method, Hillis and his colleagues drew a tree connecting the nine strains. The cladogram they produced matched the actual phylogeny they had produced in their experiment. In fact, the other four methods worked as well. Out of the 135,135 possibilities, the scientists were able to find the right one.

Species Trees and Gene Trees

Today, scientists can look at far more information in DNA than Hillis and his colleagues could. They can read individual bases of DNA. The human genome contains 3.5 billion base pairs, each of which can potentially hold a clue about our ancient history.

But scientists also have to cope with some serious shortcomings with DNA. DNA decays after death, and, even under the best circumstances, it disappears from fossils within a million years. So the vast majority of DNA that scientists study for clues to evolution comes from living species, not extinct ones. Yet living species represent only the youngest tips on the tree of life. Some 99% of all species that ever evolved have become extinct. The species alive today—all 10 million or more of them, depending on the estimate—are just a tiny remnant of the full diversity of life over its entire history.

Birds, for example, don't look much like their closest living relatives, alligators and crocodiles. The two groups of animals share a common ancestor that lived about 250 million years ago, and, once their lineages diverged, they underwent profound evolutionary changes. A DNA-based tree that includes only living species cannot reveal many details about how birds evolved from featherless reptiles.

The long branches that join living species can also lead scientists to draw the tree of life incorrectly. Over hundreds of millions of years, many mutations arise in an evolutionary lineage. In some cases, two lineages will acquire some of the same mutations independently. DNA is made up of only four bases, and so a single point mutation may well change a base to the same base more than once. This repeated mutation can make two species appear to be more closely related to each other than they really are.

Another challenge is the fact that the history of a gene does not always match the history of the species that carry it. Consider the evolution of humans, chimpanzees, and gorillas from a common ancestor. Over the course of that evolution, new alleles of genes evolved. When new species evolved, they inherited some of the alleles of their common ancestor, and not others. As those new species evolved, some of the alleles dwindled and disappeared. Others gave rise to new alleles.

Figure 7.2 shows some of the patterns that can be produced as genes and species diverge. In Figure 7.2A, a single allele in the ancestor of humans, chimpanzees, and gorillas is inherited by all three species. The gene's evolution is simple: it tracks the species tree. Things get a little more complicated in Figure 7.2B. In the common ancestors of gorillas, chimpanzees, and humans, a mutation produced a new allele of the gene. Both copies are inherited by all three living species. Despite this extra complexity, the history of the alleles parallels the history of the three species.

In Figure 7.2C the ancestral ape has two alleles, and both are passed down when the lineage splits. But by the time the living species of apes evolve, some of the alleles have been lost from each one. The surviving allele in humans is more closely related to the one in gorillas, not chimpanzees. If scientists were to analyze only these alleles, they might conclude that gorillas, rather than chimpanzees, are our closest living relatives.

Fortunately, scientists have found ways to clear up this sort of confusion between gene trees and species trees. Looking at the history of many genes, rather than just one, can give scientists a clear picture of the evolution of the

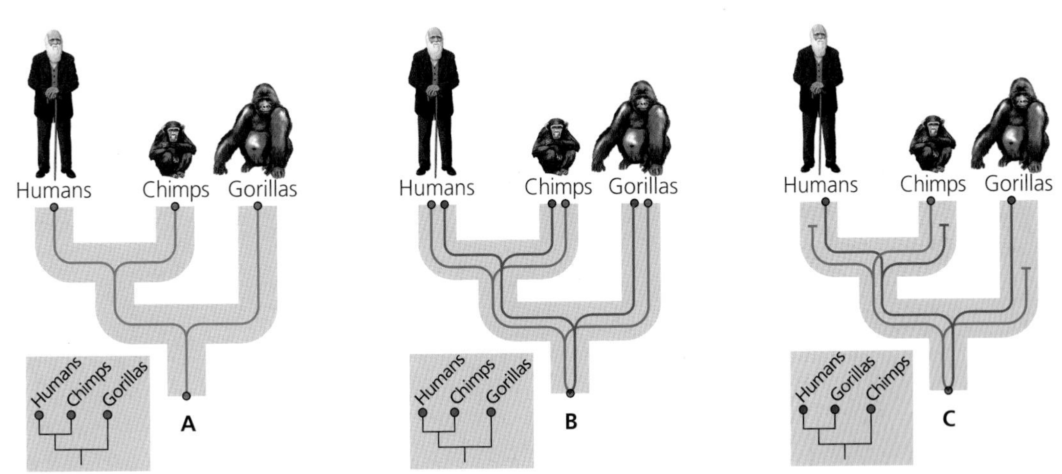

Figure 7.2 Genes do not always share the same history of the species that carry them. A: A gene diverges as apes evolve into new species. Its history is the same as the species. B: A new allele evolves in the common ancestor of African apes. Once again, the gene's history mirrors the species history. C: Due to the extinction of some alleles, the surviving version of the gene in gorillas is more closely related to the human gene than the chimpanzee gene. An analysis of this gene would suggest gorillas are more closely related to humans than chimpanzees are. One way to avoid these errors is to analyze many genes.

species that carry them. Mark Batzer, a geneticist at Louisiana State University, and his colleagues found this to be the case in their studies of a group of mobile elements called *Alu* elements (**Figure 7.3**).

These segments of DNA, measuring about 300 nucleotides long, are sometimes copied and inserted back into the genome. After a new *Alu* element arises in a single person, it may spread over generations. Some *Alu* elements are found almost exclusively only in certain ethnic groups. But some *Alu* elements are found in all people, which means that they arose in the common ancestor of living human beings. There are even some *Alu* elements shared by humans and other primate species, which must have arisen in their common ancestor, as long as 30 million years ago.

By comparing human *Alu* elements to those of other species, Batzer is reconstructing primate evolution. Batzer and his colleagues have identified 33 *Alu* elements found only in gorillas, chimpanzees, and humans, marking their common

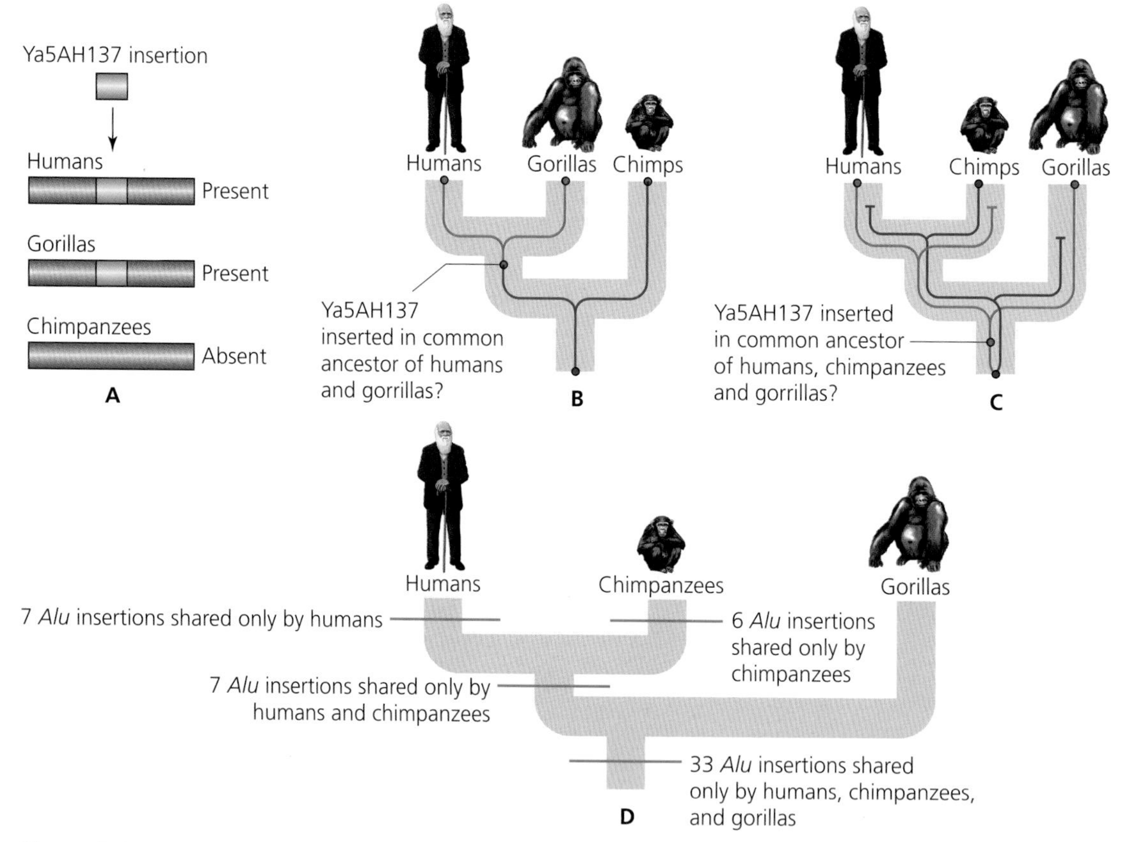

Figure 7.3 A: A mobile element called Ya5AH137 is found only in humans and gorillas. B: One possible hypothesis is that it was inserted in the common ancestor of gorillas and humans, which are more closely related to each other than to chimpanzees. C: An alternative hypothesis is that Ya5AH137 was inserted in the common ancestor of humans, chimpanzees, and gorillas. Chimpanzees, which are the closest living relatives of humans, did not inherit it. D: One way to test these alternatives is to search for similar pieces of DNA (known collectively as *Alu* elements) that arose after the common ancestor of humans, chimpanzees, and gorillas branched off from other apes. All of them point to chimpanzees being the closest living relatives of humans, showing that the hypothesis in C is the most likely.

ancestry. They also found seven *Alu* elements that are shared by humans and chimpanzees but not by gorillas. It's almost certain that this pattern could have emerged only if humans and chimpanzees are more closely related to each other than either species is to gorillas.

Looking at a single *Alu* element might have led Batzer astray. One *Alu* element in the human genome, called Ya5AH137, is found in the gorilla genome, but not in the genome of chimpanzees (Figure 7.3A). If you had only this particular bit of DNA on which to base your evolutionary tree, you might conclude that chimpanzees had branched off first, and that Ya5AH137 was inserted into the genome of the common ancestor of humans and gorillas (Figure 7.3 B). But all of the evidence from *Alu* elements points to a different conclusion: Ya5AH137 was inserted in the common ancestor of all three species. Some apes carried the insertion, while others did not. Chimpanzees ended up with the allele without Ya5AH137, while humans and gorillas still carry it. And chimpanzees, rather than gorillas, are our closest living relatives (Figure 7.3C).

Molecules and Morphology

As scientists gained the power to build evolutionary trees from DNA, they decided to test trees based on morphology alone. Paleontologists, for example, have long argued that the closest living relatives of tetrapods are lobe-fins, which today include lungfishes and coelacanths (page 64). That's a fairly precise prediction, since there are around 30,000 species of fishes alive today. Of all those fishes, paleontologists predicted that only a half dozen should share a close common ancestry with tetrapods.

In the 1990s, scientists began to gather fragments of DNA from tetrapods and fishes to test this hypothesis. The latest test was published in 2008 by Björn M. Hallström and Axel Janke of the Lund University in Sweden (**Figure 7.4**). They analyzed hundreds of segments of DNA from 12 species spanning the diversity of vertebrates, from lampreys to mammals. Their data showed that lungfishes were the closest relatives of tetrapods, and that coelacanths shared a common ancestry with lungfishes and tetrapods. The fossils made a prediction, in effect, and the DNA supported it.

Hallström and Janke were using molecular evidence to test a hypothesis based on morphology. But DNA also allows scientists to study how organisms are related to each other when morphology offers few clear clues. By comparing the DNA of living things, scientists can now explore parts of the tree of life that were once off limits. Here are three examples:

Humans: In Chapter 4, we saw how scientists use fossils to learn how humans evolved from extinct hominids. Paleontologists have unearthed many fossils of humans and their ancient hominid relatives. The oldest fossils that show clear signs of belonging to our own species, *Homo sapiens*, come from Africa and date back 200,000 years (page 56). The oldest fossils of our species beyond Africa are

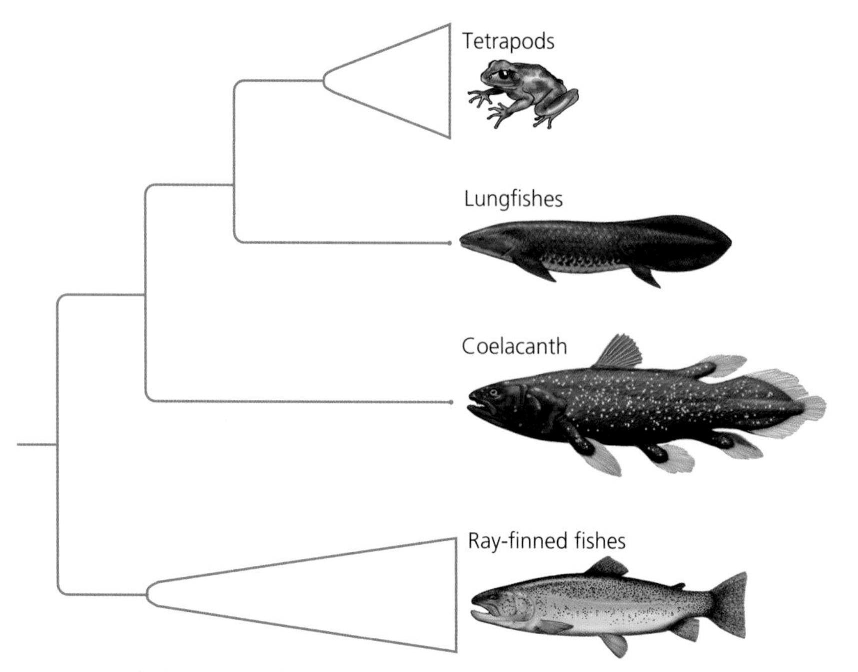

Figure 7.4 Evolutionary trees based on DNA and on fossils both support lobe-fins as the close relatives of tetrapods.

some remains in Israel dating back to about 100,000 years ago. Younger still are the first fossils of humans in Europe, Asia, and Australia. In the 1980s, Chris Stringer, a paleoanthropologist at the Natural History Museum in London, argued that these fossils were evidence that humans arose in Africa and then spread out to the other continents. Other lineages of hominids became extinct.

Geneticists have looked to human DNA to test this hypothesis. Sarah Tishkoff is in a particularly good position to do so, because she has gathered so much information about people in Africa, where Stringer proposed humans originated. Tishkoff and her colleagues analyzed DNA from Africans and compared their genetic sequences with those of people from other parts of the world. The scientists represented each person as the tip of a branch, and searched for each person's closest relative on the tree.

The results of one study are shown in **Figure 7.5**. It's based on DNA from mitochondria, tiny factories that generate energy in the cell. Mitochondrial DNA is useful to scientists like Tishkoff because it doesn't get shuffled together during sex, like the DNA in the nucleus. Instead, children inherit mitochondrial DNA only from their mothers. The sequence of DNA in each person's mitochondria is nearly identical with that of his or her mother, and her mother in turn, and so on, back through time.

When Tishkoff and her colleagues drew a tree from mitochondrial DNA, they found that Africans carry the biggest diversity of mitochondrial DNA. They also belong to branches that split off very early from those of other humans. The sequences of mitochondrial DNA of Asians, Europeans, and people of the New World are much more closely related to one another.

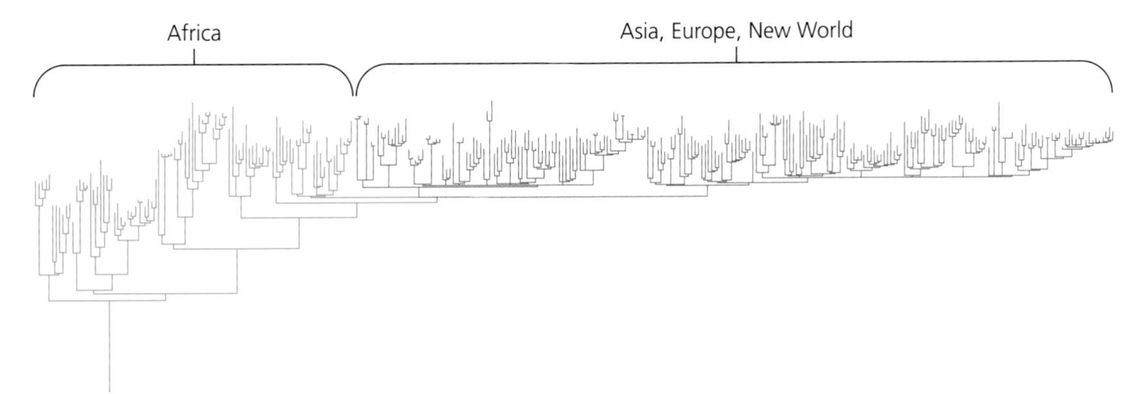

Figure 7.5 By comparing DNA from the mitochondria of humans, Sarah Tishkoff and her colleagues produced an evolutionary tree of our species. All mitochondrial DNA descends from African ancestors. Because humans have been in Africa much longer than in other parts of the world, Africans today are much more genetically diverse than other humans. A small group of Africans migrated out of the continent and became the ancestors of today's Europeans, Asians, and people of the New World. (Adapted from Gonder et al, 2005)

Tishkoff's results agree with the work of other researchers who have studied other genes. It appears that our species first evolved in Africa. Thousands of generations passed before some humans left the continent. Tishkoff's research even offers hints about where in Africa humans originated. She finds the greatest level of diversity and the deepest branches among the people of East Africa—the same region where the oldest fossils of humans have been found.

Darwin's Finches: Peter and Rosemary Grant have documented changes in the beaks of Darwin's finches through natural selection in as little as a few years (page 123). But the Grants also want to know how the birds have evolved over millions of years. Where did the birds come from, they'd like to know, and how did they evolve into 14 distinct species?

Of all Darwin's finches, only one does not live on the Galápagos Islands: the Cocos finch, which lives 800 kilometers (500 miles) away on Isla del Coco, another remote island in the east Pacific. No species of Darwin's finches lives on the mainland of South America. A few potential explanations come to mind. Perhaps the Galápagos Islands were colonized by several different species, each of which gave rise to new species on the islands. Or perhaps some Cocos finches that ended up on the Galápagos Islands produced their flock of Darwin's finches.

Bird skeletons are so delicate and the Galápagos Islands have such a harsh climate that there is no fossil record of Darwin's finches. So the Grants and their colleagues have been finding clues by grabbing the birds and drawing a few drops of blood from each one. The Grants and their colleagues then compared the DNA of Darwin's finches to that of other bird species that have been proposed as close relatives.

The scientists found that Darwin's finches are related to each other as illustrated in **Figure 7.6**. Darwin's finches are more closely related to each other than they are to any other known bird, which means that they all share a common ancestry. They did not evolve independently from different species. The tree

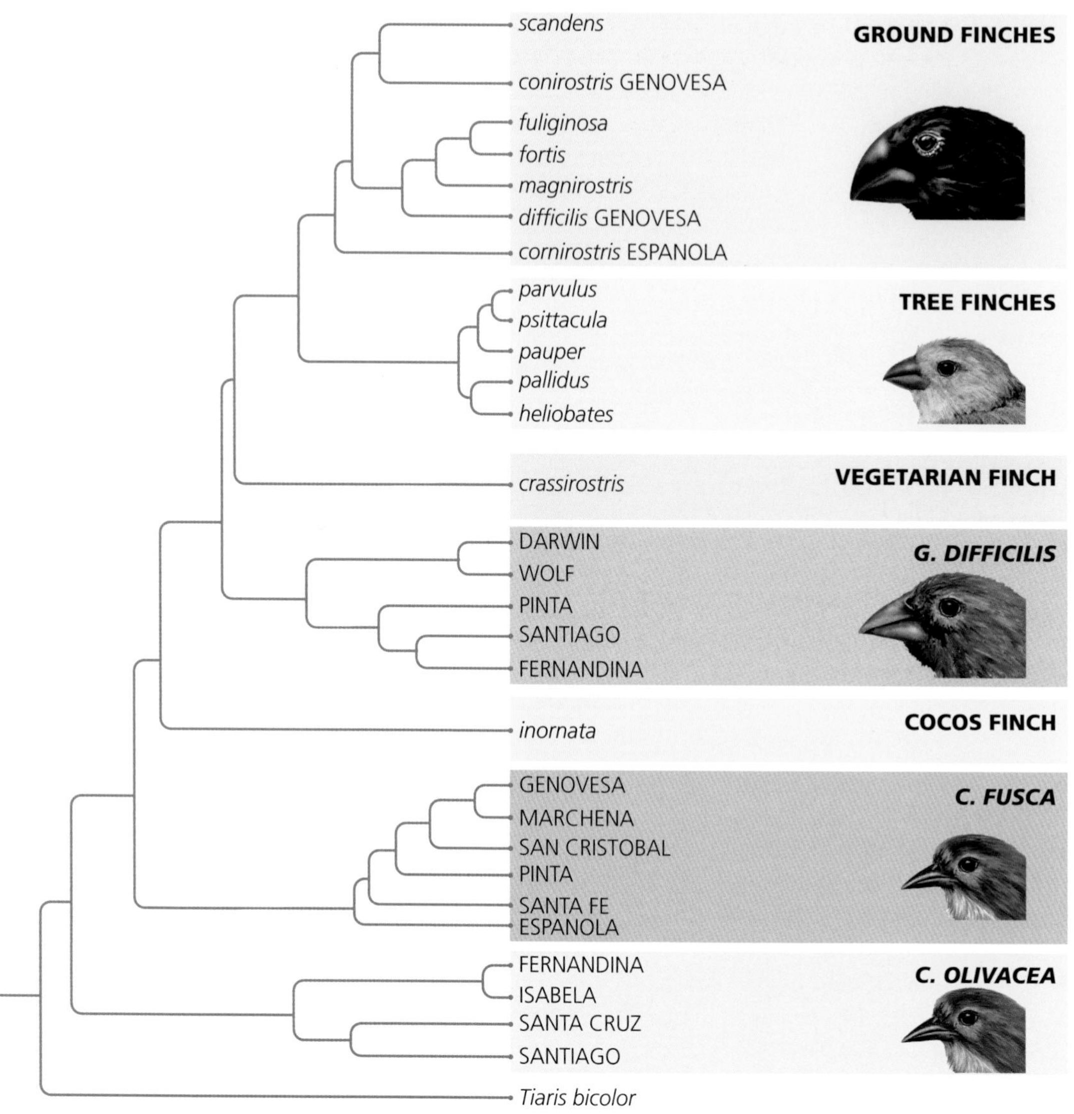

Figure 7.6 The DNA of Darwin's finches records their rapid evolution from a common ancestor that came to the Galápagos Islands. (Adapted from Grant, 2007)

also shows that Cocos finches share a closer ancestry with some species of Darwin's finches than they do with others. Darwin's finches thus could not have come from Isla del Coco. Rather, evolution went the other way. Cocos finches most likely evolved from Darwin's finches that came from the Galápagos Islands to Isla del Coco.

As for the ultimate origin of Darwin's finches, the DNA study linked Darwin's finches to a group of birds known as seed-eating tanagers that live in South America, Central America, and the Caribbean. It's not yet clear which species of that group is the closest relative of Darwin's finches, but some scientists argue that the ancestors of the dull-colored grassquit of Ecuador gave rise to them.

How did seed-eating tanagers end up thousands of kilometers from the Amazon on a lonely island in the Pacific? It's possible that a flock of birds was escaping a volcanic eruption and flew out over the ocean, to be swept away by the winds. Once the original birds settled down on the Galápagos Islands, their evolutionary tree shows, they evolved into the many different species seen today, and perhaps others that have since become extinct.

HIV (Human Immunodeficiency Virus): Today HIV is all too familiar. In 2007, an estimated 33 million people worldwide had HIV infections, and an estimated 3.1 million people were dying of AIDS-related causes every year. Yet, as diseases go, HIV is a latecomer. Scientists only became aware of it in the early 1980s, when it was still relatively rare, after which it swiftly became a global epidemic. Scientists have tried to search through medical records and blood samples for earlier cases of HIV infection that might have been overlooked. The earliest sample of HIV comes from a blood sample taken from a patient in 1959 in Kinshasa, the capital of the Democratic Republic of Congo.

The mysterious appearance of HIV led to many speculations about where it came from—including accusations that vaccination campaigns introduced it into people with vaccines contaminated with a monkey virus. But when scientists reconstructed the evolutionary tree of the virus and its relatives, they rejected those claims.

As soon as scientists discovered HIV, it was clear that it belonged to a group known as the lentiviruses. Lentiviruses are small particles with spiky knobs on their surface, and they encode their genes in RNA. They infect mammals, such as cats, horses, and primates, typically invading certain types of white blood cells. Genetic studies revealed that HIV is most closely related to strains of lentivirus that infect monkeys and apes—known as simian immunodeficiency virus, or SIV for short. HIV is not actually a single clade, like Darwin's finches. Instead, different strains have different origins.

The strain known as HIV-1, which causes the vast majority of AIDS cases, is most closely related to the SIV viruses that infect chimpanzees. HIV-2 belongs to a group of SIV strains that infect a monkey known as the sooty mangabey. A closer look at HIV-1 (**Figure 7.7**) reveals that it is actually three strains, all of which jumped from a single subspecies of chimpanzee, *Pan troglodytes troglodytes*, found in central Africa. SIV most likely evolved into HIV-1 as hunters killed apes and monkeys to sell in a growing "bush-meat" industry. Viruses in the blood of the primates could have entered cuts in the skin of the hunters, where a few of them mutated and evolved adaptations to their new host.

Knowing the structure of the HIV tree allows scientists to pinpoint those adaptations. It turns out, for example, that, as all three strains of HIV-1 evolved from chimp-virus ancestors, they all acquired the same new amino acid in the same position in the same protein. No strain of SIV in chimpanzees produces that amino acid. This mutation altered a gene encoding the shell of the virus, and experiments suggest that it was crucial to the success of the new HIV strains in humans. It's possible that the mutation allowed the virus to do a better job of manipulating its new hosts into building new copies of itself.

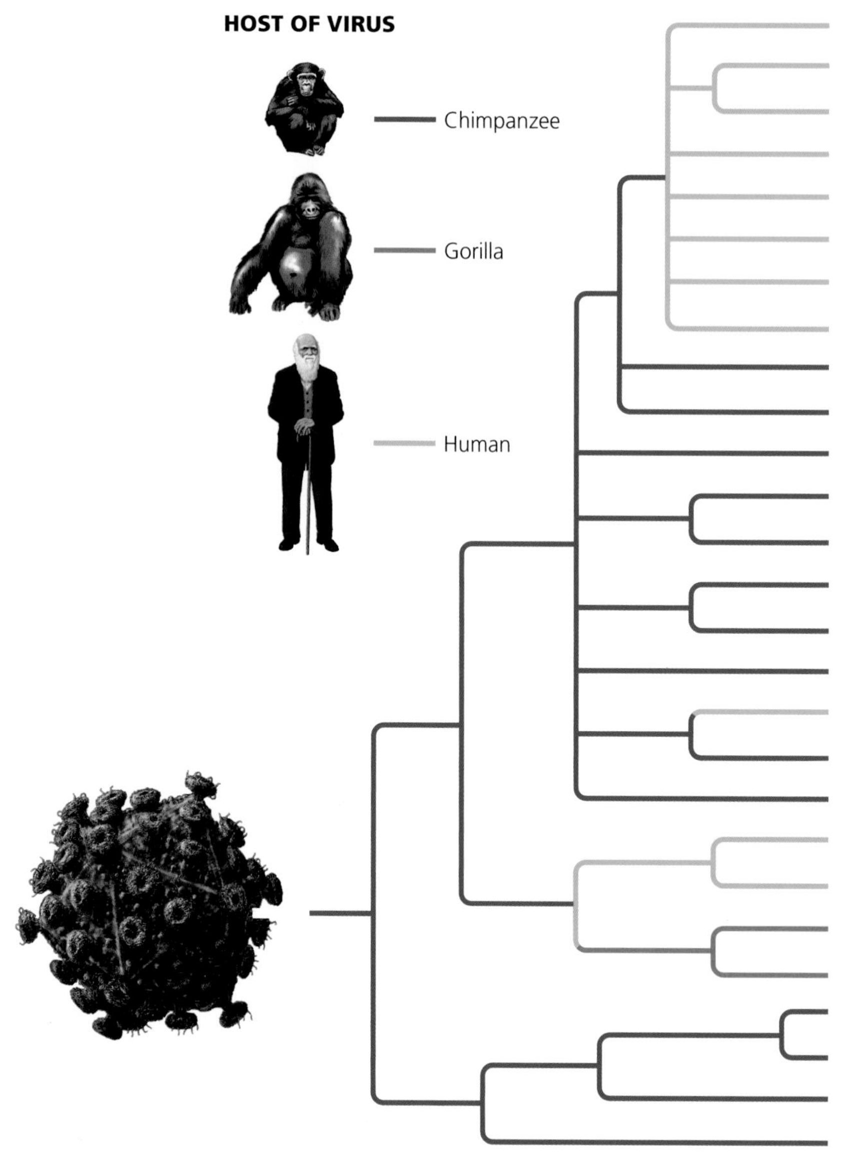

Figure 7.7 The evolutionary tree of HIV-1 reveals how the virus hopped from chimpanzee hosts to humans. (Adapted from Koella and Stearns, 2008)

Natural Selection versus Neutral Evolution

Along with solving specific mysteries about particular groups of species, molecular phylogenies also help scientists to answer broad questions about how evolution works. One of the biggest questions is how much of life's diversity can be explained by natural selection.

In the mid-1900s, researchers folded genetics and mathematics into evolutionary biology. Their work revealed how natural selection can fix alleles in populations and produce new adaptations. But other scientists began to suspect that differences in DNA sequences did not automatically translate into differences in fitness.

The first cause for doubts came from the early studies on such proteins as hemoglobin in the 1950s. Different animal species have slightly different sequences of amino acids in their hemoglobin, and yet they can all use the molecule to store oxygen. Mutations had altered the hemoglobin's structure without preventing the molecule from doing its job. These mutations were neutral, some scientists argued, and they had only become fixed in different species due to genetic drift (page 112).

Further research revealed more opportunities for neutral evolution to transform genes. Several different codons may encode the same amino acid. A mutation may alter a gene without changing the protein it makes. Scientists call this a "silent substitution."

In 1968 Motoo Kimura, a Japanese biologist, offered up the first formal neutral theory of molecular evolution. He claimed that most of the differences in genomes were the result of neutral mutations that had become fixed through genetic drift. While natural selection could produce phenotypic adaptations, Kimura argued that the patterns in genomes were mainly the result of genetic drift.

From this theory, Kimura and other researchers came up with hypotheses they could test. They predicted that neutral mutations would become fixed in populations at a roughly regular rate: the time it took for new mutations to arise and then to spread randomly through populations. When a population split into two lineages, each lineage would acquire its own unique set of neutral mutations. The more time that passed after the lineages diverged, the more neutral mutations would be fixed in each one.

Walter Fitch of the University of Wisconsin and Charles Langley of the National Institute of Environmental Health Sciences in North Carolina found some compelling evidence for neutral evolution by comparing proteins from 17 mammals. They examined one protein in particular, known as cytochrome *c*, determining its sequence in humans, horses, and other species. From these results, they determined how many mutations had arisen in each lineage. Finch and Langley then asked paleontologists to estimate when those lineages had split, based on the fossil record. They then drew a graph to compare the two sets of results. As shown in **Figure 7.8**, they discovered a striking pattern. The more distantly related two species are, the more mutations have accumulated in each lineage since they split from their common ancestor.

More evidence for neutral evolution comes from pseudogenes—genes that have mutated so that they can no longer be used to make proteins. The neutral theory predicts that pseudogenes should carry more mutations than protein-coding genes. Mutations to protein-coding genes may often be harmful, because they can change the way in which proteins work. Natural selection eliminates these harmful mutations from the population. But natural selection shouldn't weed out mutations to pseudogenes, because most of them have no function to

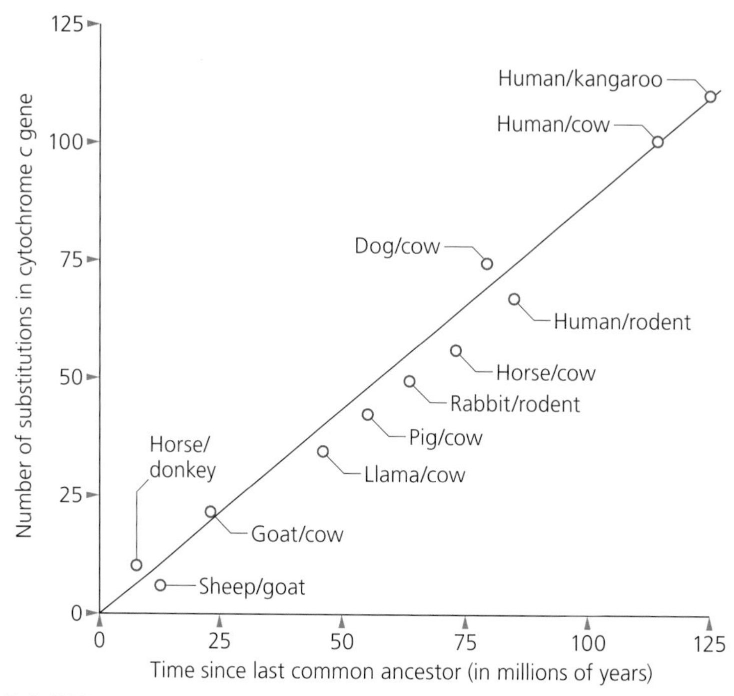

Figure 7.8 DNA mutates at a roughly clock-like rate. This graph shows how distantly related pairs of species have a large number of different substitutions in the cytochrome *c* gene. (Adapted from Moore and Moore, 2004)

begin with. So, over millions of years, more mutations should accumulate in them, while many mutations are eliminated from protein-coding genes. When scientists actually compared pseudogenes and genes in different species, they found that the pseudogenes do indeed carry more mutations.

The precise relationship between natural selection and neutral evolution is a complex one that scientists are still exploring. But it is clear that neutral evolution has had a major role in how genomes got to be the way they are today.

The Molecular Clock

As Figure 7.8 shows, neutral mutations accumulate at a roughly clock-like rate, piling up like grains of sand in the bottom of an hourglass. This discovery opened the way to using mutations to tell time. Let's say that you have sequenced the cytochrome *c* gene from another vertebrate—say a bird—and that you have calculated the number of substitutions that separate its sequence from ours. If you run your finger from that number on the *y*-axis until you reach the diagonal line and then run your finger straight down, you have an estimate for how long ago the common ancestor of both you and birds was alive. Scientists refer to this method of telling time by counting neutral mutations as the molecular clock.

Reading the molecular clock is not a simple task. Neutral evolution runs at different rates in different lineages. Even in a single lineage, different genes evolve at different rates. Over thousands or millions of years, the ticking of the molecular clock can speed up or slow down. Scientists have developed statistical methods that have allowed them to overcome some of these challenges. In 2000, for example, scientists based at Los Alamos National Laboratory set out to estimate how old the HIV-1 virus is (**Figure 7.9**). They compared the RNA from 159 HIV-1 viruses that had been isolated over the previous two decades, drawing an evolutionary tree. They then calculated the most likely rate at which the virus genes mutated, based on how different the viruses were from one another and how old they were.

Rather than requiring that the viruses all follow one clock strictly, they allowed the mutation rate to vary from branch to branch, and even from site to site within the genes. With so many possible results, this was not a calculation to undertake lightly. In fact, the scientists had to carry it out with the supercomputers at Los Alamos, which were originally built to model nuclear explosions.

The scientists estimated that the common ancestor of HIV-1 strains existed sometime between 1915 and 1941, with the most likely year being 1931. That estimate is another piece of evidence against the claim that vaccines in the 1950s were the source of HIV. To test the hypothesis that HIV-1 originated around 1931, the Los Alamos scientists examined the strain from 1959, which they had not used to make their estimate. They tallied up the mutations in the virus and used their molecular clock to estimate when it had first existed. Their estimate: 1959.

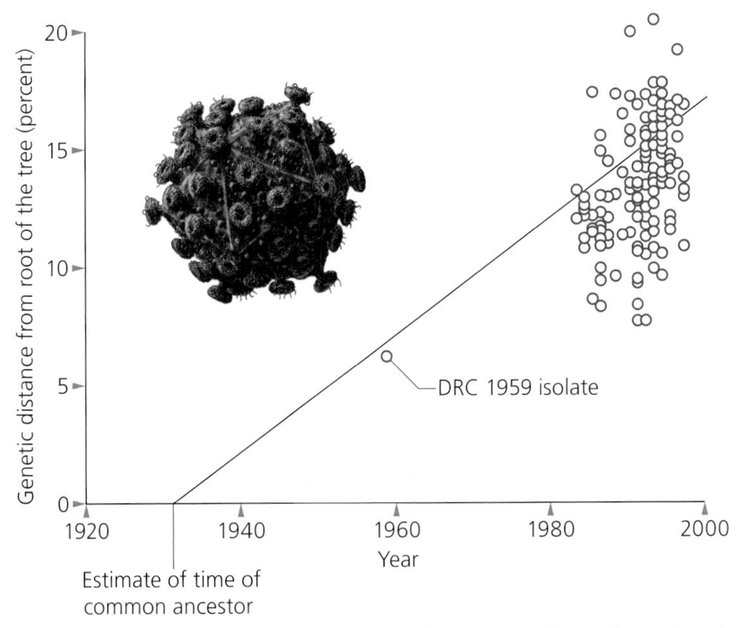

Figure 7.9 The mutations that have accumulated in HIV-1 strains indicate that their common ancestor existed in the 1930s. When scientists discovered a sample of HIV-1 from 1959, they were able to accurately predict its age by using the molecular clock. (Adapted from Rambaut et al., 2004)

The molecular clock can shed light on other chapters in evolution that took place millions of years ago. The DNA from Darwin's finches doesn't just reveal their evolutionary relationships. The mutations also allowed the Grants and their colleagues to estimate the age of the entire lineage: two to three million years. Studies on the rocks of the Galápagos Islands indicate that they began forming 10 million years ago. The crust of the Pacific seafloor has been slowly moving over an underlying blob of hot rock. The blob pushes up the crust, forming a volcanic island. Gradually, the crust moves away and the island cools and sinks, while a new island forms nearby. The original Galápagos Islands are now submerged underwater, and the biggest islands on the chain today are less than a million years old. The Grants and their colleagues hypothesize that a few birds came to one of the vanished islands two to three million years ago and then began to spread to other islands as they formed.

Ancient Selection

The way alleles are distributed in a population can reveal whether they've experienced natural selection over the past few thousand years (page 126). But once natural selection fixes an allele in a population, recombination will slowly shuffle the neighboring genes away from it. Eventually the genes become so scrambled that scientists cannot distinguish between an allele favored by natural selection and one spread by genetic drift. After tens of thousands of years, the signal fades to noise.

Fortunately, evolutionary trees give scientists other ways to search for natural selection's fingerprints, even if those fingerprints are millions of years old. And it's the neutral theory of evolution that provided scientists with one of their most important tools.

To understand how this tool works, consider a gene that evolves in three lineages for millions of years. Gene 1 encodes a protein that is very sensitive to any change to its structure. Mutations that alter the structure of its protein are fatal. As a result, natural selection eliminates most of the protein-altering mutations in gene 1. But mutations that do not alter the function of the protein aren't eliminated, and so gene 1 gradually accumulates these mutations over time. Gene 1 experiences purifying selection, so called because it purifies the gene by eliminating harmful variants.

Gene 2 experiences no natural selection. It acquires mutations, but only because those mutations are fixed by genetic drift. Gene 3, on the other hand, evolves a new sequence that changes the structure of the protein it encodes. As it experiences positive selection, it acquires a number of protein-altering mutations.

Scientists have developed statistical methods to determine how genes have evolved by comparing the silent substitutions to the nonsilent ones. They draw an evolutionary tree joining a group of species. They then examine a particular gene in each of those species, and then work backwards to determine what that gene looked like in their common ancestor. They then move forward again from the common ancestor to the living species, determining where along each branch each mutation arose.

With these results in hand, scientists can calculate the odds that a base had changed in each lineage. If a gene experiences purifying selection, the odds of a nonsilent substitution occurring are much lower than the odds of a silent substitution occurring. In a gene experiencing neutral evolution, the odds are the same for each mutation. And in a gene experiencing strong positive selection, a nonsilent substitution is more likely than a silent one (**Figure 7.10**).

This method is allowing scientists to detect natural selection that took place millions of years ago. In 2008, Marc Robinson-Rechavi, an evolutionary biologist at Lausanne University in Switzerland, and his colleagues surveyed 884 genes in a wide range of vertebrates, from zebrafish to frogs to chickens to humans—a group of animals that share a common ancestor that lived some 450 million years ago. They found evidence for positive selection in 77% of the genes. In any

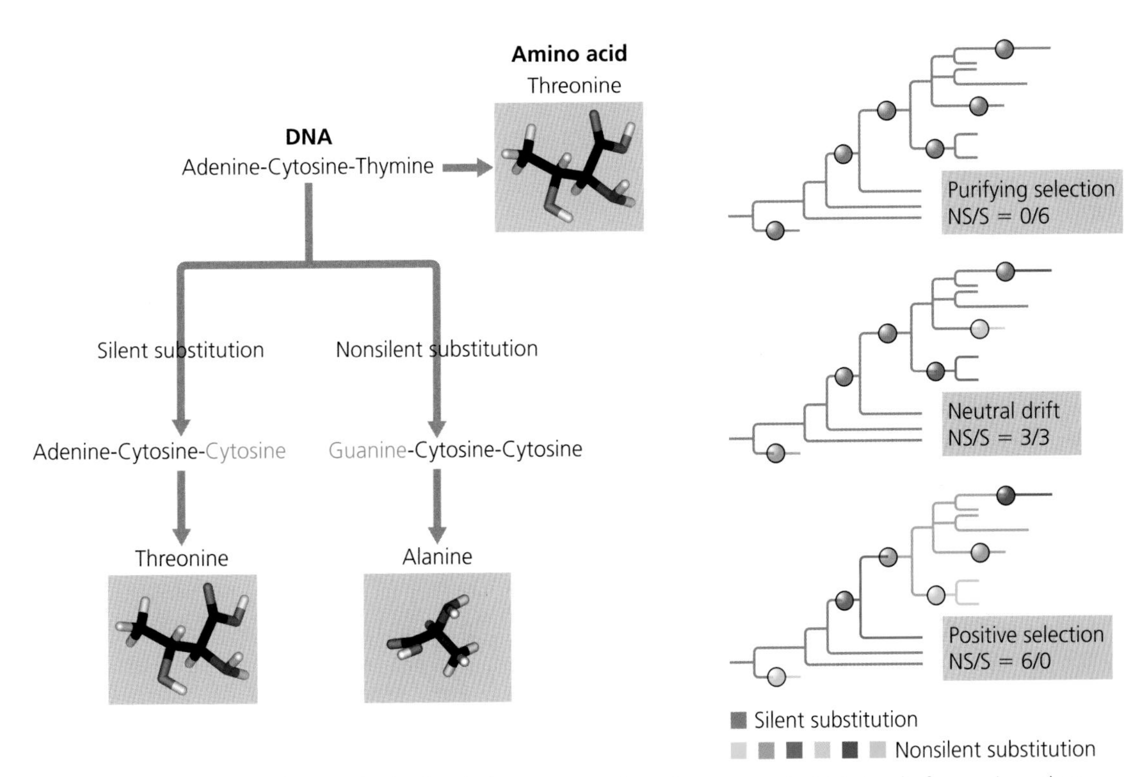

Figure 7.10 One sign of natural selection is the accumulation of an unusually high level of mutations that change the structure of proteins. Left: Silent substitutions alter a codon without changing the amino acid it encodes. Nonsilent substitutions change the amino acid, and as a result they can potentially change the way a protein functions. Right: Scientists can measure selection by comparing substitutions in a gene in different species. Top right: Nonsilent substitutions in this gene lower the fitness of organisms. Thus they are not fixed in populations. Silent substitutions are thus far more common. Middle right: Nucleotides in this gene are equally likely to acquire silent or nonsilent substitutions. This is a sign of neutral drift. Bottom right: This gene acquires nonsilent substitutions that raise fitness. They are favored by natural selection, and become fixed at a greater rate than silent substitutions.

one lineage, though, only a small percentage of the DNA in each gene had experienced positive selection. Rather than entirely overhauling the genome, natural selection precisely retooled different genes.

Robinson-Rechavi and his colleagues took a large-scale look at positive selection over the past 450 million years. Meanwhile, Richard Gibbs at Baylor College of Medicine in Houston and an international team of colleagues took a closer look at our more recent evolutionary history. Some 30 million years ago, an ancient primate gave rise to today's monkeys, apes, and humans. Gibbs and his colleagues scanned the entire genomes of rhesus macaque monkeys, chimpanzees, and humans. They could find 10,376 genes common to all three species. Of these, 9.8% had experienced no nonsilent substitutions at all. In other words, for 30 million years, these proteins have stayed the same. But 2.8% of the genes showed signs of positive selection.

By detecting natural selection this way, scientists are pinpointing some of the genes that may have helped to make us uniquely human. One of the most tantalizing of these genes is known as FOXP2, the first gene ever clearly linked to language. People with mutations to FOXP2 suffer devastating difficulties in speaking and understanding grammar.

Scientists have discovered other versions of FOXP2 in other mammals, and experiments show that it plays a role in their communication as well. Normally, mouse pups squeak to their mothers to ensure that they are properly cared for.

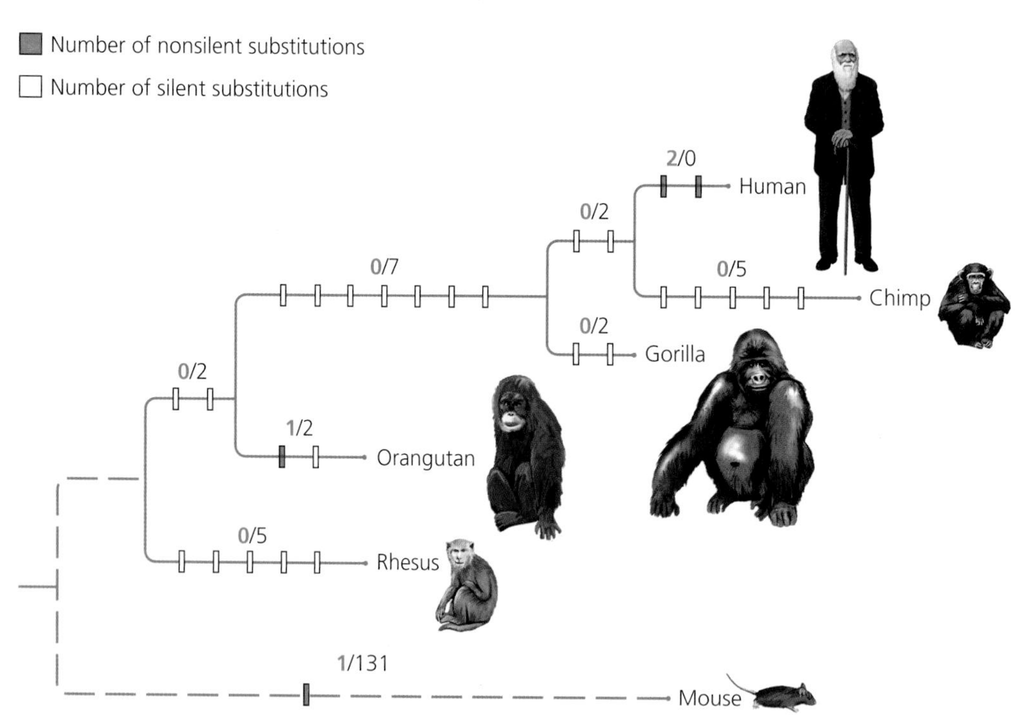

Figure 7.11 A gene linked to language, called FOXP2, has experienced strong positive selection in the human lineage. In other mammal lineages, the protein encoded by FOXP2 has changed little over millions of years. (Adapted from Enard, 2002)

When scientists shut down FOXP2 in mouse embryos, however, the mouse pups cannot squeak well, because certain neurons in their brains don't develop normally.

These results suggest that FOXP2 took on a role in animal communication at least 100 million years ago. In most of the mammals that scientists have surveyed, the amino acid sequence of the FOXP2 protein has not changed for tens of millions of years (**Figure 7.11**). The FOXP2 gene carried by a chimpanzee is practically identical with that of a mouse. In humans—and humans alone—two amino acids have changed in the protein in just the past six million years. That change represents a very powerful episode of natural selection, given how little the gene has changed in other lineages. It's possible that the transformation of FOXP2 helped to give rise to full-blown language in our species (page 346).

Deciphering the Genome

The technology for sequencing DNA is so powerful now that scientists are cataloging millions of new genes a year. But discovering just what those genes do remains a much harder puzzle to solve. One strategy that scientists use to understand the function of genes is to look at their evolutionary history.

Many genes, for example, belong to families. They evolved from an ancestral gene that duplicated many times, producing a number of related genes. While each of the new genes evolved on its own, they still bear a close relationship to other members of their family—and, fairly often, families of genes behave in the same way. So if scientists discover a new gene that belongs to a family of immune-system genes, a good working hypothesis is that the new gene serves a similar function as well.

Studying natural selection also allows scientists to learn how genes work. In recent years, a number of scientists have been trying to figure out the function of a gene called TRIM-α. Experiments have shown that it encodes a protein that helps fight against HIV-1 and related viruses. It appears to cut up the virus's RNA, so that it cannot infect cells and make new copies of itself. But, for some reason, each species of primate's TRIM-α only defends against some of the viruses. The TRIM-α of rhesus monkeys, for example, eliminates HIV-1, while human TRIM-α cannot. That's one reason why HIV is such a danger to us. If scientists can pinpoint how TRIM-α defends against certain viruses, they may get some clues to help in the fight against HIV.

To discover precisely how TRIM-α works, Sara Sawyer of the Fred Hutchinson Cancer Center in Seattle and her colleagues sequenced the gene for TRIM-α from 20 primates, including humans, gorillas, and baboons. They compared the silent and nonsilent substitutions in the gene to identify parts that experienced strong positive selection. In all the species they studied, the same stretch of 13 amino acids experienced positive selection. Subsequent experiments revealed that this little patch of TRIM-α is crucial for destroying viruses.

Sawyer created modified versions of the gene and inserted them into cells from different primate species. When a primate loses its own patch, TRIM-α loses its ability to fight off viruses. It appears that each species of primate has evolved its own version of TRIM-α to attack viruses that are specialized to infect it.

Understanding evolution doesn't just help scientists to identify the functions of genes. It also helps them to identify the genes themselves in the first place. In a long string of DNA, the segments that are protein-coding genes do not always leap out. Sometimes it is hard to distinguish them from the surrounding DNA that does not code for proteins. To flush these genes out of their genomic hiding places, scientists study their evolutionary history. Many essential genes have experienced purifying selection. Noncoding DNA that has no function can evolve neutrally, so that it varies from species to species. But the non-protein coding DNA purifying selection keeps the genes from changing much.

By comparing DNA from different species, scientists can identify these conserved genes. Ed Rubin of Lawrence Berkeley National Laboratory in California and his colleagues used this method in a study of genes that control the levels of molecules known as lipids in the blood. Lipids are an essential part of our cell membranes, but people with too many lipids floating around in their blood face a risk of heart disease. Scientists have identified some of those genes that control lipid levels by studying people with high and low levels of lipids, searching for genetic markers that tend to show up in each group. These studies revealed a cluster of genes for proteins that bind to lipids. Once scientists found these genes in humans, they were able to search for them in mice. It turned out that mice have a similar gene cluster in the same spot in their genome, which also controls their blood lipids. This gene cluster was probably already present in the common ancestor of humans and mice 100 million years ago.

Rubin and his colleagues reasoned that more lipid genes might be hiding in the noncoding DNA near the gene cluster. They decided to search the surrounding 200,000 bases. To do so, they lined up the human version of this region with the one from the mouse genome. Different nucleotides were sprinkled along each segment, the result of neutral evolution acting on noncoding DNA. But 30,000 bases away from the gene cluster, Rubin and his colleagues discovered something odd: a stretch of 1,107 bases that were remarkably similar in human and mouse DNA.

The scientists hypothesized that they had found a conserved gene. They determined what a protein encoded by the hypothetical human version of the gene looked like. It contained a series of spirals of amino acids. That discovery was encouraging because these spirals are a distinctive feature of all known lipid-binding proteins. The scientists then ran experiments on mice to test the function of the gene. They inserted extra copies of the gene into mice and found that their lipid levels dropped dramatically. When they disabled the gene in mice, the animals developed much higher lipid levels. Rubin and his colleagues then turned to humans and found polymorphisms of the gene. These polymorphisms are associated with high and low levels of heart disease. Thus, by looking back over 100 million years of evolutionary history, Rubin and his colleagues discovered a gene that may play an important role in heart disease. In Chapter 13 we'll look in more detail at how evolutionary biology sheds light on diseases.

Evolutionary trees can also shed light on the 98.8% of our DNA that does not encode proteins. Some non-protein coding elements are binding sites where proteins can attach to DNA in order to shut off the expression of genes or to speed it up. Other segments encode RNA molecules that act as sensors and can regulate the level of proteins in the cell. Functional regions of non-protein coding DNA are very hard to discover, because they are often extremely small—a few dozen bases long, in some cases—and are hidden away in vast stretches of DNA with no known function.

One way to find these segments is to take advantage of evolution. It turns out that some of these segments are conserved over the course of evolution, just as many protein-coding genes are. By aligning genomes from a number of species, scientists have identified thousands of conserved segments in the human genome. Subsequent experiments have shown that many of these conserved sequences do play important roles in the cell.

Particularly intriguing are 202 non-protein coding elements that are conserved in other vertebrates, but that have changed a great deal in the human lineage. These segments may have played a crucial role in making us uniquely human. The element that has changed the most, called HAR1, turned out to be an RNA-encoding gene that is only expressed in the brain during the development of embryos. Part of what makes us uniquely human lies in the unique anatomy of the human brain. HAR1 may help to explain how the human brain came to be so different.

TO SUM UP...

- **The relationships of species and other clades can be inferred from their DNA as well as from their morphology.**
- **The phylogeny of a single segment of DNA may be different from the phylogeny of the species that carries it.**
- **Studies on DNA show that all living humans can trace their ancestry to Africa.**
- **Darwin's finches evolved from a common ancestor that arrived on the Galápagos Islands two to three million years ago.**
- **HIV evolved from chimpanzee viruses in the early 1900s.**
- **Much of the human genome has experienced neutral evolution.**
- **Neutral evolution allows scientists to estimate the age of common ancestors by comparing the mutations in a group of organisms.**
- **The ratio of silent to nonsilent substitutions can show evidence of natural selection in the past.**
- **Detecting regions of DNA that have experienced strong selection can provide clues to the function of genes.**

Adaptations 8

From Genes to Traits

To study evolution, Bryan Fry puts his life on the line. Fry investigates the evolution of deadly snakes. He has caught venomous sea snakes on scuba dives and has trekked through the Australian outback to catch the inland taipan, the world's most toxic snake. Fry is an admitted adrenaline junkie, but he is not reckless. He prepares himself for every encounter so that he comes away safe

Snakes use a complex system of fangs, muscles, and venom glands to deliver their deadly bites. Above: Bryan Fry investigates the evolution of this complicated adaptation by analyzing the genes for venom.

and sound. It helps to know a little snake psychology. He knows, for example, that a king cobra marks its dominance over other king cobras by rearing up and touching the tops of their heads. In order to trap one of these deadly snakes, Fry first shows who's boss by tapping the snake's head. The cobra's submission gives him a chance to slip it into a bag.

Once Fry gets a snake back to his lab at the University of Melbourne, he can get a close look at the biology that makes it so deadly. A venomous snake has glands in the back of its mouth that produce venom. The cells in the glands express a set of venom genes, build the corresponding proteins, and pump the venom into the surrounding fluid. To deliver the venom to its prey, a cobra opens its jaws and stabs a pair of fangs into its victim. At the same time, muscles squeeze down on the glands, causing the venom to shoot down a pair of tubes that lead into the fangs and then to squirt out through holes at their tips.

Snakes produce complex cocktails of molecules in their venom, each of which helps them to subdue their prey. Some venoms relax the walls of the aorta, dropping the blood pressure in a snake's victim until it blacks out. Other venoms lock onto receptors in neurons, causing paralysis. Others interfere with the biochemistry inside muscle cells, causing them to break down rapidly.

A snake's ability to deliver a venomous bite is a complex adaptation. Many genes are required to produce venom molecules, the glands where the venom is made and stored, and the tubes and fangs and muscles for delivering it. The parts of the venom system depend on one another. Without venom to deliver, a cobra's fangs don't help the cobra. Without fangs, a cobra's venom cannot be effectively injected into the snake's prey.

Nature brims with such complex adaptations, from eyes to feathers to brains. In this chapter we'll explore their evolution. Natural selection is crucial for their origins, but the history of a complex adaptation is much more than the spread of a single allele in a single population. The evolution of a new complex adaptation is made possible by the evolution of new genes with new functions. Its parts must evolve into a new system that works as a unit. The evolution of complex adaptations poses one of the most interesting challenges in evolutionary biology. In this chapter, we'll see how scientists are attacking that challenge with many lines of evidence—including fossils, observations on living animals, and experiments that reveal the roles of the underlying genes.

Innovation in Our Own Time

In Chapter 6, we saw how Richard Lenski uses *E. coli* to observe natural selection taking place in his own laboratory. In recent years, he's been surprised to discover that some of the bacteria have also evolved a new adaptation: they can eat a kind of food that *E. coli* aren't supposed to be able to eat.

In 2004, Lenski and his students noticed something odd in one of the 12 flasks in which they were rearing lines of *E. coli*. The contents of the flask had become cloudy—a sign that the bacteria in the flask had gone through a population explosion. At first the scientists suspected that some other species of bacteria had slipped into the flask and was breeding quickly. But when they examined the microbes more closely, they discovered that the flask was not contaminated with another species of bacteria. It was simply packed with *E. coli*—descendants of the original ancestor with which Lenski had started the entire experiment. But somehow the bacteria in one flask had evolved a way to grow much faster than any of the others.

Careful study of the bacteria revealed their secret: they were eating a compound called citrate. Citrate, the molecule that makes lemons tart, is part of the standard nutrient broth that Lenski used to rear his *E. coli*. But *E. coli* do not eat citrate. In fact, when microbiologists try to identify what species bacteria belong to, they feed them citrate. If they can grow on citrate, then they can't be *E. coli*. And yet, strangely enough, Lenski's *E. coli* were doing just that. His students took away the glucose they normally fed the bacteria, and they discovered that this peculiar strain of *E. coli* could thrive on citrate alone.

Because these citrate eaters evolved under Lenski's watchful eye, he and his students could now begin to figure out how this new adaptation evolved. They are thawing out the ancestors of the citrate eaters to see how well they could feed on this new food. The bacteria first started feeding on citrate about 31,000 generations after the start of Lenski's experiment. But these early citrate-feeders could grow only slowly on citrate, and they made up only a small fraction of the population of *E. coli*. Over a couple thousand generations, they improved at growing on citrate, multiplying so quickly that they came to dominate their flask (**Figure 8.1**).

The gradual rise of the citrate eaters suggests that they did not evolve a new way of feeding in a single step. This ability arose through a series of mutations. In fact, Lenski's research suggests that the first mutations may not have even allowed *E. coli* to feed on citrate at all, but merely opened the way for later changes to produce this new adaptation. Nor is the bacteria done evolving this adaptation. Lenski and his students have found that the citrate eaters are acquiring new mutations that make them grow even faster on citrate than their ancestors. Lenski's team is now pinpointing the mutations that gave rise to this new adaptation. In years to come, they'll be able to run experiments to learn how each one helped make it possible.

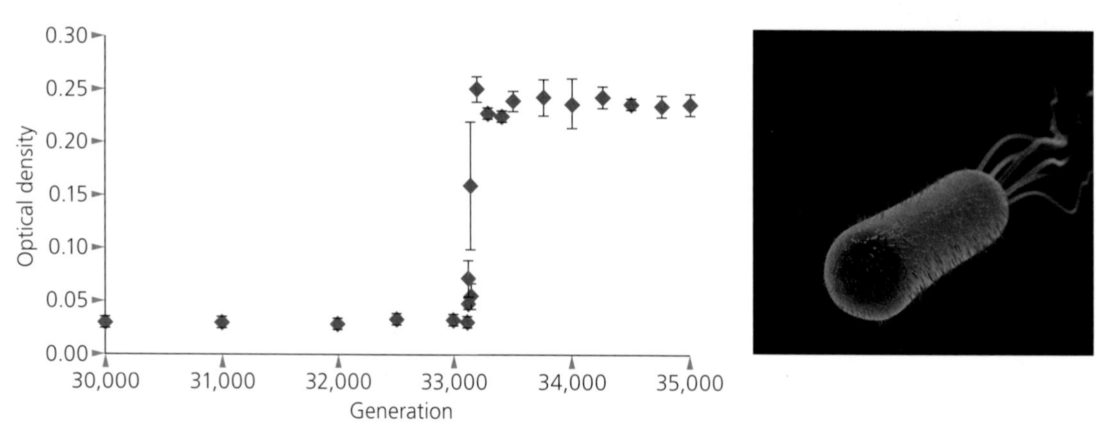

Figure 8.1 As Richard Lenski's bacteria adapted to a diet of glucose, a new trait evolved in one lineage: the ability to feed on citrate. This graph shows how this trait emerged over the course of about 2,000 generations. As the bacteria evolved to eat more citrate, they grew to greater numbers and made their flask cloudier. "Optical density" on this graph is a measure of this cloudiness. (Adapted from Blount et al., 2008)

It's not just in the lab that scientists can study bacteria evolving new adaptations. Our pollution of the environment has triggered some remarkable evolution in microbes as well. In the early 1900s chemists began developing new chlorine-based compounds to kill assorted pests. One of these compounds, called pentachlorophenol (or PCP for short), was introduced in 1936 as a way to kill the fungus that causes timber to rot. PCP eventually made its way into the soil, where it could linger for decades. At first, microbes in the soil could not break down compounds like PCP. They didn't have the right proteins for the job, simply because, before 1936, PCP did not exist in nature.

Over the years, however, scientists discovered that soil bacteria known as *Sphingobium* had evolved into PCP eaters. They could strip away PCP's five chlorine atoms and feed on the remaining carbon, hydrogen, and oxygen atoms. The bacteria needed to use five different proteins to break down PCP, each protein carrying out a small reaction such as removing a chlorine atom or rearranging the altered molecule.

How could a pathway of five proteins evolve seemingly out of nothing? To find an answer, Shelley Copley at the University of Colorado and her colleagues compared the genes for PCP-degrading proteins with other genes. It turns out that the proteins *Sphingobium* uses to break down PCP are related to proteins the bacteria use to break down other molecules, such as amino acids. Copley's research has led her to the hypothesis that the PCP-feeding pathway did not, in fact, evolve from nothing. Instead, old proteins were "recruited" from old pathways and assembled into a new one.

It may be hard to believe that a big, intricate protein well-adapted to doing one job could be able to do anything else. Yet proteins often can carry out more than one reaction, because their structure allows them to clasp onto more than one kind of molecule. One of the proteins that *Sphingobium* uses to break down PCP, known as PcpC, is an example of a promiscuous protein: not only can it break down PCP, but it can still break down amino acids.

That doesn't mean that a protein can carry out two separate reactions with equal speed. In fact, most promiscuous proteins do one job well, and another poorly. Still, even a modest ability to carry out a chemical reaction is better than none, if that reaction starts to benefit an organism. Once PCP entered the soil, for example, microbes that could break it down even at a very slow rate could enjoy an extra source of food. Natural selection then favored mutations that made the slow PCP-degrading reactions run faster. PcpC, for example, may still be able to break down amino acids, but it does so much more slowly than more specialized proteins.

There's a constraint to the evolution of new functions in proteins, however. An organism may benefit from a protein doing something new, but not if it still depends on the protein doing its old job well. It turns out there are several routes around this impasse. A multitasking gene may, for example, be accidentally copied. Now two genes can make the same protein. One of the genes is now free to evolve to make the slow reaction even faster. The other copy of the gene can continue to make proteins adapted to carry out the original reaction.

Bacteria have another way around this impasse: horizontal gene transfer. They can acquire new genes (and new copies of genes they already have) from other microbes. These imported genes can join a microbe's existing pathways, helping them to carry out new tasks.

Can I Borrow a Gene?

Gene recruitment and gene duplication are important for the evolution of new adaptations not only in bacteria, but also in multicellular organisms such as animals. In Chapter 4 we saw how scientists use fossils and the anatomy of living animals to trace the origin of adaptations such as legs and feathers. That research shows that these new adaptations do not come out of thin air. Instead, they're modified from existing adaptations often serving other functions. Evolutionary biologists can get an even deeper understanding of how such new adaptations evolve in animals by examining the genes that build them (**Figure 8.2**).

As a single fertilized egg begins to divide and give rise to an embryo, the cells in different parts of its body begin to do different things. Some begin to produce bone, while others stretch out to form neurons. Some die off, while others join together in clumps that eventually grow into brains or hearts. Yet all these cells are using the same set of genes to build themselves. One important source of the differences between these cells is a kind of protein known as a transcription factor. Transcription factors bind to specific stretches of DNA and then influence how nearby genes are expressed. Some transcription factors can switch genes on. Others switch them off. Many genes in animals can only be expressed if a series of transcription factors bind near them. As an embryo develops, cells begin to use transcription factors to turn genes on and off in a distinctive way.

The genes for transcription factors, as well as for the sites where they bind, can mutate. And sometimes these mutations can open up the opportunity for the

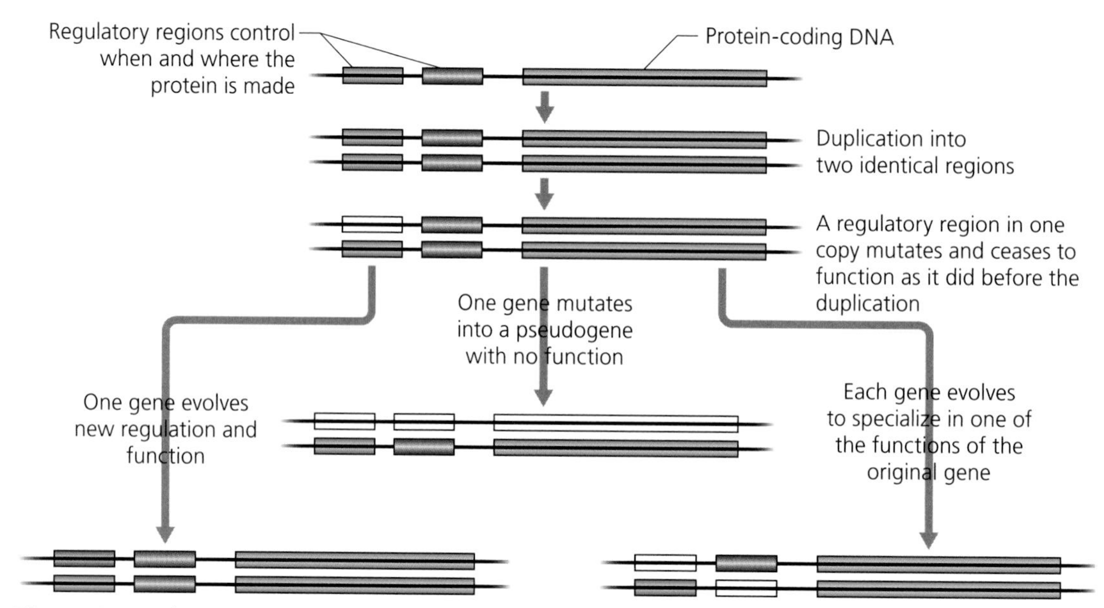

Figure 8.2 When genes duplicate, they can take on new functions.

evolution of new patterns of development. Bryan Fry and his colleagues have discovered that snake venom evolved through gene recruitment. To start their investigation, they isolated genes for venom in a wide range of snakes. Fry collected cells from the venom glands of snakes and determined which of their genes were active. About half of the active genes turned out to be ordinary housekeeping genes that carry out basic processes in all cells. The remaining active genes encoded venom molecules. Fry and his colleagues sequenced these venom genes and compared them to figure out how they evolved from an ancestral gene.

Each species produced its own distinctive cocktail of venoms, but each venom gene typically showed a close kinship with venom genes in closely related snakes. That pattern suggests that venomous snakes inherited genes for venom from a common ancestor, but, after their lineages diverged, the venom genes were shaped differently by natural selection. Green mambas and black mambas are closely related, for example, but the green mambas live in trees and the black mambas live on the ground. Experiments show that the venom of the green mamba is more effective against birds, while that of the black mamba is more effective against rodents.

Fry and his colleagues reconstructed the history of these venom genes with evolutionary trees. **Figure 8.3** shows one of these trees, for a muscle-destroying venom called crotamine. Fry compared crotamine genes with one another, as well as with other genes in snakes and other vertebrates. The tree reveals that the closest relatives of crotamine genes are genes that are expressed in the pancreas of snakes. They're known as defensins, and the snakes use them to fight infections. Pigs, mice, and humans make closely related defensins as well.

These results support the hypothesis that the defensins originally evolved in a common ancestor of snakes and mammals. As new lineages of animals split off

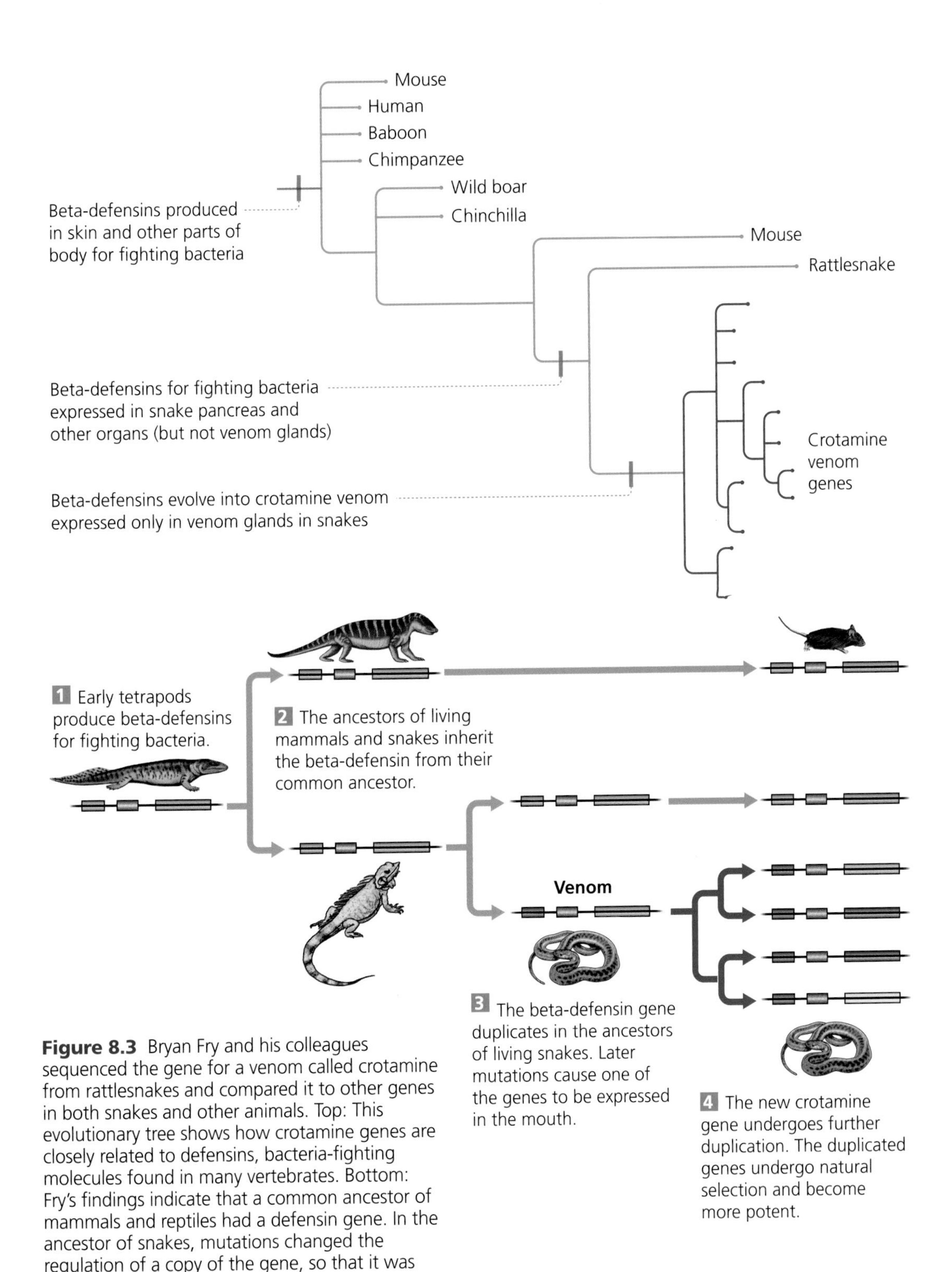

Figure 8.3 Bryan Fry and his colleagues sequenced the gene for a venom called crotamine from rattlesnakes and compared it to other genes in both snakes and other animals. Top: This evolutionary tree shows how crotamine genes are closely related to defensins, bacteria-fighting molecules found in many vertebrates. Bottom: Fry's findings indicate that a common ancestor of mammals and reptiles had a defensin gene. In the ancestor of snakes, mutations changed the regulation of a copy of the gene, so that it was expressed in the mouth. Later, the venom gene was duplicated in snakes and evolved different structures. (Adapted from Fry, 2006)

from that ancestor, they inherited that defensin gene. Accidental mutations produced extra copies of the gene, which became specialized for attacking different pathogens. The ancestors of today's snakes inherited those genes, but one of them experienced a peculiar mutation. The mutation didn't change how the protein functioned, but it did change where the protein was made. Instead of being produced in the pancreas, snakes began to produce it in their mouths.When the snakes bit their prey, they now released this defensin into the wound. Fry proposes that further mutations to the defensin gene itself changed its shape, so that it began to take on a new function. Instead of fighting pathogens, it began to damage muscles. Further mutations made it increasingly deadly, and gene duplication gave rise to an entire family of these venom genes.

Fry and his colleagues have drawn similar trees for more than two dozen venom genes. Some evolved from proteins produced in the heart, in the brain, in white blood cells, and in many other places in the snake body (**Figure 8.4**). They found, for example, the origin of a venom used by the inland taipan to make its victims black out. It evolved from proteins that slightly relax the muscles around the heart. Once these proteins evolved into venom, this slight relaxation became a rapid drop in blood pressure. In each case, venoms evolved through gene duplication, gene recruitment, and fine-tuning mutations to the genes themselves.

The venom-delivery system in a rattlesnake is, of course, more than just the venom genes. It also includes the venom glands and the fangs for delivering the venom deep into the flesh of its prey. It may be hard to imagine how this complex system evolved, since all the parts seem to depend on each other. But this kind of difficulty is no reason to discard a hypothesis. Rather than throw up his hands, Fry looked closer at the venom system and made a major discovery about how it evolved.

Fry wondered if the gene recruitment took place independently in each lineage of venomous snakes, or if some venom genes evolved before their lineages split. He discovered that some venom genes were very ancient, having evolved in the common ancestor of all snakes—even snakes not previously recognized as venomous. Garter snakes, for example, lack hollow fangs and high-pressure delivery systems. But Fry discovered that these harmless snakes make some of the same venoms that are found in rattlesnakes, and those venoms are just as potent molecule for molecule. While garter snakes may not have fangs, they can use their tiny teeth to puncture the delicate skin of a frog, through which the venom can enter their prey's body. The venom of garter snakes had gone undiscovered until Fry's discovery, simply because people didn't get hurt by their bite.

Fossils and studies on reptile DNA indicate that snakes evolved about 60 million years ago. Their closest living relatives include iguanas and monitor lizards, such as the Komodo dragon, the biggest lizard alive today. Given the ancient origin of some snake venoms, Fry now wondered whether snake venom evolved before there were snakes. He tracked down lizards closely related to snakes and discovered that many of them had glands on the sides of their jaws. He gathered RNA transcripts from those glands and sequenced them. Some of them were produced from genes closely related to venom genes in snakes. The same aorta-relaxing venom made by inland taipans, for example, is closely related to a protein produced in the mouth of the Komodo dragon.

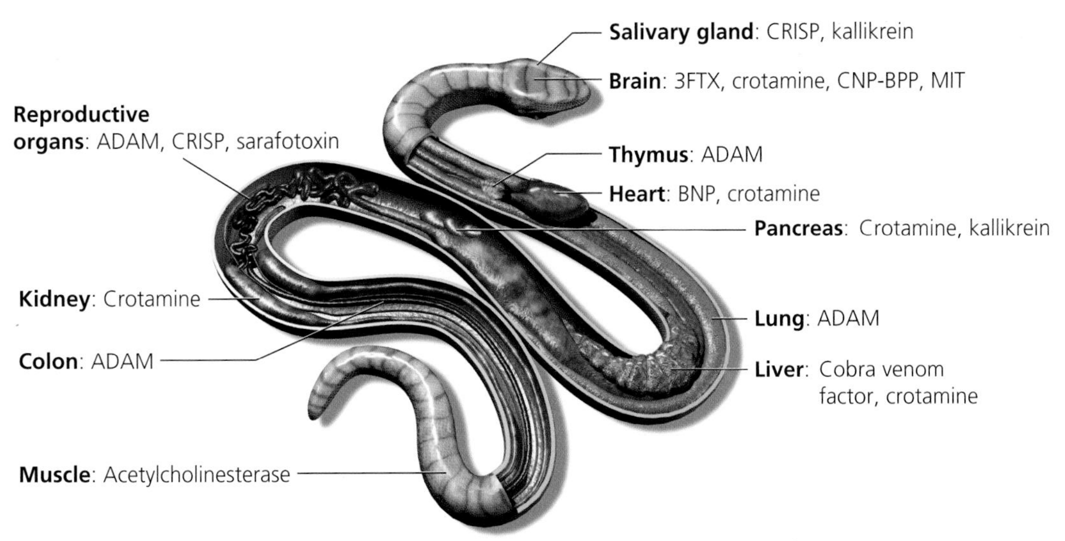

Snake toxin	Effects
3FTx	Neurotoxin
Acetylcholinesterase	Disruption of nerve impulses, causing heart and respiratory failure
ADAM	Tissue decay
BNP	Acute low blood pressure
CNP-BPP	Acute low blood pressure
Cobra venom factor	Anaphylactic shock
CRISP	Paralysis of peripheral smooth muscle, hypothermia
Crotamine	Muscle decay and neurotoxicity
Kallikrein	Acute low blood pressure, shock, destruction of blood clotting factors
MIT	Constriction of intestinal muscles, resulting in cramping, increased perception of pain
Sarafotoxin	Acute high blood pressure

Figure 8.4 Venom genes have been recruited from many organs in snakes.

Fry has proposed a surprising new hypothesis for the evolution of snake venom (**Figure 8.5**). Venom first evolved more than 200 million years ago in the common ancestor of snakes and their closest living relatives. (More distantly related lizards, such as geckos and skinks, produce no venom.) Genes with other functions were recruited for the first venoms, which the early lizards produced in mucus glands in their mouths. These early venoms were not fatal. Some of them were able to slow down the prey of lizards, while others may have caused their wounds to bleed more. When snakes lost their legs about 60 million years ago, they already had a number of venom genes. Later, some lineages of snakes evolved stronger venoms, as well as hollow fangs to improve the delivery of their venom. Rather than simply slowing down their prey, these snakes could now use venom to kill them outright. What looks to us today like a complex adaptation made up of parts that can't work on their own was actually assembled, bit by bit, over many millions of years.

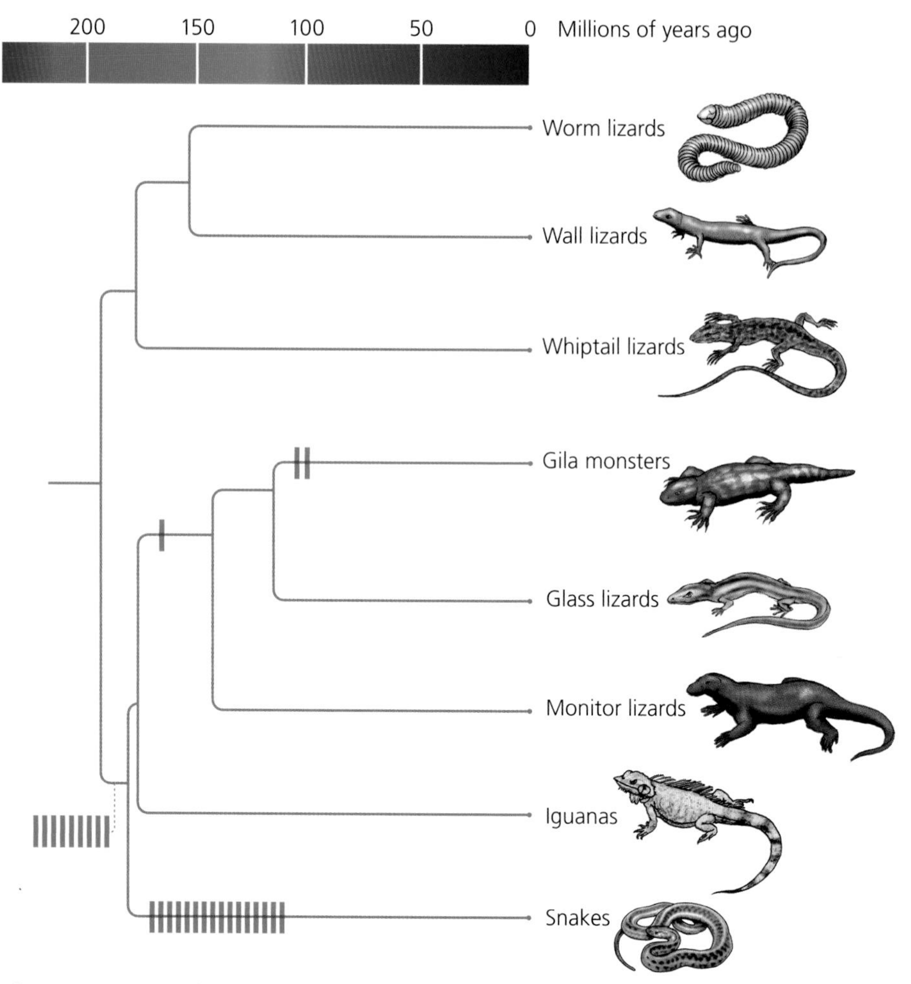

Figure 8.5 Fry and his colleagues discovered that some lizards have genes closely related to snake venom genes. Each pink bar shows when one of these genes evolved. This discovery suggests that the complex venom system in snakes began evolving millions of years before snakes evolved. (Adapted from Vidal et al., 2006)

If Fry is right, there's an unexpected answer to the question, "Which came first, the venom or the fang?" The answer is the venom. In fact, the venom may have come even before the snake.

Sculpting a Beak

Transcription factors are essential not just for making the right proteins in the right places, such as venom in venom glands; they also shape an animal's entire anatomy. Until the 1970s, biologists knew almost nothing about how genes build animal embryos. Now scientists can reconstruct the development of embryos with exquisite precision. They can, for example, manufacture dyes that turn animal embryos into rainbows of color, as each binds to a particular protein produced during development. Scientists can use RNA molecules to silence those genes to observe how development goes awry without them. Gradually, scientists are building up intricate models of development, which resemble electronic circuits (**Figure 8.6**).

By comparing the circuits in different species, scientists are reconstructing how animal shapes evolved. Despite their radically different morphologies as adults, animals share many of the same fundamental gene networks for building embryos—sometimes called the "genetic tool kit." Gene duplication, gene recruit-

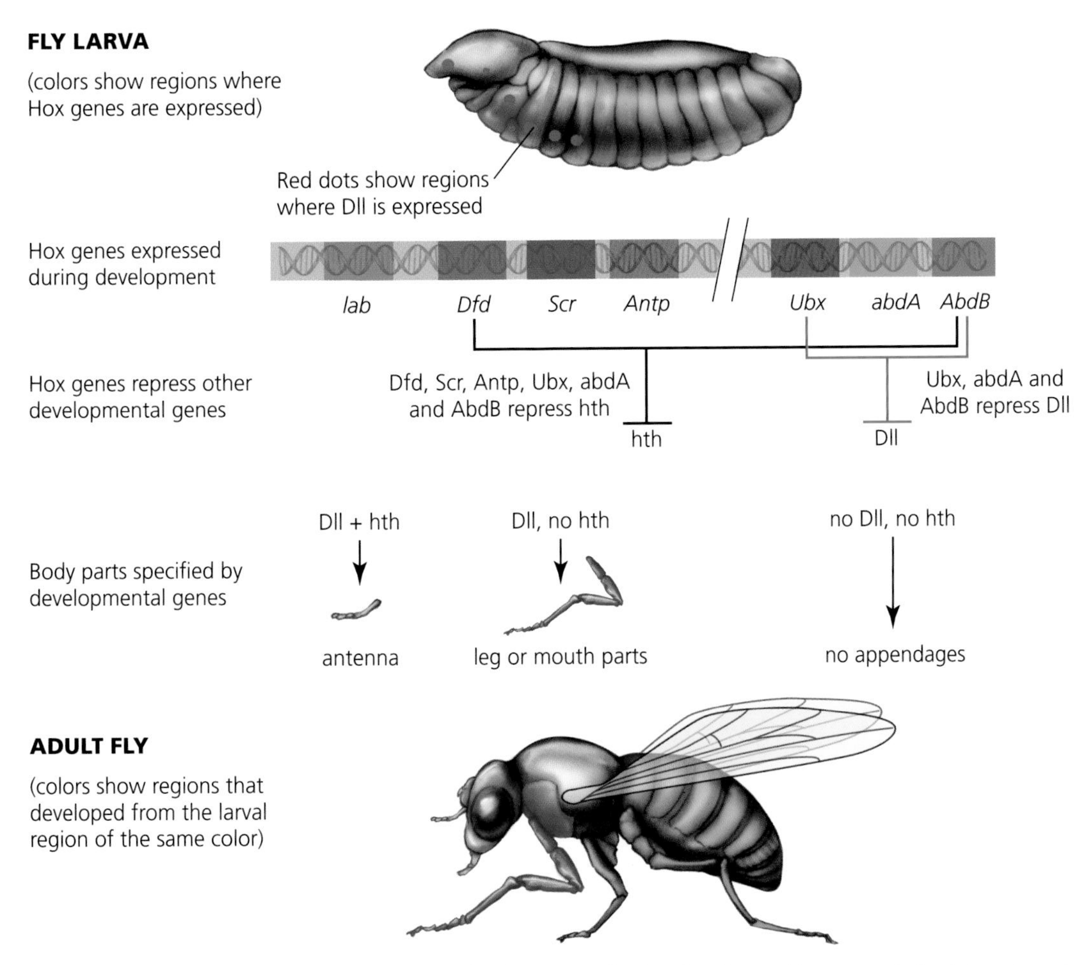

Figure 8.6 Networks of genes control the development of animal embryos. Hox genes help determine what different segments in a fly embryo will become in the adult. Remarkably, the genes are arranged in the same order in which they are expressed in the embryo. Some genes that are expressed during development shut other developmental genes down. For example, fly larva cells express a gene called Distalless or Dll. But the Hox genes produced at the back of the larva (Ubx, abdA, and AbdB) all repress the Distalless gene. Another gene, called homothorax (or hth), is repressed by all the Hox genes shown here, and so it is only produced in the fly's head. As a result, different segments make different combinations of proteins. Those combinations determine which body parts grow. For example, Distalless and homothorax together trigger the growth of antennae. Distalless without homothorax is a recipe for legs or mouthparts. No Distalless or homothorax means no appendages whatsoever. (Adapted from Gilbert, 2007 and Hueber, 2008)

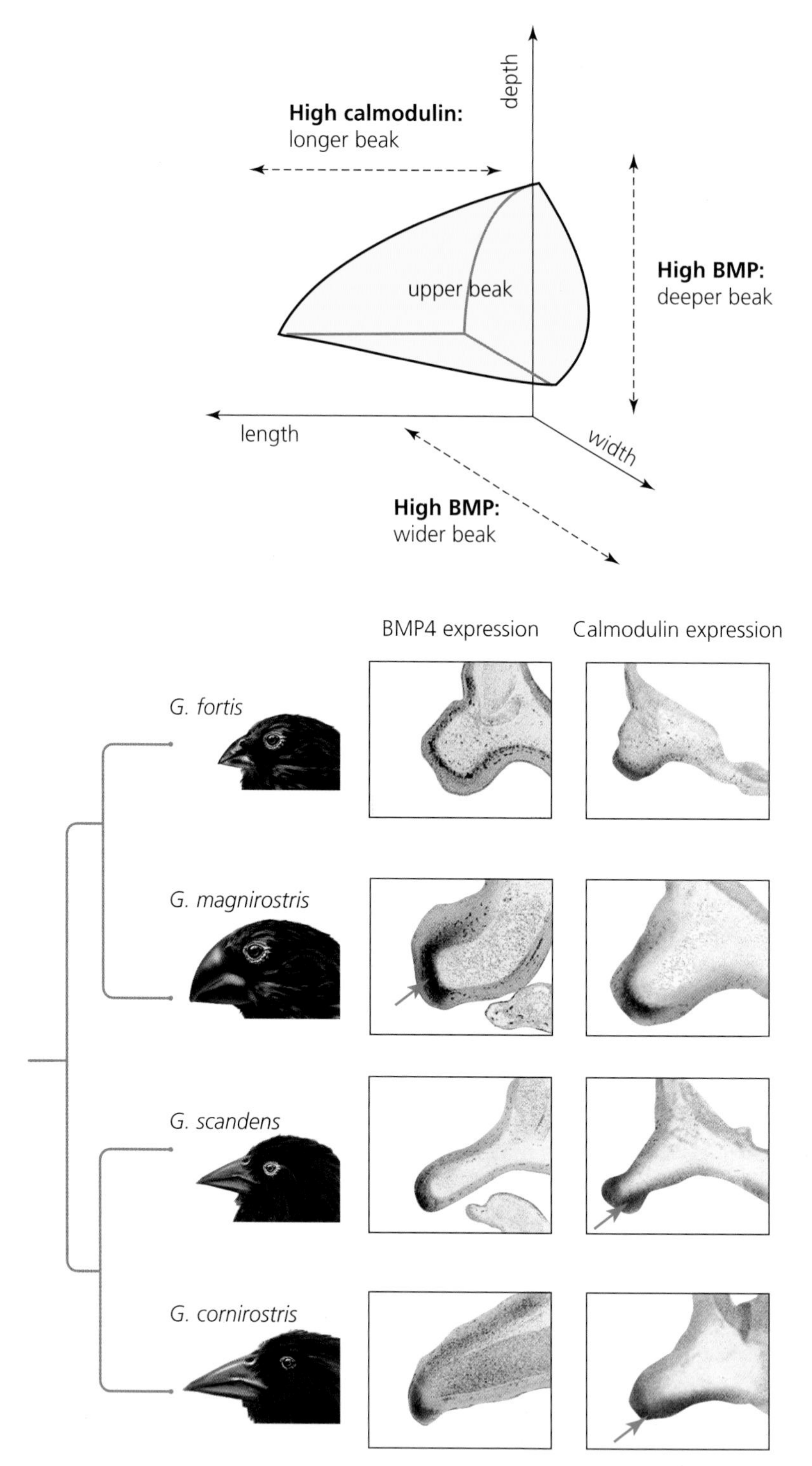

Figure 8.7 Different species of finches grow dramatically different beaks. Much of that variation is due to different levels of two proteins, known as calmodulin and BMP4. The second column shows developing beaks of bird embryos, with dark regions indicating cells expressing BMP4. The third column shows the pattern of calmodulin expression. The red arrows mark regions of very high expression. (Adapted from Grant, 2007)

ment, and shifting patterns of gene regulation have allowed new uses of the genetic tool kit, giving rise to what Darwin called "endless forms most beautiful."

To understand how development evolves, let's begin with those familiar birds, Darwin's finches. In just three million years, an ancestral finch gave rise to 14 species with an impressive range of beak shapes, each adapted to a different kind of feeding. Ground finches have deep, wide beaks for crushing hard seeds. Cactus finches have narrow beaks for snatching insects and flowers from cactuses without getting stabbed by the cactus spines. Peter Grant and Rosemary Grant teamed up with Harvard developmental biologist Cliff Tabin and his colleagues to discover what genetic changes made those new beak shapes possible.

Peter and Rosemary Grant have been investigating how changes in the development of Darwin's finches have produced dramatically different beaks.

The evolution of the beaks of Darwin's finches turns out to be, in some ways, surprisingly simple (**Figure 8.7**). Just two proteins have a powerful effect on the shapes of the beaks. One of the proteins is called BMP4 (BMP stands for bone morphological protein). Beaks that produced low levels of BMP4 become narrow, while beaks that produced lots of BMP4 become wide and deep. The other protein, known as calmodulin, controlled the length of the beaks. Low levels of calmodulin were found in the embryos of short-beaked finches, and high levels were produced in long beaks. Both BMP4 and calmodulin are known to control many other genes during development. It appears that simply changing the level of each of these master regulators plays a major role in producing the diversity of beaks found in Darwin's finches. The fact that such a simple change in gene activity can lead to the evolution of such a big phenotypic change may help explain how the birds adapted so quickly to so many different niches over the past few million years.

Recycled Feathers

It's crucial for cells in a developing embryo to have a sense of where they are. It's not enough to for a limb bud to grow into an arm, for example. To work properly, the arm has to grow from the shoulder, rather than from the neck or the base of the spine. Some developmental genes lay down the geography of an embryo, assigning cells to different parts of the body. It turns out that the same "geography genes" have been recycled again and again, to lay down the coordinates for new structures. Feathers, recent studies suggest, were produced through this evolutionary recycling.

The fossil record shows how birds evolved from dinosaurs (page 71). The evidence indicates that feathers evolved on dinosaurs long before their descendants could fly, gradually evolving from simple bristles to downy feathers and finally to feathers that could support an animal in flight. The closest relatives of birds and dinosaurs are alligators and other reptiles. Instead of feathers, they have scales on their skin. Birds have scales as well, but only on their legs. These lines of evidence point to a hypothesis: the genes that build feathers originally built scales, and were later co-opted in some parts of the skin to produce feathers. In recent years, Richard Prum, an ornithologist at Yale University, has been gathering evidence that supports this hypothesis.

Scientists have long known that both feathers and scales develop from disks of cells known as placodes in the skin of bird embryos. When a scale placode develops into a scale, Prum and his colleagues have found, its cells first express two key "geography genes" (**Figure 8.8**). One gene, *BMP2*, is expressed at the front of the placode, and the other, *Shh*, is expressed at the back. The back cells later grow faster than the front, forming a flap of hard tissue that extends over the front of the scale behind it. Both *BMP2* and *Shh* are also expressed in placodes that develop into feathers, with *BMP2* in the front and *Shh* in the back. Instead of extending backwards like a scale, however, a feather placode forms a ring that grows straight up, forming a shaft. Early dinosaur feathers may have stopped at this point, forming simple bristles. A few living birds, like the flightless kiwi of New Zealand, have bristle-like feathers as well.

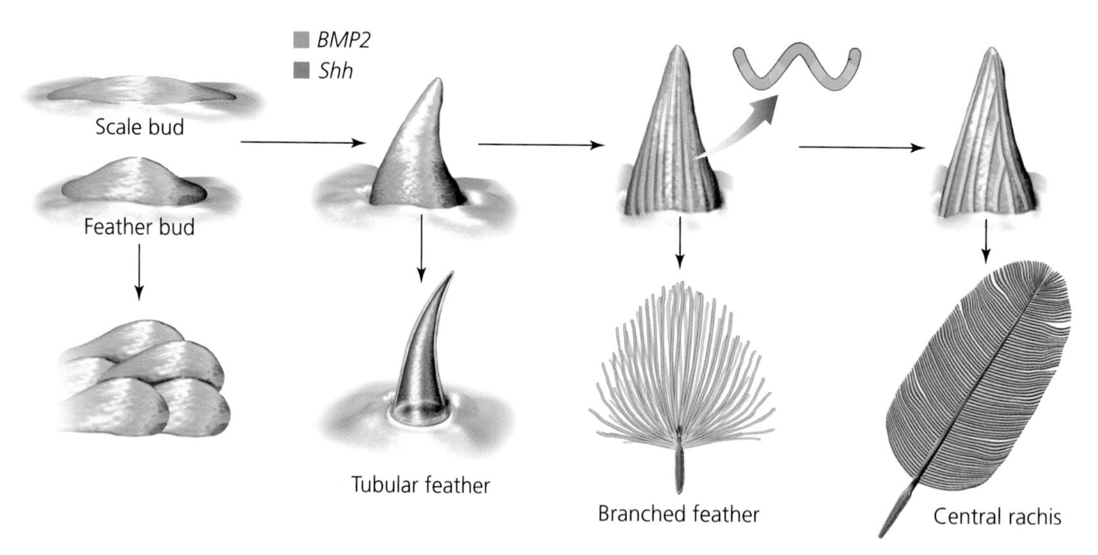

Figure 8.8 Studies on the development of feathers and on fossils of feathered dinosaurs have led to a hypothesis for the evolution of feathers. Left: Scales on birds and other reptiles develop from structures called placodes. *BMP2* and *Shh* are expressed on each side of both kinds of placodes. Early feathers may have been simple tubes growing out of placodes. Slits then evolved around the growing tube, allowing the feather to peel open into branches. *BMP2* and *Shh* were recruited again, this time to specify where the slits form. Finally, a vane (known as a central rachis) evolved, opening the way for using feathers for flight. (Adapted from Gilbert, 2007)

The next stage in feather evolution in Prum's model saw the emergence of downy feathers. A downy feather on a living bird develops from the simple shaft as parallel slits form along its sides. The location of these slits is determined by cells around the shaft, which produce BMP2 and Shh again. These genes were recruited to help lay out the geography of the feather in a new way, in other words. Once slits formed according to the BMP2-Shh map, they caused the feather to peel apart to form a plume.

It took a minor modification to change the developmental recipe for plume-shaped feathers to produce feathers with vanes. One section of the ring stopped growing slits. Instead, it developed straight up, becoming the vane. The slits, meanwhile, now extended at an angle from the vane around the rest of the ring. They now caused the feather to open up, creating barbs. Once again, Prum and his colleagues have found, BMP2 and Shh were recruited to establish the location of the vane at one end of the ring.

Feathers, according to Prum, did not evolve from dinosaur scales. Instead, the genetic circuitry that dinosaurs used to develop scales was modified to produce a new structure. And some of the same genes in that circuit could be used to build newer and newer variations on that structure, so that, ultimately, reptiles could use their skin to fly.

Flies, Mice, and Genetic Tool Kits

So far we've considered the evolution of traits limited to certain groups of animals—feathers in birds, for example, and venom in lizards and snakes. But evolutionary biologists are also discovering the common ancestry of genes that control the overall body plans of animals—even animals that look fundamentally different from one another.

At first glance, our bodies seem a far cry different from a fly's. A fly (or any other arthropod) has a segmented exoskeleton surrounding a soft interior. The main nerve of its body runs along the bottom of its abdomen, while the main structures of its digestive system run along the top of its back. It grows segmented legs, along with wings and a pair of flight-controlling clubs called halteres. Its mouth consists of leglike mandibles, and its eyes are made up of many hexagonal columns, each capturing an image of a tiny fragment of its surroundings.

A vertebrate, such as a human or mouse, has a profoundly different anatomy. Instead of an external skeleton, its skeleton grows inside its body and is surrounded by muscle and skin. Its spinal cord runs down its back, and its digestive system runs down along the abdomen. It grows four limbs; its jaws are internal bones rather than leglike structures. Its eyes are like little cameras, each able to form a detailed image.

The differences between vertebrates and arthropods—along with other major groups of animals—seemed so vast to early naturalists that they doubted that these creatures could have evolved from a common ancestor. Their vertebrate and arthropod body plans seemed to be separated by a vast gulf. But when scientists uncovered the genes that built those body plans, they realized there were homologies hidden underneath the outward differences. These different animal body plans were built with the same basic genetic tool kit.

The head-to-tail axis of a mouse, for example, is marked off by a set of genes known as Hox genes. Mice and most other vertebrates (including humans) carry

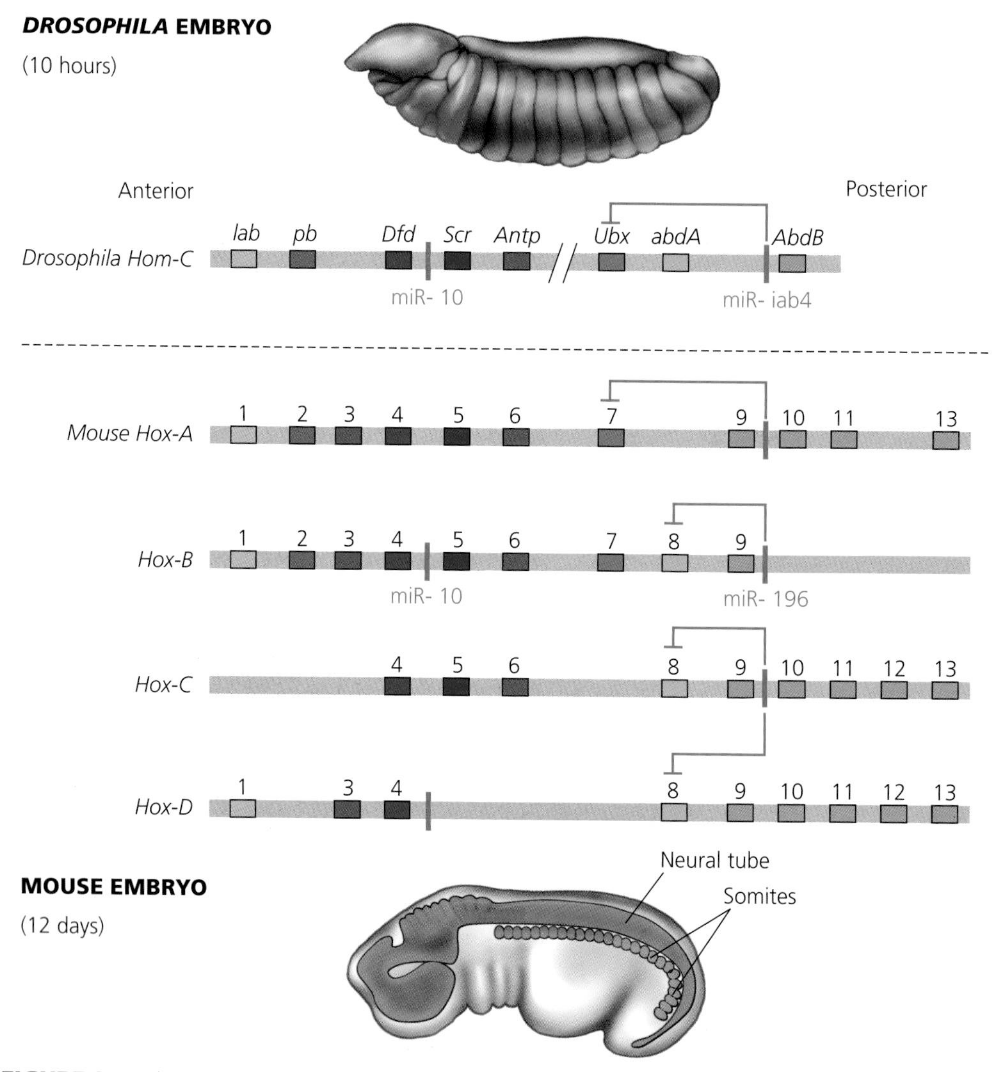

FIGURE 8.9 Flies and mice are separated by 600 million years of evolution. And yet the development of each species is controlled by homologous Hox genes. In the vertebrate lineage, the entire Hox gene cluster was duplicated twice, producing four sets of genes. Some of these genes were later lost. Yet the overall similarity of Hox genes in mice and flies is still clear. Both animals even have homologous genes in the same location in the Hox cluster that encode RNA molecules that regulate other Hox genes (marked here as miR). The best explanation for all this evidence is that the common ancestor of flies and mammals already had a set of Hox genes that controlled development. (Adapted from Gilbert, 2007, and DeRobertis, 2008)

four sets of Hox genes, and each set is arrayed along a chromosome in the same order in which it is expressed from head to tail in the body. Flies use a related set of genes to determine their own head-to-tail axis. Like the Hox genes in vertebrates, these genes are arrayed in the same order on fly chromosomes as they are expressed in a fly embryo (**Figure 8.9**).

The similarities between the Hox genes in flies and in mice has grown greater the longer scientists have studied them. In 2004, for example, David Bartel of MIT and his colleagues discovered RNA molecules that map to the Hox cluster of genes in mice. These RNA molecules keep some of the other Hox genes shut down in certain regions of the developing embryo. Flies, it turns out, have nearly identical RNA molecules. The genes that encode them are located in the same location in the fly Hox cluster. And, like the mouse RNAs, the fly RNAs silence other Hox genes. The Hox genes are so similar in mice and in flies, in fact, that they are literally interchangeable. If a scientist shuts down a Hox gene in a fly and inserts the corresponding gene from a mouse, the fly will develop normally.

Flies and mice use similar genes not only to build their bodies from head to tail, but also from front to back (**Figure 8.10**). In a developing mouse, cells along the belly (the ventral side) produce BMP4. Flies express a homologous protein called Dpp, but they express it along the back (the dorsal side). These genes determine on which side of the body the digestive system will develop. On the opposite side of each embryo, the nervous system will develop.

Flies and mice are not the only animals that share this homology in their development. So do octopuses, starfish, oysters, and earthworms. In fact, millions of species of animals use the same system of genes to determine the coordinates of their bodies. They all belong to the same lineage of animals, known as bilaterians. (The name means "two sides," which refers to the symmetry of the left and right sides of their bodies.) The common ancestor of all living bilaterians (urbilateria) lived some 570 million years ago. It had already evolved the genetic tool kit shared by bilaterians today, including a Hox gene cluster for establishing its front-to-back anatomy and a Dpp-BMP4 network to determine its front and

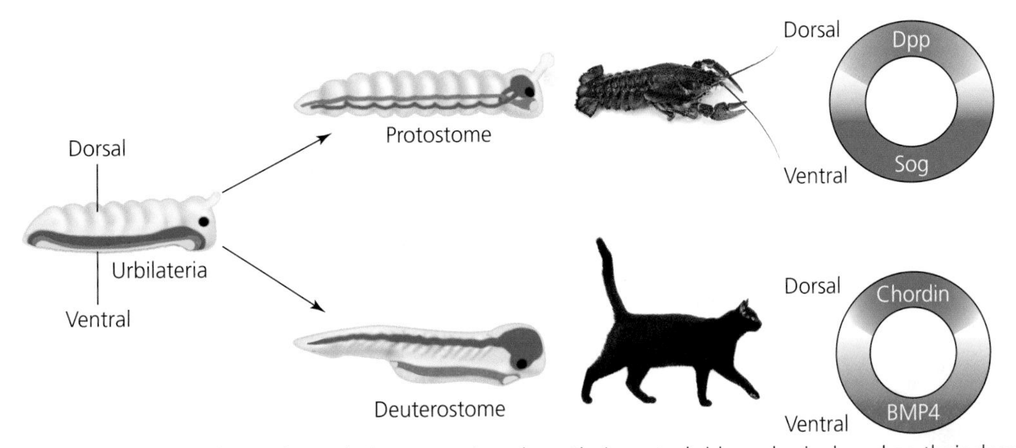

Figure 8.10 Vertebrates have their gut running along their ventral side and spinal cord on their dorsal side. Insects and many other invertebrates have the reverse arrangement. But they use homologous genes to mark where these structures will grow. (Adapted from DeRobertis, 2008)

back. Other studies suggest that the common ancestor of bilaterians also had other developmental genes for building organs such as eyes.

The descendants of that first bilaterian inherited this genetic tool kit, but they used it to build a wide diversity of bodies. Over time, Hox genes and other developmental genes came to control new genetic programs—tentacles in some cases, antennae in others, arms and legs in still others.

In the lineage leading to vertebrates, the nervous system appears to have flipped from the front to the back side. Our entire Hox cluster was duplicated twice. Shifts in the pattern of expression of Hox genes helped give rise to new variations on the vertebrate body plan. Different combinations of Hox genes are expressed from the head to tail in vertebrates, defining different parts of the spine. In a mouse embryo, for example, scientists can clearly see a distinct Hox pattern where the neck vertebrae, the lumbar vertebrae, and the thoracic vertebrae will form. In snake embryos, the Hox pattern defining the future thorax and abdomen expanded, while the neck region all but disappeared, as did the tailward region. As a result, the snake skeleton is far more uniform than that of other reptiles.

Evolving Eyes

"The eye to this day gives me a cold shudder," Charles Darwin once wrote to a friend. If his theory of evolution was everything he thought it was, a complex organ such as the human eye could not lie beyond its reach. And no one appreciated the beautiful construction of the eye more than Darwin—from the way the lens was perfectly positioned to focus light onto the retina to the way the iris adjusted the amount of light that could enter the eye. In *The Origin of Species,* he wrote that the idea of natural selection producing the eye "seems, I freely confess, absurd in the highest possible degree."

For Darwin, the key word in that sentence was *seems.* If you look at the different sort of eyes out in the natural world and consider the ways in which they could have evolved, Darwin realized, the absurdity disappears. The objection that the human eye couldn't possibly have evolved, he wrote, "can hardly be considered real."

Today evolutionary biologists are deciphering the origins not just of our eyes but of the dozens of different kinds of eyes that animals use (**Figure 8.11**). Fly eyes are built out of columns. Scallops have a delicate chain of eyes peeking out from their shells. Flatworms have simple light-sensitive spots.

Octopuses and squids have camera eyes like we do, but with some major differences. The photoreceptors of octopuses and squids point out from the retina, towards the pupil. Our own eyes have the reverse arrangement. Our photoreceptors are pointed back at the wall of the retina, away from the pupil.

Scientists have long believed that these different eyes evolved independently. Yet they share an underlying unity in the genes used to build them. By tracing the history of these shared genes, scientists are testing Darwin's hypothesis that a complex eye could have evolved through a series of intermediate steps.

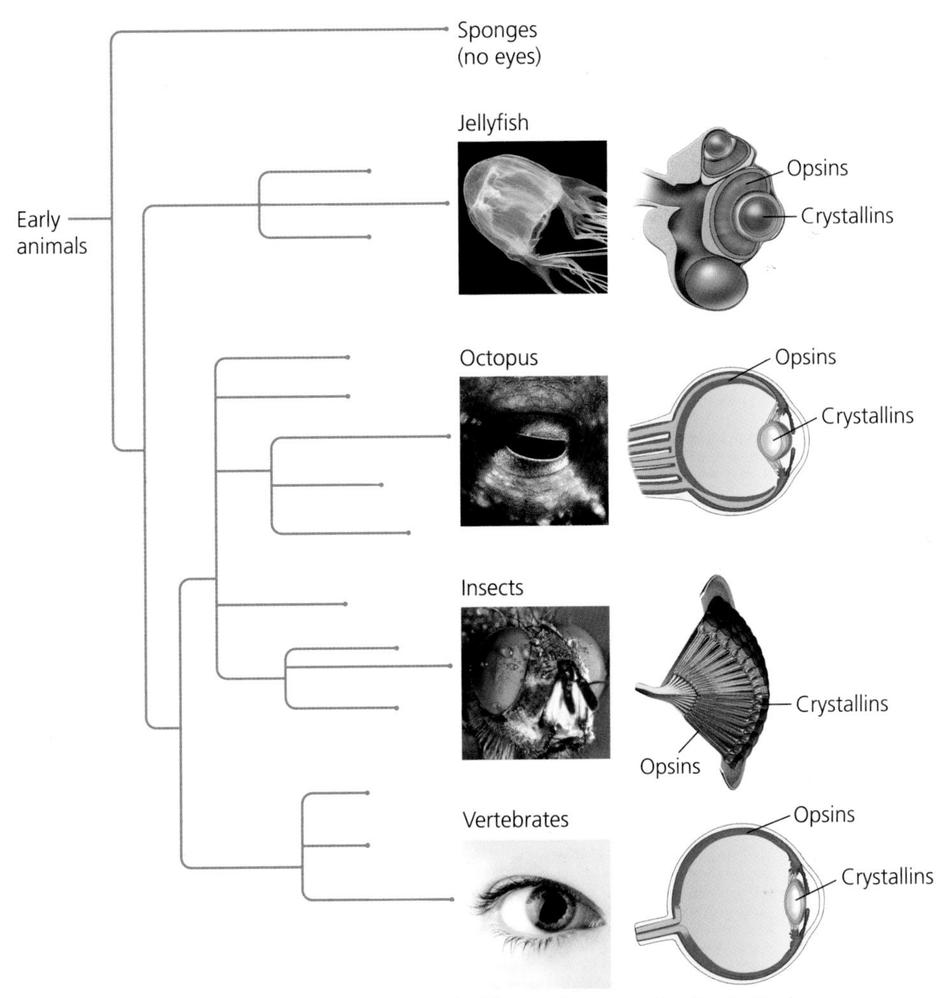

Figure 8.11 Complex eyes have evolved in several different lineages of animals. Each kind of eye contains crystallins for directing incoming light and opsins for capturing it. But the particular molecules each kind of eye uses as opsins and crystallins are different from the others.

When light enters your eye, it strikes a molecule known as an opsin. Opsins sit on the surface of photoreceptor cells, and, when they catch photons, they trigger a series of chemical reactions that causes the photoreceptor to send an electrical message towards the brain.

Biologists have long known that all vertebrates carry the same basic kind of opsin in their eyes, known as a c-opsin. All c-opsins have the same basic molecular shape, whether they're in the eye of a shark or the eye of a hummingbird. All c-opsins are stored in a stack of disks, each of which grows out of a hairlike extension of the retina called a cilium. In all vertebrates, c-opsins relay their signal from the stack of disks through a pathway of proteins called the phosphodiesterase pathway. All of these homologies suggest that c-opsins were present in the common ancestor of all living vertebrates.

How *Not* to Study Evolution

The idea of evolution caused an uproar through much of the 1800s, starting with naturalists such as Lamarck. When Charles Darwin first published *The Origin of Species*, scientists engaged in fierce public debates over his theory. Scientists would not come to a consensus about some of the most important questions about the theory—such as how heredity makes natural selection possible—for more than 60 years after Darwin's book was first published.

As these issues were settled, evolutionary biologists moved on to deeper questions and continued to debate each other. Today, when scientists meet at conferences to present their research on evolution, the disputes can get as fierce as they do at any scientific meeting. Evolutionary biologists argue about the best way to reconstruct evolutionary trees. They argue about what caused mass extinctions. They argue about the relative importance of natural selection, sexual selection, and other processes in evolution. These debates can get spirited, and sometimes downright rough. But evolutionary biologists do not rehash long-settled subjects, such as the fact that life has evolved over billions of years, or that complex traits evolved in a step-wise fashion from earlier traits. It would be just as pointless for astronomers to revive debates about whether the Earth revolves around the Sun, or vice versa.

Today a number of organizations and individuals claim that evolution is fundamentally untrue. They hold a wide range of views. Some claim that the Earth is only 6,000 years old, and that God created all life pretty much in its current form at the beginning of that time. Other people don't object to the notion that life is in fact billions of years old. Instead, they claim that major features of biology are the result of something that they call "intelligent design"—a planned creation by some kind of intelligent agent they claim not to be able to identify.

These claims are known collectively as creationism, because they all explain the diversity of life as the result of direct creation, rather than the result of natural processes such as natural selection. Evolutionary biologists reject creationism, both because its objections to evolution have no basis in the evidence and because it is ultimately a nonscientific view of nature.

Many arguments for creationism begin with the construction of a straw-man version of evolution. The straw man is knocked down, and its fall is supposed to be taken as evidence that evolution is false and creationism is true. For example, evolution is caricatured as being purely random. Since organisms have many genes that encode many interacting proteins, the argument goes, the odds are astronomically tiny that pure randomness

Other bilaterians—such as insects, octopuses, and scallops—don't have c-opsins in their eyes. Instead, they build another molecule, known as an r-opsin. Instead of keeping r-opsins in a stack of disks, they store r-opsins in foldings in the membranes of photoreceptors. R-opsins all send their signals through the same pathway of proteins (not the same pathway as c-opsins send signals in vertebrates). Again, the homologies in r-opsins suggest they evolved in the common ancestor of insects, scallops, octopuses, and other invertebrates that have r-opsins in their eyes.

These findings suggested that r-opsins and c-opsins evolved only after bilaterians had branched off into two lineages. In recent years, however, evolution-

could produce a complex trait like the eye. But natural selection is not a random process. Whether a particular organism gets a particular mutation is a matter of chance. Natural selection only spreads mutations through an entire population if they have some beneficial effect. Many lines of evidence show that natural selection can favor an entire series of mutations, which together give rise to new traits.

Another straw-man argument is the supposed absence of fossils that document major transitions in evolution. Because fossils have not been found to document every intermediate stage in evolution, creationists claim, evolutionary biologists cannot claim that new species evolved from old ones. But the process by which fossils form makes it inevitable that the fossil record is incomplete. What matters is the fact that the fossil record is consistent with the theory of evolution. Before the 1980s, for example, paleontologists had found no fossils of whales that could shed light on the early stages of their evolution. Based on their study of living whales, biologists predicted that whales descended from terrestrial mammals, most likely artiodactyls (a group that includes such hoofed animals as pigs, camels, and hippopotamuses—for more, see page 8). Over the past three decades, scientists have found not just one of these transitional fossils, but fossils from about 30 different species of early whales, documenting a 10-million-year transition. And these early whales show clear links to artiodactyls.

Many arguments for creationism begin with the construction of a straw-man version of evolution.

Understanding evolution helps us make sense of the mechanisms underlying the patterns of fossils and living whales, and it guides us towards new evidence. Creationism in its various forms lacks this sort of explanatory power. Creationists seem to believe that things are the way they are simply because they are. What explains the presence of 13 distinct species of closely related finches with different beaks on the Galápagos Islands? They were created that way.

To say that new species were created directly by God, as many creationists do, is to make a religious claim, not a scientific one. Scientists cannot evaluate a theory about a supernatural agent, since scientific theories explain phenomena that follow natural, repeatable patterns. To invoke an intelligent designer instead does not make creationism any more scientific. The designer is either supernatural or natural. If the designer is supernatural, the claim of intelligent design is not scientific. If the designer is natural, it should be possible to make predictions about how it has produced life's diversity. But since we can't know the nature of the designer, we're told, this claim reaches a scientific dead end.

ary biologists have discovered opsins where they weren't supposed to be. It turns out, for example, that humans also make r-opsins. We just don't make them on the surfaces of photoreceptors where they can catch light. Instead, r-opsins help to process images captured by the retina before they're transmitted to the brain.

In 2004, Detlev Arendt of the European Molecular Biology Laboratory and his colleagues also found c-opsins where they weren't supposed to be. They were probing the nervous system of an animal known as a ragworm, which captures light with r-opsins. Arendt and his colleagues discovered a pair of organs atop the ragworm's brain that grew photoreceptors packed with c-opsins. Arendt sequenced the gene for the ragworm c-opsins and compared it with genes for other opsins. He found that it is more closely related to the genes for c-opsins in our own eyes than it is to the genes for r-opsins in the ragworm's own eyes. These findings have led Arendt and other researchers to revise their hypothesis about the origin of opsins: the common ancestor of all bilaterians must already have had both kinds of opsins.

But Todd Oakley, a biologist at the University of California at Santa Barbara, wondered if opsins might be even older. To find out, Oakley and his colleagues turned to the closest living relatives of bilaterians. Known as the cnidarians, this lineage includes jellyfish, sea anemone, and corals.

Biologists have long known that some cnidarians can sense light. Some jellyfish even have eye-like organs that can form crude images. In other ways, though, cnidarians are radically different from bilaterians. They have no brain or even a central nerve cord, for example. Instead, they have only a loose net of nerves. These dramatic differences had led some researchers to hypothesize that bilaterians and cnidarians had evolved eyes independently. In other words, the common ancestor of cnidarians and bilaterians did not have eyes.

In recent years, scientists have sequenced the entire genomes of two species of cnidarians, the starlet sea anemone (*Nematostella vectensis*) and a freshwater hydra (*Hydra magnipapillata*). Scanning their genomes, Oakley and his colleagues discovered that both species of cnidarians have genes for opsins—the first time opsin genes had ever been found in a nonbilaterian. The scientists carried out experiments on some of these genes and discovered that they are expressed in the sensory neurons of the cnidarians. Oakley's research suggests that, as he had suspected, opsins evolved much earlier than bilaterians.

With discoveries from scientists such as Oakley and Arendt, we can start to get a sense of how opsins evolved. Opsins belong to a family of proteins called G-protein coupled receptors (GPCRs). They're also known as serpentine proteins, for the way they snake in and out of cell membranes. Serpentine proteins relay many different kinds of signals in the cells of eukaryotes. Yeast cells use them to detect odorlike molecules called pheromones released by other yeast cells. Early in the evolution of animals, a serpentine protein mutated so that it could pick up a new kind of signal: light. (**Figure 8.12**)

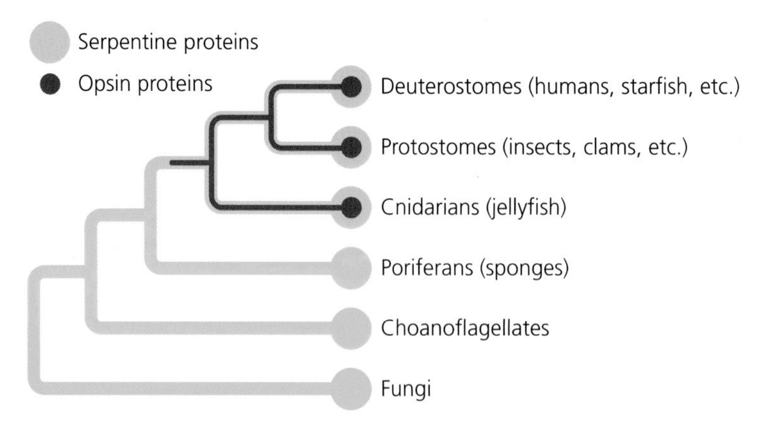

Figure 8.12 The opsins in animal eyes all share a common ancestry, having evolved in the common ancestor of jellyfish, insects, and humans. They evolved from a family of proteins known as serpentine proteins, which are carried by all animals, as well as related organisms such as fungi. (Adapted from Oakley, 2008)

At some point, the original opsin gene was duplicated (**Figure 8.13**). The two kinds of opsins may have carried out different tasks. One may have been sensitive to a certain wavelength of light, for example, while the other tracked the cycle of night and day. When cnidarians and bilaterians diverged, perhaps 620 million years ago, they each inherited both kinds of opsins. In each lineage, the opsins were further duplicated and evolved into new forms. And thus, from a single opsin early in the history of animals, a diversity of light-sensing molecules has evolved.

The earliest eyes were probably just simple eyespots that could only tell the difference between light and dark. Only later did some animals evolve spherical eyes that could focus light into images. Crucial to these image-forming eyes was the evolution of lenses that could focus light. Lenses are made of remarkable molecules called crystallins, which are among the most specialized proteins in the body. They are transparent, and yet can alter the path of incoming light so as to focus an image on the retina. Crystallins are also the most stable proteins in the

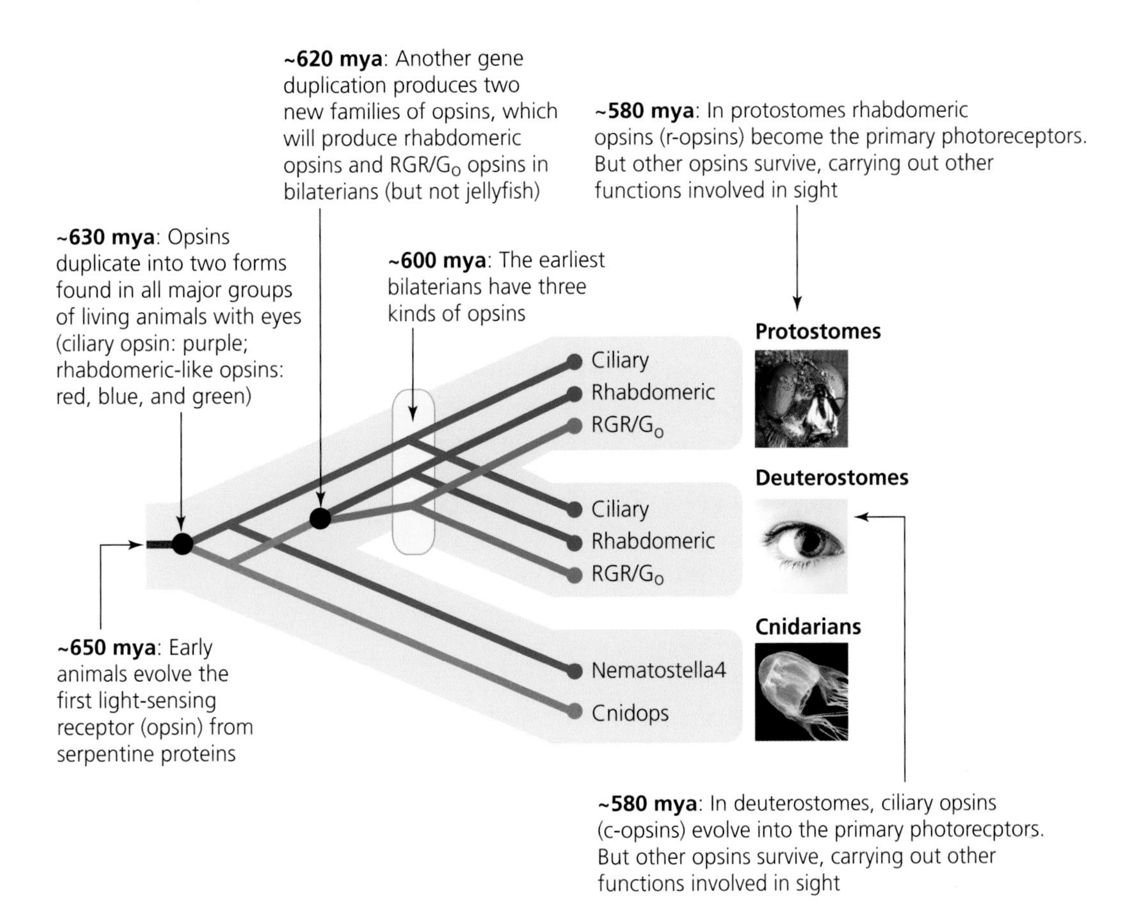

Figure 8.13 Recent studies on opsin evolution show that the common ancestor of cnidarians and bilaterians already had two kinds of opsins. Over the past 600 million years, these opsin genes have duplicated. Some of these opsins capture light in animal eyes. But other opsins carry out other functions. (Adapted from Plachetzki et al, 2007)

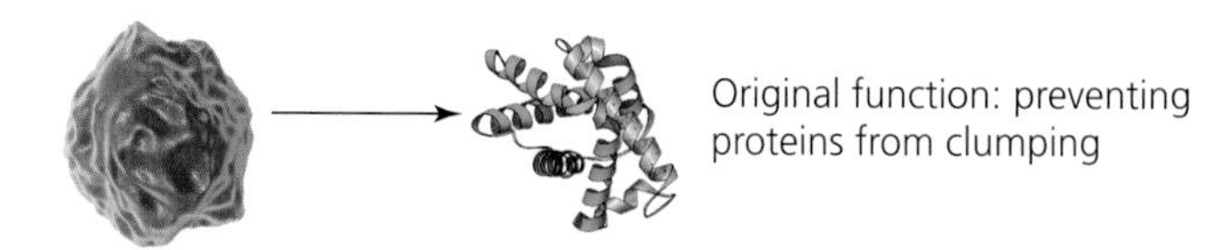

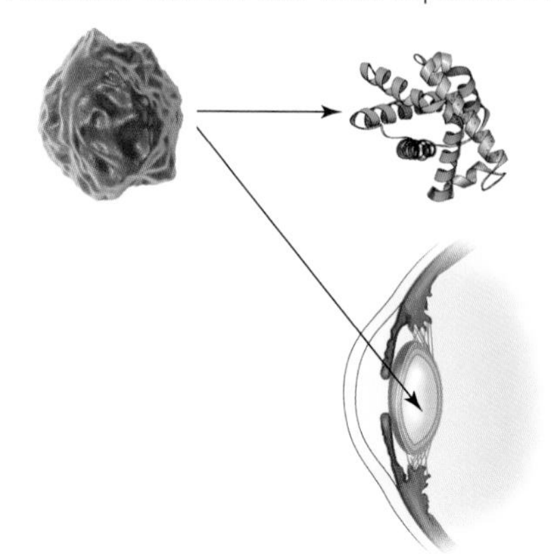

Figure 8.14 Crystallins in the lens of the human eye evolved through gene recruitment. They started out carrying out other functions in the body, such as preventing proteins from clumping. Mutations caused these proteins to be produced in the eye as well, where they helped focus images.

body, keeping their structure for decades. (Cataracts are caused by crystallins clumping late in life.)

It turns out that crystallins also evolved from recruited genes. All vertebrates, for example, have crystallins in their lenses known as α-crystallins. They started out not as light-focusing molecules, however, but as a kind of first aid for cells (**Figure 8.14**). When cells get hot, their proteins lose their shape. They use so-called heat-shock proteins to cradle overheated proteins so that they can still carry out their jobs. Scientists have found that α-crystallins not only serve to focus light in the eye, but also act as heat-shock proteins in other parts of the body. This evidence indicates that in an early vertebrate, a mutation caused α-crystallins to be produced on the surface of their eyes. It turned out to have the right optical properties for bending light. Later mutations fine-tuned α-crystallins, making them better at their new job.

Vertebrates also produce other crystallins in their eyes, and some crystallins are limited to only certain groups, such as birds or lizards. And invertebrates with eyes, such as insects and squid, make crystallins of their own. Scientists are gradually discovering the origins of all these crystallins. It turns out that many different kinds of proteins have been recruited, and they all proved to be good for bending light.

In 2007, Trevor Lamb of Australian National University and his colleagues synthesized these studies and many others to produce a detailed hypothesis about the evolution of the vertebrate eye (**Figure 8.15**). The forerunners of vertebrates produced light-sensitive eyespots on their brains that were packed with

1. Early chordates with light-sensitive eyespots expressing photoreceptor genes

2. Light-sensitive regions bulge outwards to the sides of the head

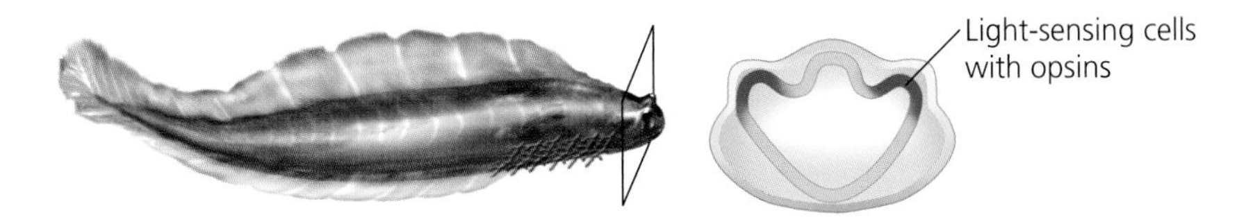

3. Patch folds inward into a cup, beneath unpigmented skin (lens placode)

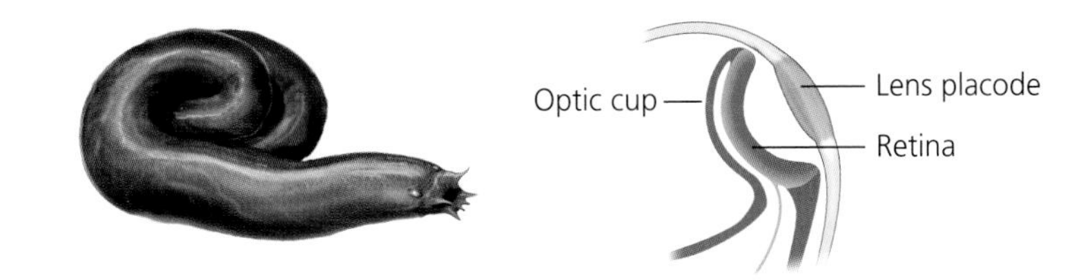

4. Surface becomes transparent, and lens evolves ability to focus an image

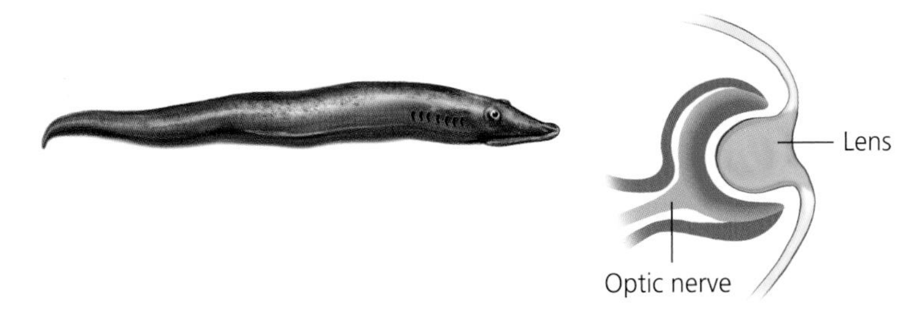

5. Eyes become spherical, evolve greater acuity

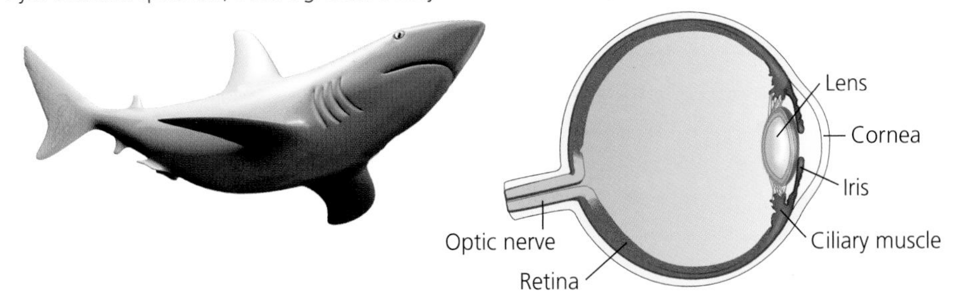

Figure 8.15 This figure illustrates one recent hypothesis for how the vertebrate eye evolved. It started out as a simple light sensor and then gradually evolved into precise image-forming organs. Living vertebrates and relatives of vertebrates offer clues to how this transformation took place. (Adapted from Lamb et al, 2008)

photoreceptors carrying c-opsins. These light-sensitive regions ballooned out to either side of the head, and later evolved an inward folding to form a cup. Early vertebrates could then do more than merely detect light: they could get clues about where the light was coming from. The ancestors of hagfish branched off at this stage of vertebrate eye evolution, and today their eyes offer some clues to what the eyes of our own early ancestors would have looked like.

After hagfish diverged from the other vertebrates, Lamb and his colleagues argue, a thin patch of tissue evolved on the surface of the eye. Light could pass through the patch, and crystallins were recruited into it, leading to the evolution of a lens. At first the lens probably only focused light crudely. But even a crude image was better than none. A predator could follow the fuzzy outline of its prey, and its prey could flee at the fuzzy sight of its attackers. Mutations that improved the focusing power of the lens were favored by natural selection, leading to the evolution of a spherical eye that could produce a crisp image.

The evolution of the vertebrate eye did not stop there. Some fish evolved double lenses, which allowed them to see above and below the water's surface at the same time. Birds evolved the ability to see in ultraviolet light. But all subsequent vertebrate eyes would be variations on the basic theme established half a billion years ago.

Constraining Evolution

In this chapter, we've been considering some of the many remarkable adaptations that have evolved over the history of life. But many evolutionary biologists want to understand why more kinds of adaptation *haven't* evolved. Why aren't there any hawk-sized dragonflies? Why are there no nine-toed tetrapods? It's clear that there are constraints on evolution, and scientists are investigating their causes.

Some clues about constraints come from fossils. While insects are small today, 300 million years ago they reached tremendous sizes. The wingspan of some ancient dragonflies reached almost a meter across. Some scientists have argued that insects are prevented from evolving to such great sizes again because the chemistry of the planet has changed. Insects get oxygen through tiny tubes penetrating their exoskeletons. As the insects become bigger, they need more tubes to supply enough oxygen to support their larger bodies. These tubes are an inefficient way to get oxygen into a big animal. In bigger insects, these tubes occupy a bigger proportion of their bodies, particularly in their legs. They become weaker and weaker even as the animal becomes bigger and bigger. It may be no coincidence that 300 million years ago, when insects got big, oxygen levels in the atmosphere rose very high for a time. That pulse of oxygen may have temporarily lifted the constraints on insect size.

The changing atmosphere constrains evolution from the outside. Genes themselves can constrain evolution from within, making it difficult for certain developmental programs to evolve. Michael Coates, a paleontologist at the University of Chicago, has argued that such developmental constraints are behind the lack of nine-toed tetrapods today. The earliest tetrapods had an abundance of toes. Some of them had six, others seven, and still others eight. Yet, by 340 million years ago, the only tetrapods still alive had five digits on each foot, and that limit has not been broken since. Paleontologists have not found any fossils of a tetrapod species from the past 340 million years with six or more digits on each limb. They've only found tetrapods that lost one or more of their original five digits. The ancestors of horses, for example, started out with five toes. Over millions of years the three middle digits fused into a single hoof. The outside two digits are now tiny and vestigial.

It is unlikely that five digits are the upper limit today because they're somehow more adaptive than six. After all, the fossil record shows that early tetrapods did just fine with as many as eight digits. Some living tetrapods spend much of their time in the same shallow-water habitats as early tetrapods, and yet they have not re-evolved any extra digits.

Coates suspects that the answer lies in some of the genes that help build toes and fingers. Hox genes play a central role in their development, while also guiding the development of limbs, the head-to-tail axis of the entire body, and even sexual organs. In other words, Hox genes are pleiotropic: they have many effects on an organism. As a result, a mutation to a single Hox gene may do more than just change the number of fingers on a hand. In fact, there is a disease that is caused by just such a mutation, called hand-foot-genital syndrome (page 98). It's possible that the pleiotropy of Hox genes constrains tetrapod digits. Even if an extra digit would be favored by natural selection, the mutation that caused it would also cause harmful effects elsewhere in the body.

The biggest constraint on evolution, however, may be history itself. New adaptations do not evolve from scratch. They are modifications of previously existing structures, pathways, and other traits. Even traits that are impressive in their complexity have deep flaws. The vertebrate eye, for example, has photoreceptors pointing away from the light. The photoreceptors all send projections to the optic nerve on the surface of the retina, and the nerve actually plunges back through the retina in order to travel back to the brain. The place where the optic nerve passes through the retina forms a blind spot in our vision. In order to get a complete picture of what we see, our eyes constantly dart around so that our brains can fill in the blind spot with extra visual information.

Developmental constraints help to make sense of some of the weirder examples of animal anatomy. Fish, for example, grow a series of nerve branches from their spinal cord that extend into their gill pouches. Tetrapods evolved from lobefins, inheriting this basic arrangement of nerves. But, as they evolved heads and necks, the ancestral gill arches shifted their positions and their sizes. The nerves migrated with the arches, stretching into peculiar paths. The most spectacular of these, called the recurrent layryngeal nerve, extends down the neck to the chest, loops around a lung ligament, and then runs back up the neck to the larynx. In long-necked giraffes, the nerve grows to a length of 20 feet in order to make this U-turn, when one foot of nerve would have done quite nicely.

Convergent Evolution

The mountain lions and wolves that roam North America are placental mammals. In other words, female lions and wolves carry their developing fetuses in their uterus, where an organ called the placenta grows to help the fetuses grow. Lions and wolves also roamed Australia, but they were of a different sort. They were marsupial mammals. In other words, they do not grow placentas. Instead, their young crawl out of the uterus and into a pouch to finish developing. Marsupial and placental mammals diverged from a common ancestor about 130 million years ago, and both lineages eventually produced species that looked remarkably similar and occupied the same ecological niches (**Figure 8.16**).

The evolution of two lineages into a similar form is known as convergent evolution. Two distantly related lineages will often evolve into the same form by different routes. A dolphin, for example, has a tapered body and a fluke at the end of its tail that can generate force much like a bird wing in the air. Sharks and tuna have similar bodies, allowing them to move efficiently in water as well. But dol-

Figure 8.16 Top: *Smilodon* lived in North and South America between 2.5 million and 10,000 years ago. Bottom: *Thylacosmilus* lived in South America from 10 million to 3 million years ago. They looked nearly identical, but their common ancestor lived 130 million years ago. *Smilodon* was a placental mammal closely related to cats, while *Thylacosmilus* was marsupial more closely related to an opossum. They are a striking example of convergent evolution.

phins evolved their fishlike bodies through the loss of their hind legs and many other modifications to a terrestrial-mammal body (page 8). While sharks and tuna swim from side to side, dolphins swim by moving their tails up and down.

What makes cases of convergent evolution like those of dolphins and fishes so intriguing is the combination of similarities and differences. In many cases, organisms can't evolve into identical forms because of developmental constraints. Many species of fishes, for example, grow flattened bodies that they use to swim close to the seafloor. One group includes rays and skates, which are closely related to sharks. They evolved a flat shape by acquiring "wings" of muscle that stretched out to either side of their bodies. Another group, known as flatfish, evolved from ray-finned fishes (**Figure 8.17**). They also have flat bodies, but evolved them along a very different path. The ancestors of flatfishes had typical ray-finned fish bodies that were flattened from side to side. They swim across the seafloor by turning onto one side, rather than swimming on their bellies like a ray.

This raises an evolutionary challenge for the flatfishes: What to do about their eyes? With one side facing down, one of their eyes will be staring at the seafloor. It turns out that, over millions of years, the down-facing eye of the flatfish gradually migrated to the up-facing side of the head. **Figure 8.17** shows an evolutionary tree that includes a remarkable flatfish fossil that had an eye midway through this transition. Today, the development of flatfish embryos still reflects this history. The eyes first develop on both sides of the head, and then one of the eyes gradually moves from one side of the head to the other.

If you compare the genes of flatfishes and rays, you'll find deeply different genetic programs producing the same flattened body. But in other cases, two lineages have evolved identical phenotypes by independently acquiring mutations

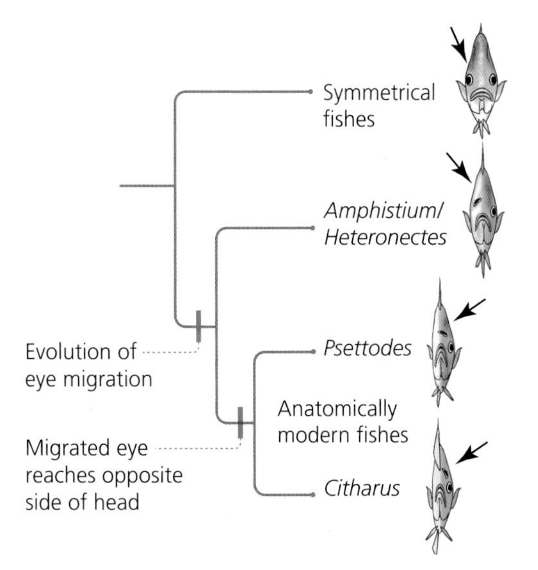

Figure 8.17 Flatfishes lie flat on the seafloor, with two eyes looking up. Fossils of extinct relatives of flatfishes show how they evolved from ordinary fish: one eye migrated up to the top of their head and moved to the other side. (Adapted from Friedman et al., 2008)

on the same genes. This kind of evolution is known as parallelism. One of the best documented cases of parallelism involves the stickleback fish. Populations of sticklebacks that live in the ocean typically produce armor plates with sharp spines on their bodies to ward off predators. Towards the end of the last Ice Age, about 15,000 years ago, some populations of sticklebacks moved up rivers and into lakes. There they enjoyed a predator-free life. Each population of freshwater sticklebacks independently evolved less armor, and some have entirely lost their spines.

David Kingsley, a biologist at Stanford University, and his colleagues have uncovered the genetic changes that produced this parallelism. They gathered sticklebacks from lakes in California, Washington, British Columbia, and the Northwest Territories of Canada, as well as in Iceland and Scotland. They bred these lake sticklebacks with their heavily armored and spiked cousins from the ocean. Some of their offspring developed armor and spikes; some didn't. By comparing the DNA of these hybrids, Kingsley and his colleagues discovered that, in every lake they studied, mutations to the same gene (called *Eda*) had led to the reduction of armor.

The researchers also discovered that all the sticklebacks that grew spines inherited the same segment of DNA from their marine parent. Fish lacking spines had inherited the same segment from the freshwater parent. When they examined that segment closely, they discovered that a gene called *Pitx1* was essential for the development of the spines. Yet the sequence of *Pitx1* was identical in both the marine and freshwater sticklebacks. The only difference between the two kinds of fish is where *Pitx1* is expressed. In freshwater sticklebacks, *Pitx1* is expressed in the developing nose, thymus gland, and sensory neurons. Marine sticklebacks express the same gene in all those cells, as well as in cells that will develop into spines. It thus appears that in freshwater sticklebacks a mutation has disabled one binding site near *Pitx1* that makes it become active in the spines, while the other binding sites and the gene itself are unchanged.

As scientists probe deeper into the developmental programs that have evolved in animals, it gets harder to draw a clean line between convergence and parallelism. Eyes, once again, provide an excellent example. Complex eyes can be found in many animals. In each case, they contain lenses to focus light on photoreceptors. But that does not mean that these animals share a common ancestor that also had a complex eye. The octopus eye, for example, is radically different from our own. For one thing, its receptors face forward, not backward. The crystallins that make up our lenses were borrowed from different genes than those in octopuses. So, on one level, the complex eye is a case of convergent evolution.

But all complex eyes also share a deep ancestry. Their opsins evolved from the same ancestral opsin. The development of all complex eyes in bilaterians is controlled in part by the same gene, known as *Pax-6*. When scientists inserted the *Pax-6* gene from a mouse into a fly's genome, the fly sprouted extra eyes over its body.

Is the fly eye homologous to the mouse eye, or is it convergent? The answer is both, depending on the level at which you look at it.

TO SUM UP...

- Gene duplication is an important kind of mutation for the evolution of new functions.
- Genes are sometimes recruited to become active in new organs or in new gene networks.
- Some genes influence many other genes involved in development. Mutations in a few of these regulatory genes can produce far-reaching changes in development.
- Animals as different as mammals and insects use the same "tool kit" of regulatory genes during development.
- During the evolution of animal eyes, genes for photoreceptors and crystallins were recruited and duplicated. Simple light-sensing eyes gradually evolved into complex camera eyes.
- Pleiotropy and physical constraints can restrict the possible forms into which a species can evolve.
- In convergent evolution, two separate lineages evolve similar adaptations.

The Origin of Species 9

Rick Brenneman loaded his rifle and quietly took aim. He had come thousands of kilometers from his home in Omaha, Nebraska, to stalk giraffes on the Hoanib River, which flows through a remote corner of Namibia in southern Africa. He selected a tall bull from a herd and fired. He made a direct hit. But the bull giraffe did not collapse and die. Instead, it staggered a little and then ran away.

Above: Rick Brenneman shoots needles into giraffes to collect their DNA. He and his colleagues have challenged the conventional hypothesis that giraffes belong to one species. They argue that there are actually six species of giraffes (pictured on the oppostie page).

Brenneman had fired a carbon dioxide–powered rifle, sending a needle, rather than a bullet, into the giraffe's side. The needle fell out of the giraffe with a tiny chunk of skin lodged in its tip. Brenneman and his colleagues walked across the savanna and picked up the needle. He was not hunting giraffes; he was hunting giraffe DNA.

Brenneman is a conservation geneticist at the Henry Doorly Zoo in Omaha. He and his colleagues have been firing darts at giraffes across Africa to help save them from extinction. Although giraffes live throughout much of Africa, their population has dropped by 30% in just the past decade, leaving fewer than 100,000 individuals in the wild.

You might think that there wasn't much left to learn about giraffes. After all, they're about as easy to spot as any animal on Earth, with their tall necks towering above the African plains. They don't lurk among the microbes hidden in the soil, or swim in some deep underwater canyon. Zoologists have studied giraffes carefully for well over a century. And yet Brenneman and his colleagues discovered something astonishing when they took a look at the DNA of 266 giraffes from Namibia, Kenya, Niger, Uganda, Zimbabwe, Tanzania, and South Africa. If you consult a standard field guide to African wildlife, you'll find a single species of giraffe on Earth: *Giraffa camelopardis*. But now that Brenneman and his colleagues have analyzed the DNA, they argue that there are in fact six species of giraffes rather than one.

How could six species of giraffes be hiding in plain sight? The answer takes us to the heart of what it means to be a species, and the process by which a new species evolves.

Species Before Evolution

Long before the dawn of science, humans were naming species. To be able to hunt animals and gather plants, people had to know what they were talking about. Taxonomy, the modern science of naming species, emerged in the 1600s and came into its own in the next century, thanks largely to the work of Swedish naturalist Carl Linnaeus. Linnaeus invented a system to sort living things into groups, inside of which were smaller groups. Every member of a particular group shared certain key traits. Humans belonged to the mammal class, and within that class the primate order, and within that order the genus *Homo*, and within that genus the species *Homo sapiens*. Linnaeus declared that each species had existed since creation. "There are as many species as the Infinite Being produced diverse forms in the beginning," he wrote.

Linnaeus's new order made the work of taxonomists much easier, but trying to draw the lines between species often proved frustrating. Two species of mice might interbreed where their ranges overlapped, raising the question of what name to give to the hybrids. Within species there was confusion as well. The willow ptarmigans in Ireland, for example, have slightly different plumage than the willow ptarmigans in Norway, which differ in turn from the ones in Finland. Naturalists could not agree about whether they belonged to different ptarmigan species or were just varieties—subsets, in other words—of a single species.

Charles Darwin, for one, was amused by these struggles. "It is really laughable to see what different ideas are prominent in various naturalists' minds, when they speak of 'species,'" he wrote in 1856. "It all comes, I believe, from trying to define the indefinable."

Species, Darwin argued, were not fixed since creation. They had evolved. Each group of organisms that we call a species starts out as a variety of an older species. Over time, natural selection transforms them as they adapt to their environment. Meanwhile, other varieties become extinct. An old variety ends up being markedly different from all other organisms—what we see as a species in its own right.

"I look at the term 'species' as one arbitrarily given, for the sake of convenience, to a set of individuals closely resembling each other," Darwin declared.

Good Barriers Make Good Species

There are trillions of animals and plants on Earth, and, at any moment, a vast number of them are mating or producing offspring. But who mates with whom is far from a random process. Bald eagles, for example, do not mate with dolphins or sunflowers. They mate only with other bald eagles, and so their offspring have only bald eagle DNA. The reproductive barriers that surround bald eagles include their geography. In the wild, bald eagles live in North America; the Atlantic Ocean prevents them from mating with eagles in Africa. Other kinds of reproductive barriers prevent bald eagles from mating with animals with overlapping ranges. Some of those reproductive barriers separate the bald eagles from other species by time, rather than space. Bald eagles sleep at night, and so they don't have the opportunity to mate with nocturnal animals. Other reproductive barriers are ecological. A bald eagle may catch fish from a lake at the bottom of which live mud-dwelling nematode worms. But since the eagle lives in the air and the nematodes in the mud, they never encounter each other. Of course, if a bald eagle and a mud-dwelling nematode did ever meet each

This pair of white-tailed sea eagles is engaging in an aerial courtship dance. Their ritual helps keep their species reproductively isolated from other bird species.

other, mating would be difficult, to say the least. Their sexual organs simply would not fit together. Thus anatomy can also create reproductive barriers.

Yet there are many species that have overlapping ranges, occupy similar niches in their ecosystems, and even have the right anatomy to mate—and yet they remain distinct from each other. The reproductive barriers between these species are subtle, but no less important than mountains or anatomy. In many species, for example, males and females only mate after an elaborate courtship ritual. Among bald eagles and related species, a male and female perform an aerial dance, darting and diving at each other, locking talons, falling from the sky, and then separating at the last possible moment. These aerial dances can lead to a lifelong bond between a male and female eagle; they will mate only with each other and work together to raise their young.

All of these reproductive barriers block reproduction at some point *before* a sperm from a male fertilizes an egg from a female. They create prezygotic isolation, named for the zygote (a cell that is the result of fertilization). Postzygotic isolation, on the other hand, is created by reproductive barriers that come into play after fertilization. When some species interbreed, the hybrid embryo fails to develop. In other cases, the hybrid offspring manages to survive to adulthood, but is sterile. When male donkeys and female horses interbreed, for example, they produce mules, which cannot interbreed with either parent species. In still other cases, hybrids can reproduce, but their offspring have lower fitness than their purebreed grandparents. As a result, there is relatively little flow of genes from one species to the other and back, and so the species don't merge.

These reproductive barriers help keep sexually reproducing species distinct. But these barriers were not always in place. Like the rest of biology, they also evolved. The origin of species, biologists now know, is, to a great extent, the origin of reproductive barriers (**Figure 9.1**).

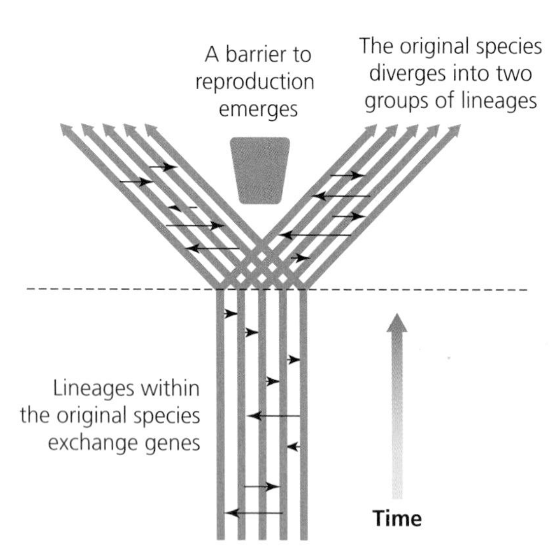

Figure 9.1 A key step in the evolution of new species is the splitting in two of an original population. A geographical barrier can divide it, but scientists have also identified other factors that can act as barriers. (Adapted from Doolittle, 2008)

Splitting Species

The origin of new species can take far longer than a human lifetime. But we humans can get some clues to how it happens by running laboratory experiments. In the 1970s, for example, Larry Hurd and Robert Eisenberg of Cornell University tried to produce reproductive isolation in a swarm of houseflies. They put a thousand flies into a chamber, with traps at the top and the bottom. They plucked the first 50 insects to fly up into the top trap, and the first 50 to fly into the bottom one. The flies in each set were then allowed to mate, but only with other flies that had flown in the same direction.

Hurd and Eisenberg then repeated the experiment for the next generation of each set. They put the offspring of the upward-flying flies in the chamber and collected the first 50 flies to breed a new generation; they did the same with the downward-flying ones.

This process produced a dramatic change in the behavior of the houseflies. In the first round, it took three or four hours to capture 50 flies in each trap. After 16 generations, it took only 10 minutes. After 16 generations, Hurd and Eisenberg mixed the top-flying flies and the bottom-flying flies together to let them mate. In every run of the experiment, nearly all of the flies chose mates from their own breed.

In the late 1980s, William Rice, then at the National Autonomous University of Mexico, and George Salt of the University of California, Davis, ran a more elaborate version of this experiment. They wanted to re-create the decisions that *Drosophila* flies make as they search for fruits, which serve both as their source of food and as the place where they mate and lay eggs. Rice and Salt fitted together a series of tubes to build a maze for the flies. The flies flew and crawled their way down the tubes. They had to choose which direction to turn at a series of forks. The first choice was whether to turn down a lit tube or down a dark one. The flies then reached a fork where one tube went up and the other went down. At the third fork, they could move towards either of two odors (ethanol or acetaldehyde, produced by different kinds of fruit). Finally, the flies ended up in one of eight vials at the end of the maze, where they mated and laid eggs.

Rice and Salt picked out the eggs from two of the vials. They reared the flies until they emerged from their pupae, and then put them into the same maze. The flies evolved a stronger and stronger preference for the same path as their parents. The preference became complete within about 30 generations. Both of these experiments demonstrate that the selection for traits, such as an attraction to ethanol, can produce reproductive isolation as a by-product.

Biologists use the insights they get from these kinds of experiments to test hypotheses about how new species evolve in the wild. In the mid-1900s, as the science of population genetics was coming into its own, biologists offered the first detailed account of how reproductive barriers could evolve in nature. Ernst Mayr, an ornithologist, and Theodosius Dobzhansky, a geneticist who studied *Drosophilia*, offered different versions of the same basic idea: new species evolve when old ones are geographically split.

A population can be geographically isolated from the rest of its species in many ways. Sometimes birds get swept from their mainland home to a remote island. Glaciers can slice down through the range of a salamander, leaving a population on either side. Some species shift their range as the climate changes, such as when the planet warmed at the end of the last Ice Age 12,000 years ago. Some animal species in the southwest United States lived in continuous belts of lowlands during the Ice Age. To find suitably cool temperatures after the Ice Age, they shifted up the sides of mountains. Eventually each mountain population became isolated from the others, because the animals no longer ventured down to the lowlands.

Once a population becomes isolated, it continues to evolve. It adapts through natural selection to its own habitat, and genetic drift fixes alleles that have little to do with fitness. Along the way, reproductive barriers arise as a side effect of this local evolution.

After thousands of years, separated populations may come back into contact. The glacier may melt, letting salamanders mingle once more. The birds may return from their island to the mainland. What happens next depends on how effective the barriers have become. If the reproductive barriers are low, the population will interbreed easily with the rest of its species. In other cases, however, reproductive barriers reduce the flow of genes. Sometimes two populations that come back into contact will be able to produce hybrids, but the hybrids are either sterile or less fit than their parents. In such cases, a so-called hybrid zone may emerge where the two populations meet. If the population can remain reproductively isolated even after it makes contact with its old species, it can be considered a new species.

The evolution of new species through geographic isolation is now known as allopatric speciation (allopatric means "in another place"). Allopatric speciation takes place across vast stretches of time and space, and so scientists cannot reproduce its full scope with laboratory experiments. But experiments can allow them to test some of the proposed steps in the process. To test these hypotheses further, scientists must search for patterns in nature that they would predict if allopatric speciation were indeed an important force in evolution.

Some of the most potent evidence for allopatric speciation comes from species that live in the Pacific and Atlantic oceans on either side of Panama (**Figure 9.2**). The Isthmus of Panama only began to emerge from the sea about 15 million years ago, and it did not finish forming until about three million years ago. Before then, North America and South America were isolated from each other by water, and so marine animals could swim easily between the Atlantic and Pacific oceans, mixing their genes in their offspring. Once the Isthmus of Panama arose, however, species began to split into Atlantic and Pacific populations. In many cases, the closest relative of a species that lives in the Atlantic now lives in the Pacific, and vice versa. Scientists brought some closely related Atlantic and Pacific species of shrimp into laboratories and placed them in the same tank. They discovered that only 1% of the matings between the shrimp from different oceans produced viable offspring. A geographical barrier—in this case the Isthmus of Panama—had thus allowed reproductive barriers to evolve.

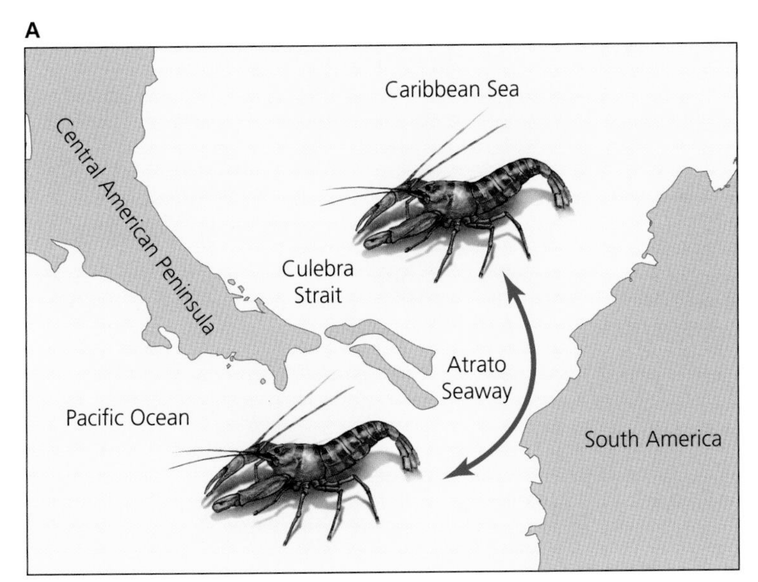

Figure 9.2 A: Before about three million years ago, the Atlantic and Pacific Oceans were joined by the Atrato Seaway. Some species of shrimp had ranges that extended into both oceans. B: When the Isthmus of Panama formed, it isolated Atlantic and Pacific populations of some shrimp. Genetic studies show that the closest relatives of some shrimp species in the Atlantic are not other Atlantic shrimp, but Pacific species. These studies are evidence of allopatric speciation, or the formation of new species by geographic barriers.

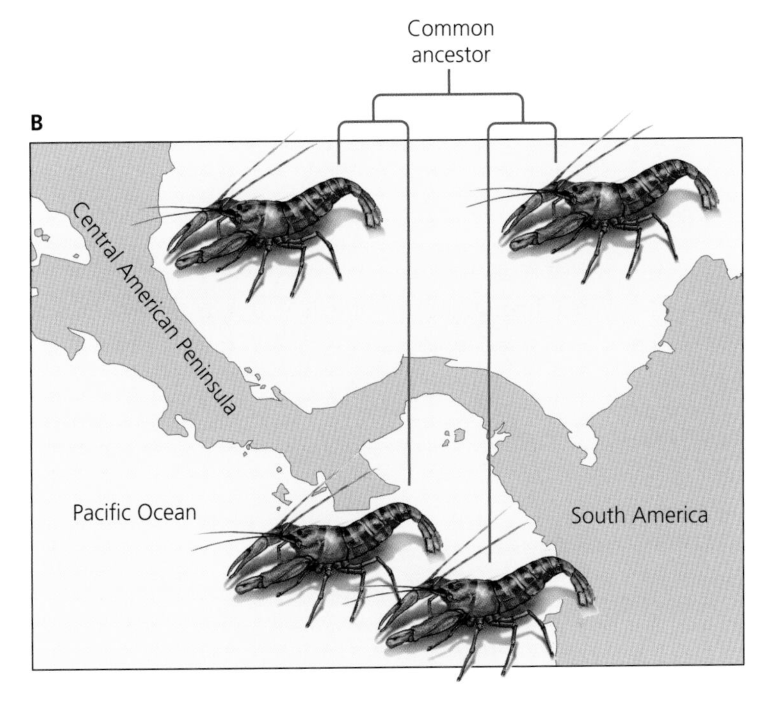

Rings of Species

Over the past few decades, scientists have identified many species that have experienced allopatric speciation. But in at least a few cases, populations have managed to diverge without being geographically cut off from each other.

One of the most striking examples of speciation without total geographical isolation is that of the greenish warbler, a small bird that lives in forests across

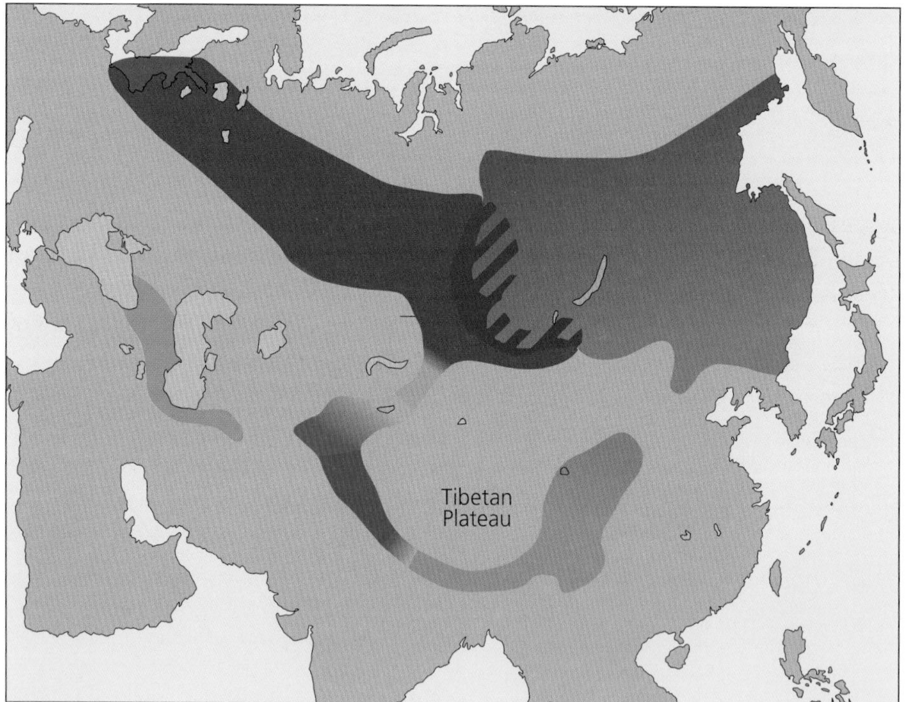

Figure 9.3 Greenish warblers originally evolved south of the Himalayas. They later spread eastward and westward along the mountain range and then expanded their range north. Today, ornithologists recognize six subspecies of greenish warblers (shown here in different colors). In Siberia, the two arms of the species range overlap. But the western Siberian warblers (left) and the eastern ones (right) rarely interbreed. That's because they sing different courtship songs and have different patterns on their feathers. They behave like two separate species. (Adapted from Irwin, 2005)

much of northern and central Asia. Darren Irwin, a biologist at the University of British Columbia, has been traveling to Russia, Mongolia, China, and Nepal to study the birds, observing their colors, listening to their songs, and sampling their DNA. By analyzing their genes, Irwin has found that the oldest populations are located in the southern end of their range, along the southern face of the Himalayas (**Figure 9.3**). It was here that the birds probably survived during the last Ice Age in one of the few regions in Asia where suitable forests still grew. As the glaciers retreated about 20,000 years ago and forests expanded northward, the range of the greenish warbler expanded as well. Some birds expanded their range around the eastern side of the Himalayas, up through China into eastern Siberia. Another population expanded northward around the west side, into Central Asia. The Tibetan Plateau to the north of the Himalayas has remained dry and treeless, and so the greenish warblers did not colonize that region. Instead, they spread like two outstretched arms, their hands meeting finally in central Siberia.

It was there in Siberia that Irwin discovered something very strange. The eastern and western populations of greenish warblers are, in effect, two separate species.

The greenish warblers in western Siberia have a single bar on their wings, while those in eastern Siberia have two. As subtle as this difference may seem to us humans, it's very important to greenish warblers, because they use their wing bars to communicate with each other. Male warblers from the two ends of the bird's range also sing different songs. If an eastern Siberian male greenish warbler hears another eastern Siberian male greenish warbler singing, it will respond aggressively. If it hears a western Siberian male, it makes no response. Because of these differences in their songs and plumage, the eastern and western Siberian greenish warblers almost never inbreed even though their ranges overlap.

The greenish warbler is an example of what's known as a ring species: a series of neighboring populations that can interbreed with their neighbors, but whose two overlapping "end" populations do not. If the only greenish warblers Irwin knew about were the two "end" populations that he encountered in central Siberia, he would have concluded they were two distinct species. In fact, they are at the opposite ends of an almost continuous belt of greenish warblers.

Species Side by Side

Some animals and plants appear to have diverged while living side by side—a process called sympatric speciation. One of the best examples of sympatric speciation can be found in Lake Apoyo, which formed in the crater of an extinct volcano in Nicaragua. Axel Meyer, a biologist at the University of Konstanz in Germany, and his colleagues examined two species of fish that live in the lake. One species, the Midas cichlid (*Amphilophus citrinellus*), has a big body and uses powerful jaws to crush snails at the lake bottom. The slender arrow cichlid (*A. zaliosus*) lives in the open water, where it eats insect larvae.

Lake Apoyo formed less than 23,000 years ago when its volcano became extinct and filled with rainwater. Meyer's team studied the DNA of the two cichlids and compared it with that of fish in neighboring lakes. They concluded that the Midas cichlid originally invaded the lake, perhaps swept in during a hurricane. The arrow cichlids then branched off from the Midas cichlids, evolving a distinctive body shape and no longer breeding with their parent species.

Meyer's results suggest that the arrow cichlid evolved from relatively slender Midas cichlids, shifting from a diet of snails to a diet of insect larvae. They enjoyed more reproductive success if they mated with other slender cichlids, because their slender offspring could swim more efficiently in the open water. Over time, the fish evolved the mating preferences that now help keep the two populations distinct.

Cichlid fishes colonized Lake Apoyo in Nicaragua less than 23,000 years ago and have since diverged into two distinct forms. Large cichlids (lower left) crush snails on the lake bottom. Slender ones (lower right) eat insect larvae in the open water. These two forms evolved without the help of a geographical barrier.

The Speed of Speciation

In the mid-1990s, Jerry Coyne of the University of Chicago and H. Allen Orr of Rochester University decided to see how long it takes for *Drosophila* flies to become reproductively isolated. First they came up with a scale for scoring how much two populations of flies could interbreed. Populations with no reproductive barriers scored a zero, while completely isolated populations got a score of one. Coyne and Orr found that sympatric species of *Drosophila* flies—flies that were isolated enough to live side by side without merging into a single species—scored .94 on average.

Coyne and Orr then gathered data on how well 171 species of *Drosophila* could interbreed and gave each pair of species a score of their own. With the help of a molecular clock, they also estimated how much time had passed since each pair of species had diverged from a common ancestor. The result of their study (**Figure 9.4**) reveals that reproductive isolation takes time to evolve. Pairs of *Drosophila* species typically needed a few hundred thousand years to become as reproductively isolated as sympatric species are.

While reproductive isolation may take hundreds of thousands or even millions of years to evolve in animals, plants sometimes evolve new species in just a few generations. This rapid speciation begins when pollen grains from one

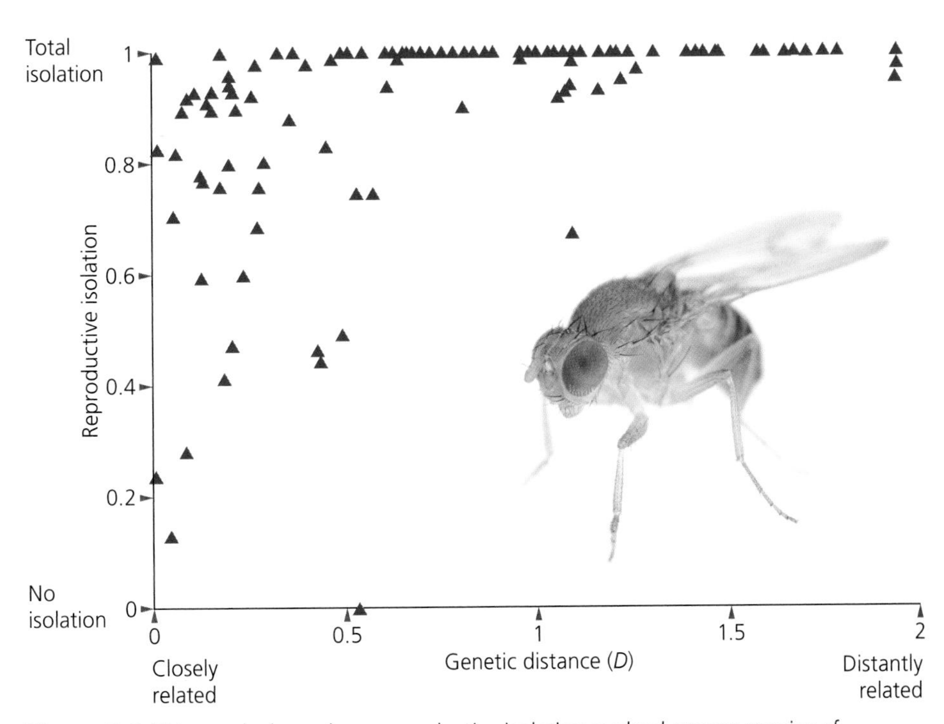

Figure 9.4 This graph shows how reproductive isolation evolved among species of *Drosophila*. The genetic distance (*D*) between two species increases with time. It takes roughly a million years for *D* to reach a value of one. By then, a typical pair of *Drosophila* species no longer interbreed. (Adapted from Coyne and Orr, 2004)

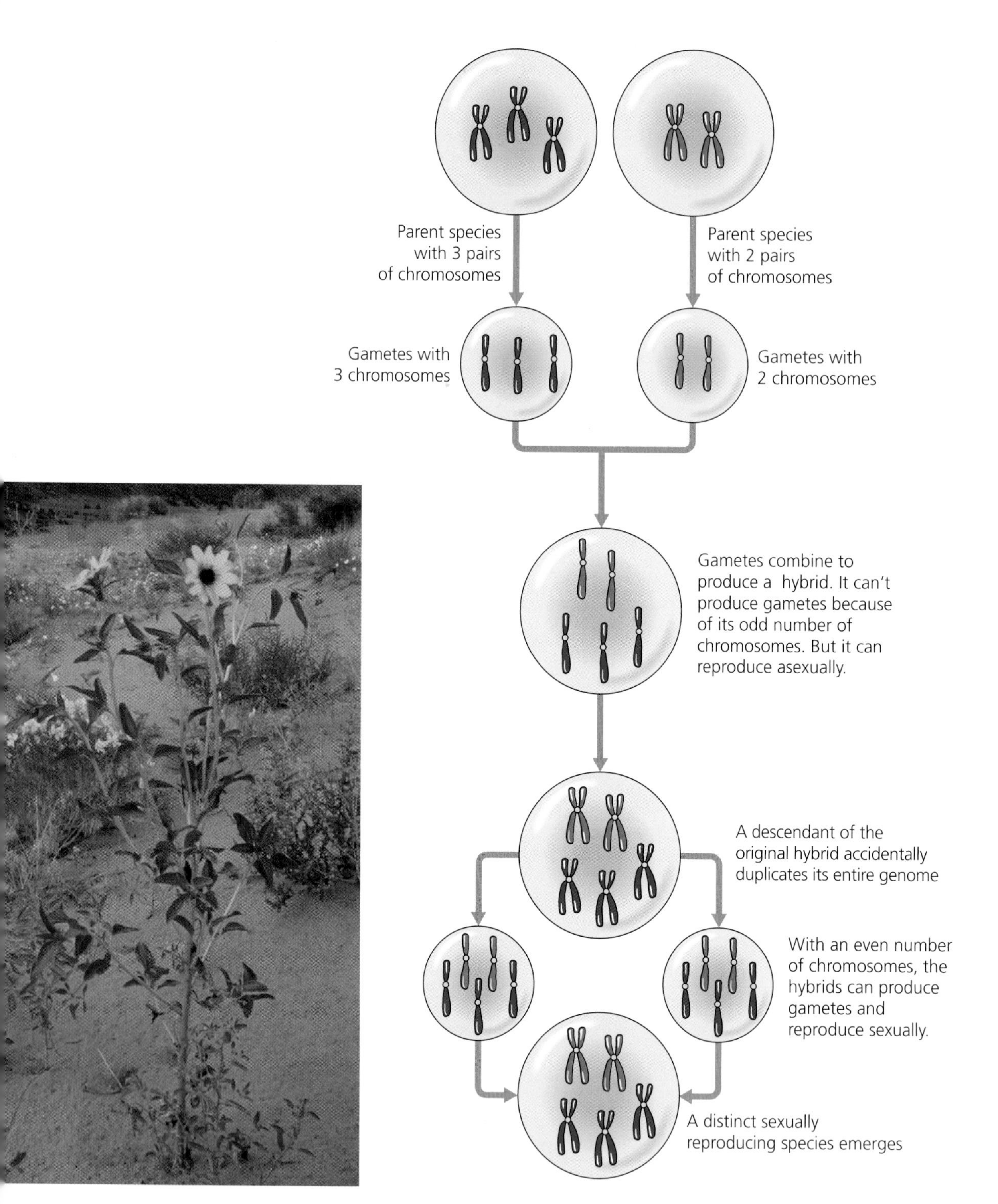

Figure 9.5 Left: The sunflower *Helianthus anomalus* evolved when two sunflower species formed hybrids. Right: One of several possible routes by which hybrid plants can become new species.

species land on another species and fertilize its ovules, producing hybrid seeds. Often these hybrid seeds fail to grow. In other cases, the seeds develop into plants, but they cannot produce viable seeds of their own. And in still other cases, hybrids can successfully breed with one or both of their parental species. But on rare occasion, these hybrids become a species of their own.

Figure 9.5 shows how this speciation takes place. A species with two pairs of chromosomes mates with a species with three. The hybrid cannot produce viable gametes, because it cannot divide its five chromosomes evenly. But, in some cases, hybrids can reproduce asexually. On rare occasions, the hybrid's offspring accidentally duplicate their entire genomes. Now those offspring can pair off their chromosomes and produce viable gametes. Those gametes can come together to produce new offspring. Now the hybrid can reproduce sexually, but it can't interbreed easily with the two species that produced it. In other words, it has become a distinct, sexually reproducing species.

This process is called allopolyploidy. In some cases a single hybrid can produce a new species through allopolyploidy, but often the process is more complex. Two plant species may produce many hybrid individuals, which then begin to interbreed among themselves. These multiple origins give the new allopolyploid species more genetic variation, which can allow it to adapt faster through natural selection.

While there's much left to learn about how allopolyploidy produces new species, one thing is clear: it is surprisingly common. Perhaps half of all flowering plant species evolved this way.

Uncovering Hidden Species

Over the past century, biologists have learned a great deal about the genetic diversity within and between species. They have learned about the processes that drive that diversity, such as natural selection and the evolution of reproduction. And along the way, they've come up with new concepts for what it means to be a species.

In 1942, Ernst Mayr laid out one of the most influential of those concepts, which later came to be known as the biological species concept. He defined a species as a group of "actually or potentially interbreeding populations which are reproductively isolated from other such groups." It's now clear that reproductive isolation does indeed evolve among species, and many biologists consider Mayr's concept to be very useful. But some critics have pointed out that the biological species concept doesn't always match reality very well.

Many species that seem quite distinct have turned out to be able to hybridize with other "good" species. In Europe, for example, 16% of all the butterfly species produce viable hybrids. The biological species concept is also difficult to apply to many populations. When the last Ice Age ended 12,000 years ago, mountain-dwelling populations of animals and plants became isolated on "sky islands." Should the shrews on two neighboring mountains be considered two separate species because they don't reproduce? It's hard to say, based on the biological species concept, because the shrews never make the journey from one mountain to the other and thus never have the chance to interbreed.

Some scientists now favor looking at the history of a population to decide whether it's a species or not. This view is known as the phylogenetic species concept. It holds that a species is a group of populations that has been evolving independently of other groups of populations. According to the phylogenetic species concept, scientists can identify a species by finding unique traits that a group of populations share, but that are not found in other closely related populations.

The phylogenetic species concept is very practical. Scientists don't have to establish that a species is reproductively isolated to show that it is indeed a species; instead, they can rely on traits that are easier to identify, such as unique mutations, color patterns, or shapes. But the phylogenetic species concept has its share of detractors, too. Its critics argue that it will lead scientists to split species down into smaller and smaller species whenever they discover a tiny difference between two populations.

Some scientists are finding that the best solution is to draw on different concepts, so that they can test hypotheses about species with different lines of evidence. That's the strategy Rick Brenneman and his colleagues used to map the diversity of giraffes in Africa. One line of evidence they gathered was giraffe DNA: they collected tissue from 266 giraffes across Africa and sequenced 1,707 nucleotides from the DNA of each animal. (The scientists also sequenced the same DNA sequence from an okapi, a short-necked relative of the giraffes.) From the giraffe DNA, the scientists produced an evolutionary tree, which is illustrated in **Figure 9.6**. Their results indicate that the giraffes evolved from a

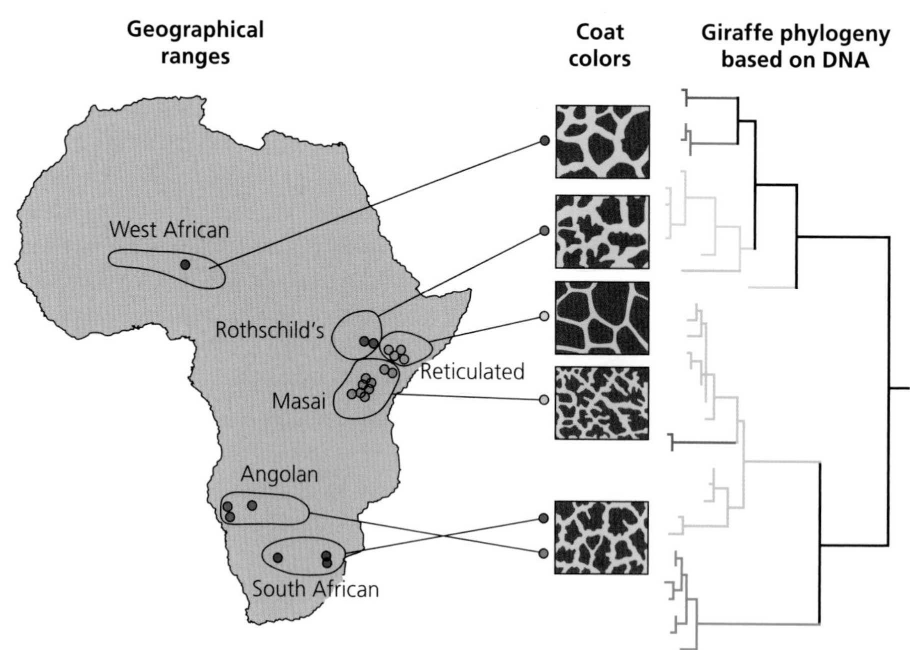

Figure 9.6 Rick Brenneman and his colleagues have reconstructed the evolution of giraffes from a common ancestor. As this map shows, they have diverged into several lineages, each of which has a distinct range and coat color. The scientists propose that they are six different species. (Adapted from Brown et al., 2007)

common ancestor about a million years ago. Its descendants diverged into six lineages. Each lineage has evolved a distinct pattern of brown patches on its skin.

The traits in each lineage are consistent with there being six different species of giraffes, according to the phylogenetic species concept. But Brenneman and his colleagues also looked at how reproductively isolated the lineages are. The West African, Angolan, and South African lineages are too geographically isolated to interbreed with the other giraffes, but the three lineages in East Africa live close to each other and could, at least theoretically, produce hybrids. But Brenneman and his colleagues found only three hybrid giraffes—less than 1% of the animals they darted. For the most part, the lineages of giraffes had not mixed their DNA together. The giraffes may be so picky about the coat pattern of their mates that they avoid interbreeding.

Based on these lines of evidence, the scientists proposed that the giraffes belong to six separate species. If they're right, it makes a big difference to how conservation biologists try to save giraffes from extinction. Rather than trying to preserve a single species spread across most of Africa, they may need to come up with strategies tailored for six different species adapted to six different habitats.

Species Beyond Barriers

When it comes to genetic diversity, the best place to look is not in the animal kingdom. In the 1990s, as scientists developed techniques for studying molecular phylogeny, they began to compare DNA from as wide a range of living things as they could find. **Figure 9.7** illustrates a sampling of this diversity. The lengths of the branches show how genetically distinct each lineage is from the others.

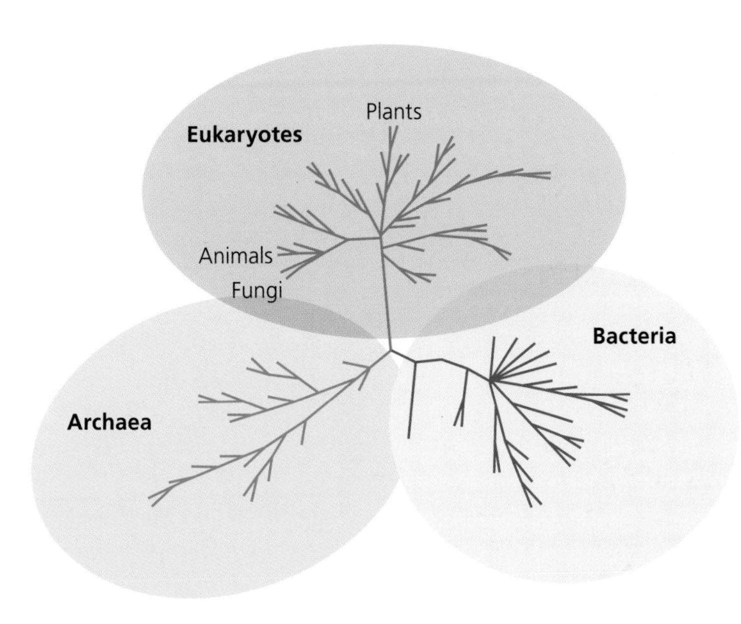

Figure 9.7 Many studies on DNA have yielded a tree of life made up of three main branches, known as bacteria, archaea, and eukaryotes. The number of mutations in each branch is represented by its length. This tree shows how animals represent a very small part of the genetic diversity of life on Earth. (Adapted from Barton et al., 2007)

The animal kingdom is just a small tuft of twigs on this tree of life. In fact, all multicellular organisms—animals, plants, and fungi—represent only a narrow slice of life's genetic diversity.

Single-celled organisms, by contrast, are vastly more diverse. Many studies find that they belong to three main lineages (known as domains). In one of these domains are the eukaryotes, which include not only animals, plants, and fungi, but also a huge range of single-celled protozoans. Eukaryotes share many traits in common, such as keeping their DNA in a sac known as the nucleus. The other two domains are made up only of single-celled organisms. Known as bacteria and archaea, they look identical to the untrained eye, but microbiologists have found a number of differences between them, such as the molecules they use to build their membranes.

This deep diversity shouldn't come as too much of a surprise, inasmuch as life on Earth was single-celled for billions of years before animals and plants emerged. The diversity of animals and plants may also be constrained by the ways they get energy they need to grow. Animals generally just eat other organisms, while plants generally just harness sunlight. Microbes, on the other hand, do both—and many other things as well. Some feed on methane gas; some feed on rocks. Microbes can also be much more robust than animals and plants; you can find them far underneath the seafloor, at the bottom of acid-drenched mine shafts, and even trapped in salt crystals. While zoologists such as Rick Brenneman are discovering hidden diversity among animals, microbiologists are discovering far more. There may be 5,000 different species living in the human gut. A single teaspoonful of soil may contain 10,000 bacterial species.

But what exactly is a microbial species? Bacteria and archaea reproduce asexually, dividing into genetically identical copies. So the concept of reproductive barriers between them is meaningless. In a flask of bacteria, there may be a new mutant in every generation; that mutant will pass down its mutation to its offspring, marking them as a distinct lineage. Does that single mutation qualify the entire lineage as a new species?

Some researchers have pointed out that, while bacteria do not mate like animals do, they do trade genes. Viruses may carry genes from one host to another, or bacteria may simply slurp up naked DNA, which then slips into their genome. There is some evidence that closely related strains trade more genes than distantly related ones—a microbial version of the reproductive barriers between animal species.

But critics have pointed to some problems with the analogy. Although animals and plants can trade genes every time they reproduce, microbes may do so very rarely. On the other hand, microbes don't just accept genes from their close relatives; scientists have documented cases of horizontal gene transfer between microbes separated by billions of years of evolution. Over millions of years, genomes of microbes can come to look like mosaics (**Figure 9.8**).

To gauge how much horizontal gene transfer has taken place among microbes, Tal Dagan and William Martin of the University of Düsseldorf analyzed the genomes of 181 bacteria. In 2008 they reported that, on average, 81%

of the genes in a single genome had experienced horizontal gene transfer at some point in their history. In some respects, Dagan and Martin argue, the history of life looks more like a web than a tree (**Figure 9.9**).

If microbial genomes are mosaics, some critics argue that it's impossible to draw boundaries around microbial species. But other researchers, such as Frederick Cohan of Wesleyan University think microbial species should be taken seriously. Cohan and his colleagues study the dense ecosystem of microbes that live around hot springs in Yellowstone National Park. The researchers have found that the bacteria and archaea in that ecosystem can be sorted into genetic clusters and ecological clusters. Each genetically related group of microbes lives

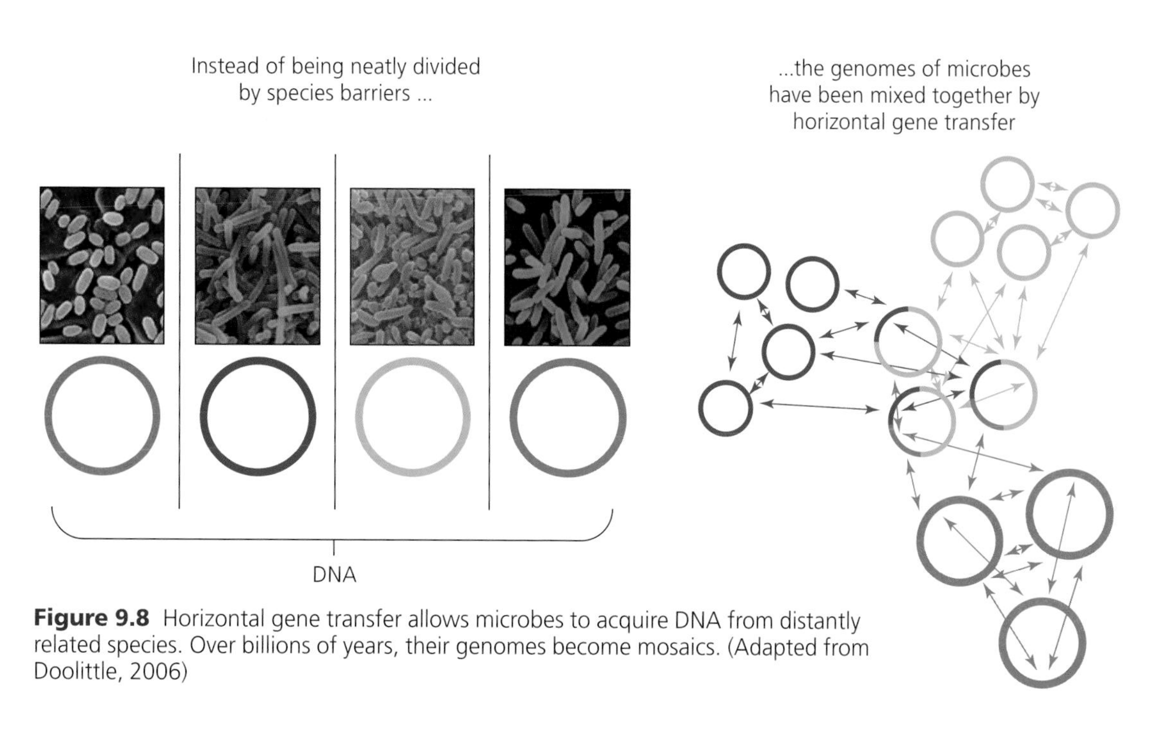

Figure 9.8 Horizontal gene transfer allows microbes to acquire DNA from distantly related species. Over billions of years, their genomes become mosaics. (Adapted from Doolittle, 2006)

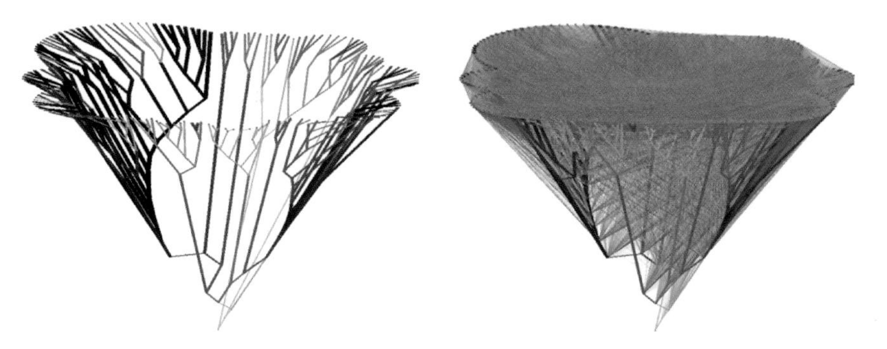

Figure 9.9 Left: An evolutionary tree of 181 species of bacteria, calculated from some of their genes. Right: Over billions of years, many genes have moved from one branch to another. Horizontal gene transfer events are indicated here by colored lines. (Adapted from Dagan and Martin, 2008)

in a certain niche in the hot springs—enjoying a certain temperature, for example, or requiring a certain amount of sunlight. Cohan argues that, when a lineage of bacteria or archaea adapts to a particular ecological niche—one that's distinct from the niches of other lineages—it's right to call it a species. Such species are like animal species in that their members are evolving together. Bald eagles evolve collectively as a species, because they share genes and are continually experiencing natural selection for the same ecological niche. A given species of bacteria is continually adapting to a niche of its own (**Figure 9.10**).

If microbiologists follow Cohan's advice, many new species of microbes will have to be named. To avoid confusion, Cohan does not want to come up with completely original names. Instead he wants to add an "ecovar" name at the end ("ecovar" stands for "ecological variant"). The bacterial strain that caused the first recorded outbreak of Legionnaires' disease in Philadelphia, for example, would be called "*Legionella pneumophila* ecovar Philadelphia."

Understanding the nature of microbial species could help public-health workers to prepare for the emergence of other diseases in the future. Disease-causing bacteria often evolve from relatively harmless microbes that dwell quietly within their hosts. It may take decades of evolution before such organisms cause an epidemic large enough for public-health workers to notice. Classifying these new species could let public-health workers anticipate outbreaks and give them time to prepare. Solving the mystery of species turns out not just to be important for understanding the history of life. Our own well-being may depend upon it.

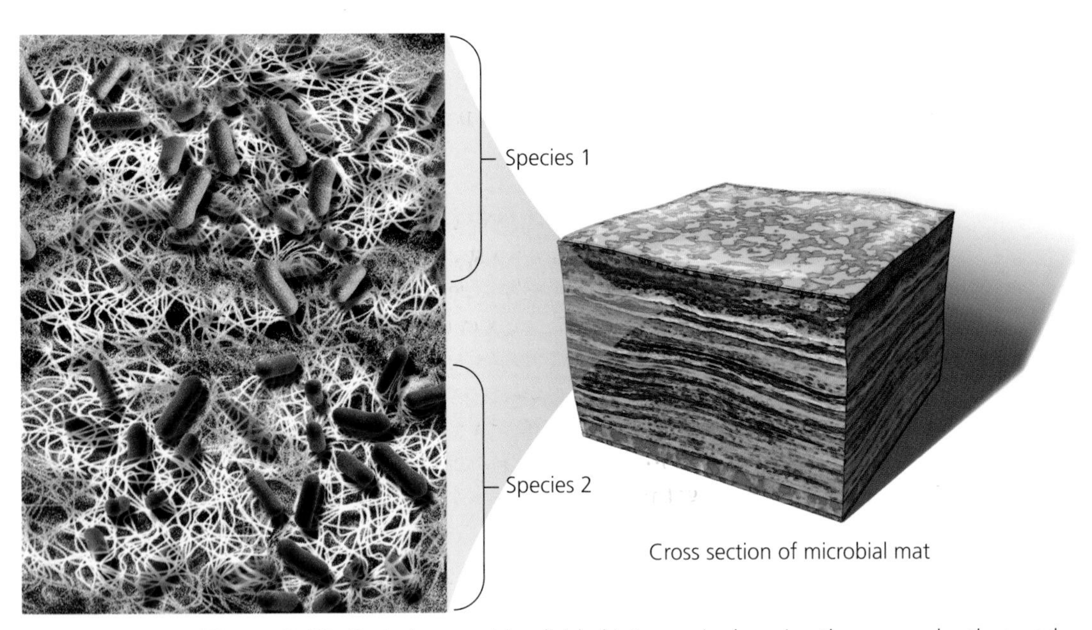

Figure 9.10 Bacteria cannot be divided into species by using the same rules that work for animals. A different way to define bacterial species is based on the way they are adapted to narrow ecological niches. This figure shows how bacteria are distributed in a hot spring in Yellowstone. Different species are adapted to particular temperatures and concentrations of different minerals and other nutrients.

On the Origin of Our Own Species

It is easy to draw a line between our own species, *Homo sapiens*, and all other species alive today. Our closest living relatives are chimpanzees and bonobos. There are no chimp–human hybrids walking among us, confusing us with human-sized brains and knuckle-walking bodies. But the line is not so easy to draw between the hominid species that have existed on Earth over the past seven million years.

When paleoanthropologists discover a new hominid fossil, they compare it with previously discovered ones to see if it belongs to the same species. Some species of hominids have proven to be clearly distinct, but others are more ambiguous. Paleoanthropologists are divided about whether certain hominid fossils belong to species of their own or are actually members of other species. Much of their trouble arises from the relatively limited number of fossils they've found of some hominid species. Our own species contains lots of variation, such as the spectacular range of height from Pygmies to basketball players. Only by comparing many humans can biologists recognize the limits to human variation, as well as the distinctive traits that all humans carry and other species lack. Paleoanthropologists don't have that luxury of data to study.

Over the past century, the uncertainty about hominid species has also helped to fuel debates about the origin of our own species. At one extreme, some researchers argued that all of the hominids in the past million years throughout the Old World belonged to a single species. While local populations evolved distinctive traits, they remained linked to the rest of the species by gene flow.

Toward the end of the 1900s, an alternative view emerged. Some researchers argued that hominids in Europe and Asia represented separate branches of hominid evolution. The direct ancestors of our own species evolved in Africa over the past million years, and *Homo sapiens* only spread out to other continents some 50,000 years ago. The other hominid species eventually became extinct, leaving *Homo sapiens* as the last hominid species on Earth.

Some of the evidence for this "out-of-Africa" hypothesis comes from studies on fossils. While the oldest hominid fossils—dating back six million years—are rare, paleoanthropologists have gathered a fairly large collection of hominid fossils from the Old World from the past one million years. It's big enough for them to carry out statistical tests on the shapes of the fossils. Paleoanthropologists have found that these fossils cluster together in distinct groups. There's little overlap between hominid fossils from Africa over the past 200,000 years and the fossils of Neanderthals, for example.

It's also possible to investigate the nature of hominid species with DNA. As we saw in Chapter 7, Sarah Tishkoff and other researchers have traced the ancestry of genes in living humans to people living in Africa less than 200,000 years ago. Other clues are coming from the DNA of an extinct group of hominids—the Neanderthals.

In the mid-1990s, Svante Pääbo (a biologist now at the Max Planck Institute for Evolutionary Anthropology in Leipzig) wondered if fossils of Neanderthals might still preserve fragments of DNA. In 1997 he and his colleagues reported their first successful fishing expedition, a tiny stretch of Neanderthal DNA measuring just 379 base pairs long. Since then, they've found fragments of DNA from more than a dozen other Neanderthals, from sites ranging from Spain in the west to Siberia in the east. And in February 2009, Pääbo marked Darwin's 200th birthday by announcing that he and his colleagues had completed the rough draft of the entire genome from a single Neanderthal that had lived 30,000 years ago.

As more and more Neanderthal DNA came to light, Pääbo and other researchers compared it to human DNA. They drew evolutionary trees representing how the genetic material in human and Neanderthal individuals had evolved from a common ancestor. The trees have proven very consistent. The DNA from any given Neanderthal is more closely related to the DNA of other Neanderthals than to that of any living human. The DNA of humans and Neanderthals can be traced back to a common ancestor that lived 800,000 years ago.

These results fit the idea that humans and Neanderthals were two separate species, or at least were two subspecies that rarely interbred. The ancestors of Neanderthals may have migrated out of Africa and then gradually adapted to their habitat in Europe and the Near East. Meanwhile, the ancestors of humans continued to evolve in Africa, becoming recognizable members of *Homo sapiens* about 200,000 years ago.

When humans expanded out of Africa, they moved into territories where other hominids already lived. In Asia, they may have encountered surviving populations of *Homo erectus*. In Europe, they would have encountered Neanderthals. The evidence from Neanderthal DNA supports the idea that Neanderthals and humans did not mate indiscriminately with each other, blending their genes together. But some researchers have suggested that the two populations of hominids may have interbred at least a few times, and that humans still carry a little Neanderthal DNA today.

To understand how this could happen, imagine that Neanderthals and humans had mated and that their children were healthy enough to grow up to adulthood. If these hybrids mated with humans, the DNA from the Neanderthals could have spread into *Homo sapiens*. Over the generations, recombination would break up the Neanderthal DNA into small segments and mix it into the human gene pool. Segments of Neanderthal DNA that raised fitness might be spread through natural selection; some neutral segments might be spread by genetic drift. Neanderthal genes that lowered fitness would be eliminated. Genetic drift might also make nonhuman DNA rarer and rarer over the generations, until it finally disappeared.

Several teams of scientists have searched for DNA in humans that might have come from Neanderthals or other hominids this way. Bruce Lahn, a biologist at the University of Chicago, and his colleagues announced in 2006 that they had found evidence for interbreeding in a gene known as microcephalin, which encodes a protein involved in the development of the brain (**Figure 9.11**). Two alleles of microcephalin exist in humans, with the more common version found in two-thirds of all people. The variations of the more common form of microcephalin are nearly identical to each other. Their similarity suggests that they descended from an ancestral allele just 37,000 years ago. But the common microcephalin allele and the rare one are markedly different from each other. Based on the mutations in the alleles, Lahn and his colleagues estimate that they evolved from an ancestral gene more than a million years ago.

Lahn and his colleagues propose that the ancestral microcephalin allele existed in the common ancestor of humans and Neanderthals. The more common form of microcephalin only entered our species 37,000 years ago, when humans arrived in Europe and began to interbreed with Neanderthals. Once the Neanderthal allele entered the human gene pool it spread rapidly thanks to natural selection.

Other researchers have found other segments of human DNA that also hint at interbreeding. Michael Hammer of the University of Arizona and his colleagues examined a gene called *RRM2P4* and drew an evolutionary tree of alleles in people from Africa, Europe, and Asia. The alleles, they discovered, share a common ancestor that dates back two million years. The deepest branches belong to alleles found in Asian people. Yet two million years ago, there were no direct ances-

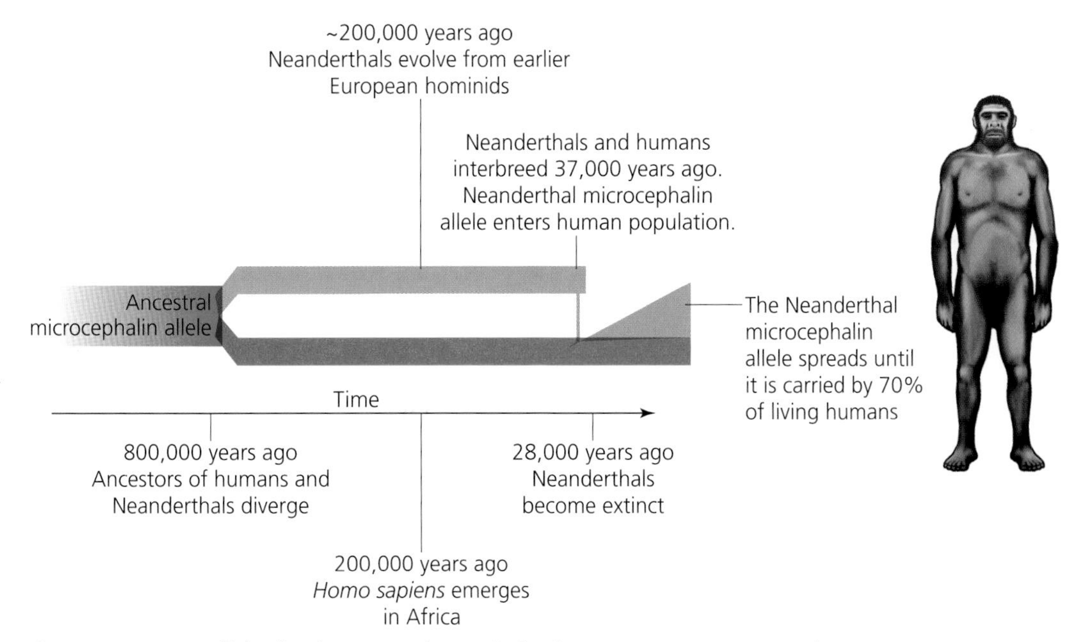

Figure 9.11 Two alleles for the gene microcephalin share a common ancestry about a million years ago. But the variants of the more common allele descend from a common ancestor that's just 37,000 years old. One explanation recently offered for this pattern is that the common allele entered our species through interbreeding with Neanderthals. (Adapted from Evans, 2007)

tors of humans in Asia—only *Homo erectus*. Hammer and his colleagues propose that *Homo sapiens* expanded into Asia 50,000 years ago and interbred with Asian populations of *Homo erectus*. The *RRM2P4* allele entered the human gene pool, and gradually spread out of Asia and into human populations on other continents.

Many scientists reject arguments for hominid interbreeding. In 2007, Laurent Excoffier, a population geneticist at the University of Bern in Switzerland, and his colleagues created mathematical models of some of the alternative hypotheses for the origin of humans. At one extreme, they built a model in which humans expanded from Africa without interbreeding at all with other hominids. At the other extreme, they built a model in which genes could flow across all of the hominid populations in the Old World. Excoffier and his colleagues then measured the probability that each model would produce the actual patterns of genetic variability that have been found in living humans.

Excoffier's group found that the model that best explained the evidence was one in which humans evolved in Africa and spread from there about 60,000 years ago with little interbreeding or none whatsoever. Ancient alleles such as the ones identified by Lahn and Hammer actually originated in Africa and survived there over the past one to two million years, Excoffier argues. They only spread to other continents when *Homo sapiens* expanded out of Africa.

In the next few years, a vast amount of data will emerge that scientists will use to test hypotheses about our speciation. It's a fair bet that researchers will be able to sequence genomes of other individual Neanderthals, and it's even possible that researchers will be able to isolate DNA from other hominid species. There's also a lot left to learn from human genetic diversity. Most studies on human genetic diversity look at relatively few snippets of DNA from a large number of people. If scientists could compare entire genomes, they could be more confident in their arguments about the history of *Homo sapiens*. However, by the end of 2008, scientists had produced complete genomes of only four people on the entire planet. Thanks to the plummeting price of DNA sequencing, that number will soon multiply many times over. And as the genomes multiply, our understanding of our own species will grow as well.

TO SUM UP...

- **Naturalists began classifying organisms into species before Darwin developed his theory of evolution.**
- **Speciation is the evolution of new species.**
- **Reproductive barriers can evolve between isolated populations.**
- **Allopatric speciation occurs when geographically isolated populations evolve reproductive barriers.**
- **Speciation can also take place without geographical isolation. Ring species, sympatric speciation, and allopolyploidy are examples.**

- **Complete reproductive isolation can take millions of years to evolve between two species.**
- **Genetic studies sometimes reveal that what appears to be a single species may actually be several related species.**
- **Species of bacteria and archaea are difficult to identify using concepts developed for animal species.**
- ***Homo sapiens* evolved about 200,000 years ago. Scientists are debating how much interbreeding took place between our species and other hominids.**

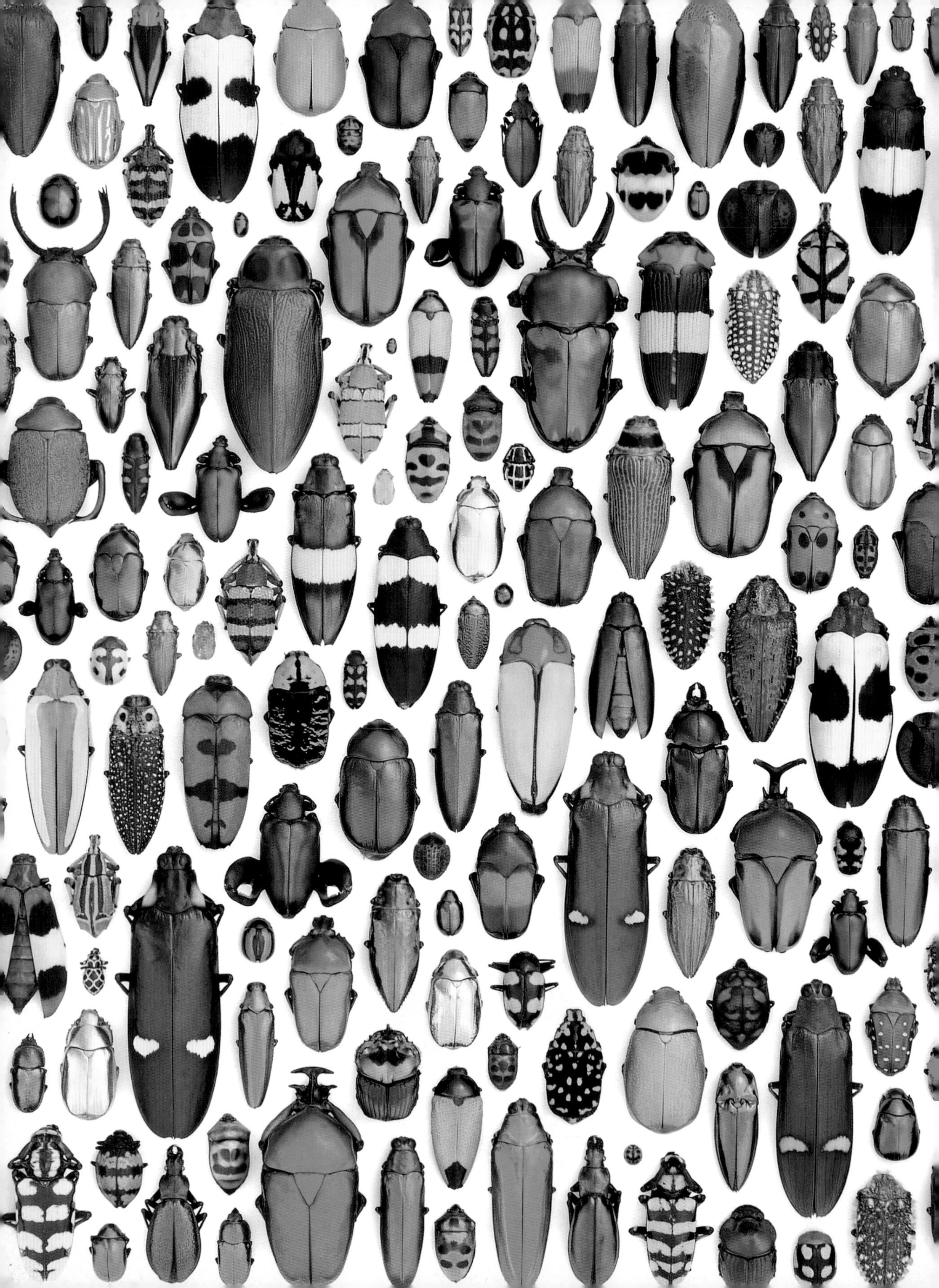

Radiations & Extinctions

10

Biodiversity Through the Ages

There's a story that scientists like to tell about the great evolutionary biologist J. B. S. Haldane. Supposedly, Haldane once found himself in the company of a group of theologians. They asked him what one could conclude about the nature of the Creator from a study of his creation. "An inordinate fondness for beetles," Haldane replied.

There are some 350,000 named species of beetles—70 times more species than all the mammal species on Earth. Insects, the lineage to which beetles belong, include a million named species, the majority of all 1.8 million species scientists have ever described.

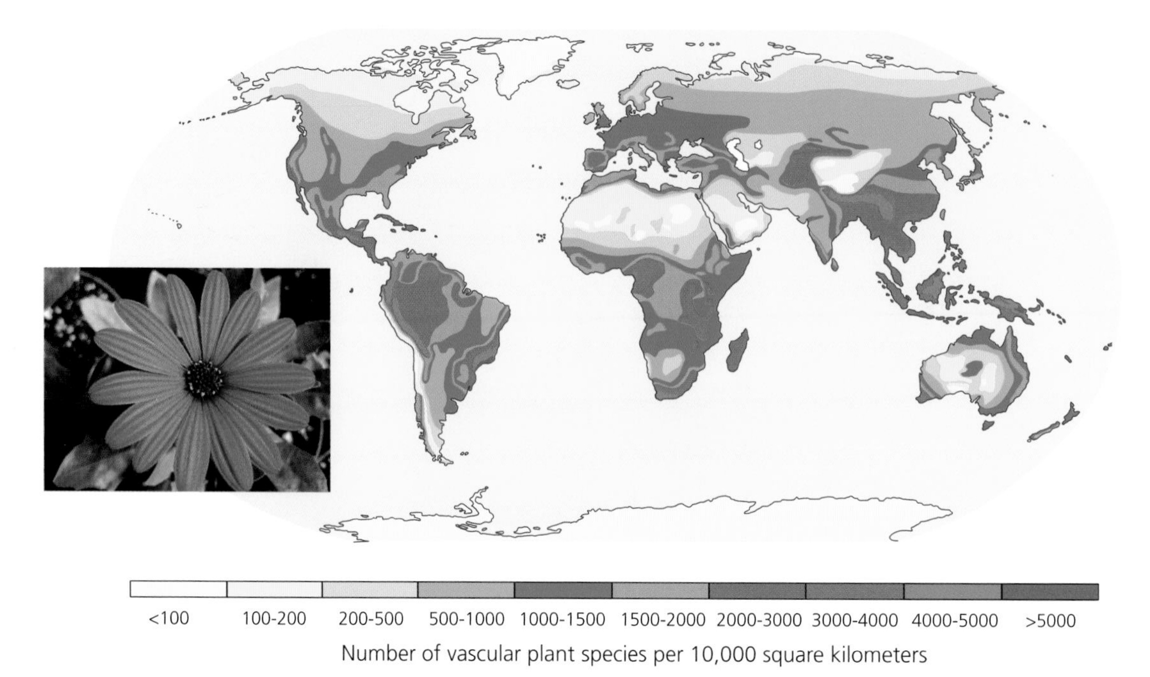

Figure 10.1 The diversity of plants is much higher in the tropics than in the regions near the poles. Animals and other groups of species show a similar pattern of diversity. (Adapted from Benton, 2008)

Biological diversity (or biodiversity for short) is one of the most intriguing features of life. Why are there so many insects on Earth and so few mammals? Why is biodiversity richest in the tropics, rather than being spread smoothly across the planet (**Figure 10.1**)? Why do different continents have different patterns of diversity? Almost everywhere on Earth, for example, placental mammals make up the vast diversity of mammal diversity. On Australia, however, there is a huge diversity of marsupial mammals.

Biodiversity has also formed striking patterns through the history of life, as illustrated in **Figure 10.2**. A large team of scientists produced this graph by analyzing records for 3.5 million fossils of marine invertebrates that lived during the past 540 million years. They divided up that time into 48 intervals and calculated how many genera were alive in each one. The graph shows that among marine invertebrates, biodiversity is higher today than it was 540 million years ago. But the pace of this rise was not steady. There were periods in which diversity rose rapidly, as well as periods in which it dropped drastically.

In this chapter we'll examine how scientists study biodiversity, analyzing patterns over space and time and then creating hypotheses they can test. We'll explore how lineages of species grow, and then how they become extinct. We may, biologists fear, be in the early stages of a catastrophic bout of extinctions on a scale not seen for millions of years. By understanding the past of biodiversity, scientists can make some predictions about the future we are creating.

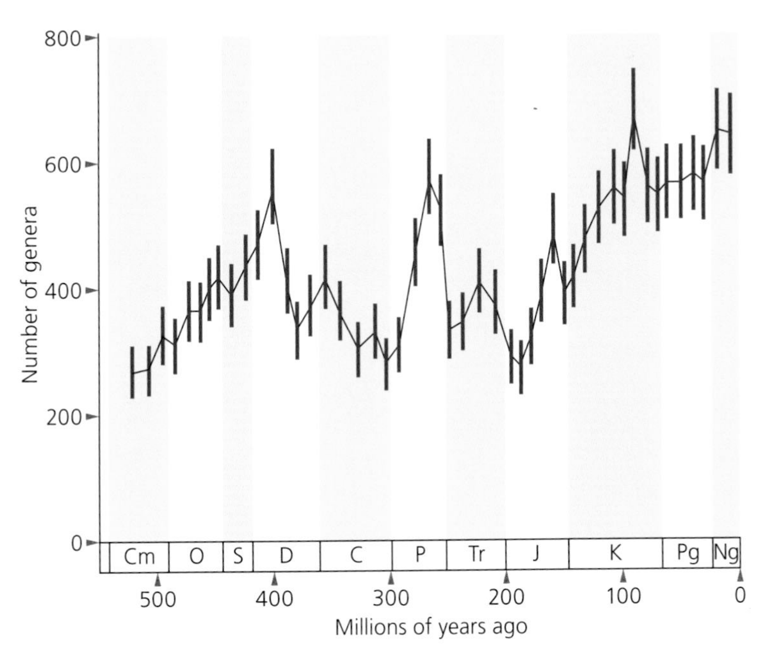

Figure 10.2 A team of paleontologists analyzed 3.5 million fossils of marine invertebrates that lived over the past 540 million years to determine the history of diversity. As this graph shows, diversity has risen and fallen several times, but today there are about twice as many genera as there were at the beginning of this period. (Adapted from Alroy et al., 2008)

Riding the Continents

Few people have heard of the mite harvestman, and fewer still would recognize it at close range. It is related to the far more familiar daddy longlegs, but its legs are stubby rather than long, and its body is about as big as a sesame seed. On the floors of the humid forests where it dwells, it looks like a speck of dirt. As unglamorous as the mite harvestman may seem, however, it has a spectacular history to unfold.

An individual mite harvestman may spend its entire life in a few square meters of forest floor. The range of an entire species may be less than 100 kilometers (60 miles) across. Yet there are 5,000 species of mite harvestman, and they can be found on five continents and a number of islands. Sarah Boyer, a biologist at Macalester College in Minnesota, and her colleagues have traveled around the world to catch mite harvestmen, and they've used the DNA of the animals to draw an evolutionary tree. At first glance, their results seem bizarre. One lineage, for example, is only found in Chile, South Africa, and Sri Lanka—countries separated by thousands of kilometers of ocean (**Figure 10.3**).

But the results of Boyer's research make sense if you remember that Chile, South Africa, and Sri Lanka have not always been where they are today. Over millions of years, continents have slowly moved across the globe. Mite harvestmen belong to an ancient lineage; fossils show that they branched off from other invertebrates at least 400 million years ago. Back then, much of the world's land

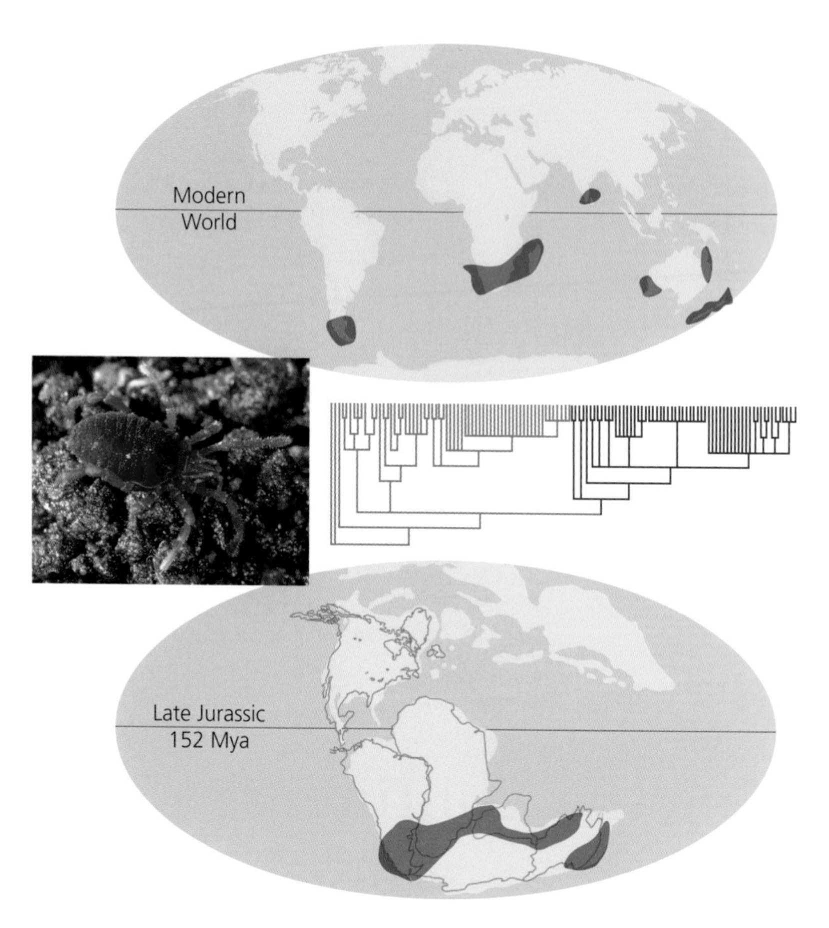

Figure 10.3 One lineage of mite harvestmen can be found on continents and islands separated by thousands of miles of ocean. They reached their present locations thanks to continental drift. Around 150 million years ago, the ranges of these invertebrates formed a continuous belt. Later, the continents broke apart and moved away, taking the mite harvestmen with them. (Adapted from Boyer et al., 2007)

was fused together in a single supercontinent. When Boyer mapped the locations of the mite harvestmen on a map of ancient Earth, she found that they were all close to each other in the Southern Hemisphere.

The study of how biodiversity is spread around the world is known as biogeography. Mite harvestmen illustrate one of the most common patterns in biogeography, called vicariance: species become separated from each other when geographical barriers emerge. Those barriers can be formed by oceans, as in the case of the mite harvestmen; they can also be separated by rising mountains, spreading deserts, and shifting rivers. The other major pattern in biogeography, known as dispersal, occurs when species themselves spread away from their place of origin. Birds can fly from one island to another, for example, and insects can float on driftwood.

The biogeography of many groups of species is the result of both dispersal and vicariance. Most living species of marsupials can be found today on Australia and its surrounding islands. But marsupials originally evolved thousands of kilometers away (**Figure 10.4**). The oldest fossils of marsupial-like mammals, dating back 150 million years, come from China. At the time, Asia was linked to North America, and by 120 million years ago marsupials had spread there as well. Many new lineages of marsupials evolved in North America over the next 55 million years. From there, some of these marsupials spread to Europe, even

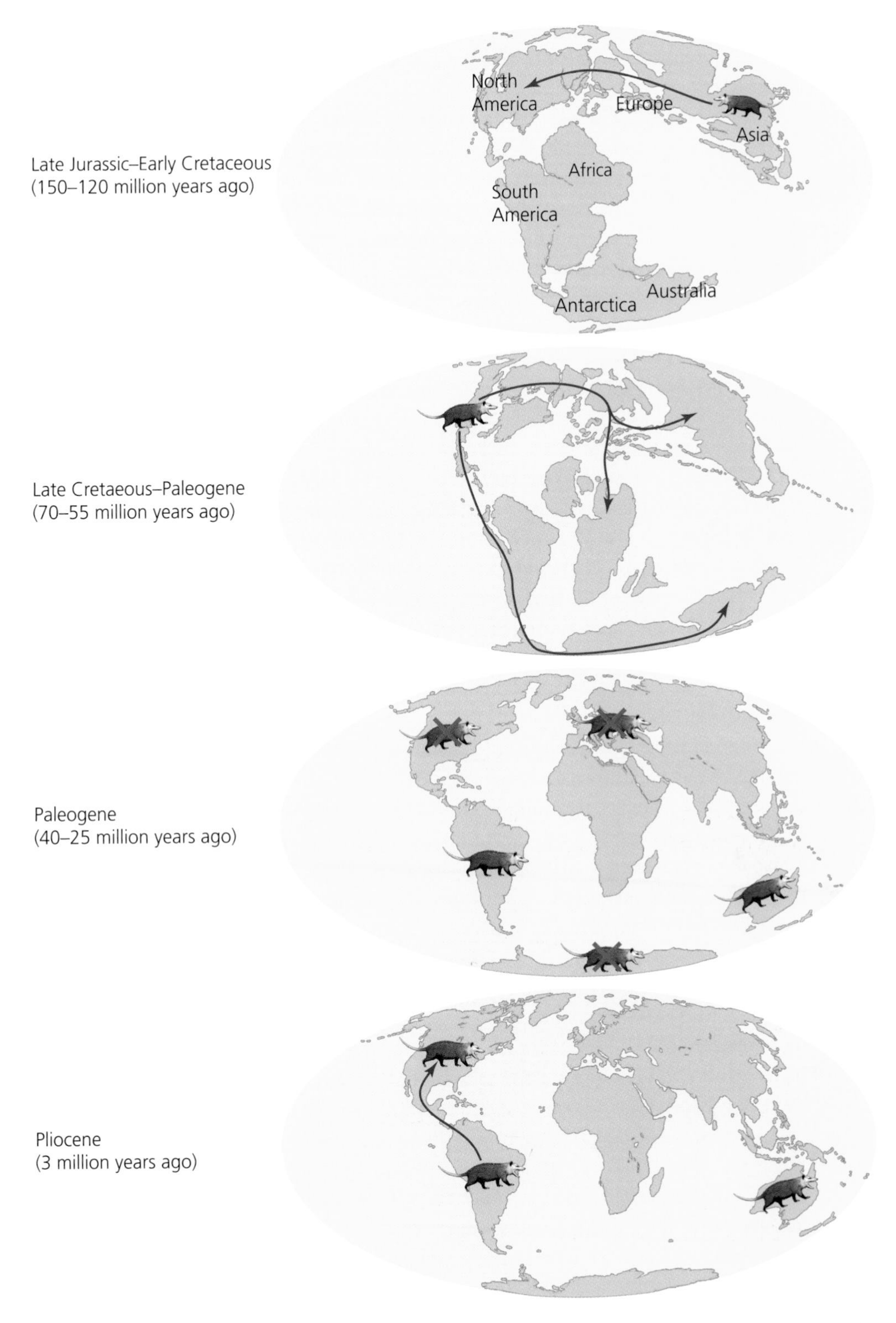

Figure 10.4 The fossil record sheds light on the spread of marsupial mammals around the world.

reaching as far as North Africa and Central Asia. All of these northern hemisphere marsupials eventually died out in a series of extinctions between 30 and 25 million years ago.

But marsupials did not die out entirely. Another group of North American marsupials dispersed to South America around 70 million years ago. From there, they expanded into Antarctica and Australia, both of which were attached to South America at the time. Marsupials arrived in Australia no later than 55 million years ago, the age of the oldest marsupial fossils found there. Later, South America, Antarctica, and Australia began to drift apart, each carrying with it a population of marsupials. The fossil record shows that marsupials were still in Antarctica 40 million years ago. But as the continent moved nearer to the South Pole and became cold, these animals became extinct.

In South America, marsupials diversified into a wide range of different forms, including cat-like marsupial sabertooths. These large carnivorous species became extinct, along with many other unique South American marsupials, when the continent reconnected to North America a few million years ago. However, there are still many different species of small and medium-sized marsupials living in South America today. One South American marsupial, the familiar Virginia opossum, even recolonized North America.

Australia, meanwhile, drifted in isolation for over 40 million years. The fossil record of Australia is too patchy for paleontologists to say whether there were any placental mammals in Australia at this time. Abundant Australian fossils date back to about 25 million years ago, at which point all the mammals in Austrlia were marsupials. They evolved into a spectacular range of forms, including kangaroos and koalas. It was not until 15 million years ago that Australia moved close enough to Asia to allow placental mammals—rats and bats—to begin to colonize the continent. These invaders diversified into many ecological niches, but they don't seem to have displaced any of the marsupial species that were already there.

Isolated islands have also allowed dispersing species to evolve into remarkable new forms. The ancestors of Darwin's finches colonized the Galápagos Islands two to three million years ago, after which they evolved into 14 species that live nowhere else on Earth. On some other islands, birds have become flightless. On the island of Mauritius in the Indian Ocean, for example, there once lived a big flightless bird called the dodo. It became extinct in the 1600s, but Beth Shapiro, a biologist now at the Pennsylvania State University, was able to extract some DNA from a dodo bone in a museum collection. Its DNA revealed that the dodo had a close kinship with species of pigeons native to southeast Asia. Only after the ancestors of the dodos diverged from flying pigeons and ended up on the island of Mauritius did they lose their wings and become huge land-dwellers. A similar transformation took place on Hawaii, where geese from Canada settled and became large and flightless.

Hawaiian geese and dodos may have lost the ability to fly for the same reason. The islands where their flying ancestors arrived lacked large predators that would have menaced them. Instead of investing energy in flight muscles that they never needed to use, the birds that had the greatest reproductive success were the ones that were better at getting energy from the food that was available on their new island homes.

The Pace of Evolution

Biodiversity forms patterns not just across space, but also across time. New species emerge, old ones become extinct; rates of diversification speed up and slow down. These long-term patterns in evolution get their start in the generation-to-generation processes of natural selection, genetic drift, and reproductive isolation.

When a lineage of organisms evolves over a few million years, these processes can potentially produce a wide range of patterns (see **Figure 10.5**). Natural selection may produce a significant change in a trait such as body size, for example. On the other hand, the average size of a species may not change significantly at all (a pattern known as stasis). Stabilizing selection can produce stasis by eliminating the genotypes that give rise to very big or very small sizes. It's also possible for a species to experience a lot of small changes that don't add up to any significant trend. (This type of pattern is known as a random walk, because it resembles the path of someone who randomly chooses where to take each new step.)

At the same time, a species can split in two. The rate at which old species in a lineage produce new ones can be fast or slow (see **Figure 10.5c**). Over millions of years, one lineage may split into a large number of new species, while a related

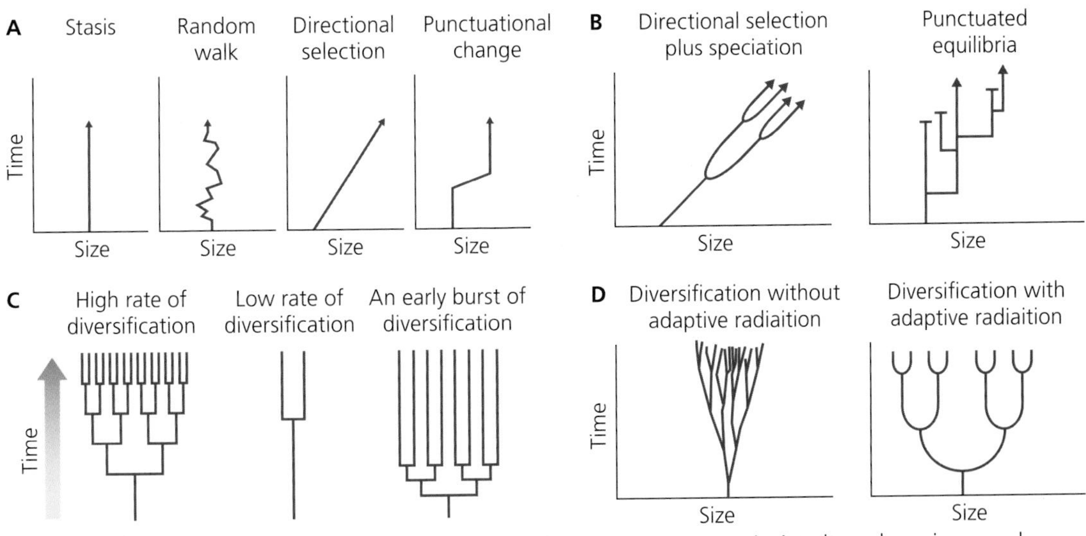

Figure 10.5 Over long periods of time, evolution can form many patterns. A: A trait, such as size, may be constrained by stabilizing selection, undergo small changes that don't add up to a significant shift, experience long-term selection in one direction, or experience a brief punctuation of change. B: A lineage may also branch into new species while experiencing different kinds of morphological change. C: The rate at which new species evolve is different in different lineages. It can also change in a single lineage. D: In an adaptive radiation, a lineage evolves new species and also evolves to occupy a wide range of niches.

lineage hardly speciates at all. It's also possible for a lineage's rate of speciation to slow down or speed up.

Even as new species are evolving, however, others may become extinct. The rate at which species become extinct may be low in one lineage and high in another. It's also possible for the rate of extinction to rise, only to drop again later.

All of these processes can also unfold at the same time, and so the range of possible long-term patterns in evolution can be enormous. A lineage with a low rate of speciation may end up enormously diverse because its rate of extinction is even lower. On the other hand, a lineage that produces new species at a rapid rate may still have relatively few species if those species become extinct quickly. Evolutionary change may happen mainly within the lifetime of species, or it may occur in bursts when new species evolve. A lineage may produce many species that are all very similar to each other, or evolve a wide range of forms.

Any one of these patterns is plausible, given what biologists know about how evolution works. Which of these patterns actually dominate the history of life is a question that they can investigate by studying both living and extinct species.

Evolutionary Fits and Starts

One of the most influential studies of the pace of evolutionary change was published in 1971 by two young paleontologists at the American Museum of Natural History named Niles Eldredge and Stephen Jay Gould. They pointed out that the fossils of a typical species showed few signs of change during its lifetime. New species branching off from old ones had small but distinctive differences. Eldredge carefully documented this stasis in trilobites, an extinct lineage of armored arthropods. He counted the rows of columns in the eyes of each subspecies. He found that they did not change over six million years.

Eldredge and Gould proposed that this pattern was the result of stasis punctuated by relatively fast evolutionary change, a combination they dubbed punctuated equilibria. They argued that natural selection might adapt populations within a species to their local conditions, but overall the species experienced very little change in its lifetime. Most change occurred when a small population became isolated and branched off as a new species. Eldredge and Gould argued that paleontologists could not find fossils from these branchings for two reasons: the populations were small, and they evolved into new species in just thousands of years—a geological blink of an eye.

This provocative argument has inspired practically an entire generation of paleontologists to test it with new evidence. But testing punctuated equilibria has turned out to be a challenge in itself. It demands dense fossil records that chronicle the rise of new species. Scientists have also had to develop sophisticated statistical tests to determine whether a pattern of change recorded in those fossils is explained best as stasis, a random walk, or directional change.

Scientists now have a number of cases in which evolution appears to unfold in fits and starts. **Figure 10.6** (top) comes from a study by Jeremy Jackson and Alan

Cheetham of bryozoans, small animals that grow in crustlike colonies on submerged rocks and reefs. On the other hand, more gradual, directional patterns of change have also emerged. **Figure 10.6** also charts the evolution of a diatom called *Rhizosolenia* that left a fairly dense fossil record over the past few million years. One structure on the diatom gradually changed shape as an ancestral species split in two.

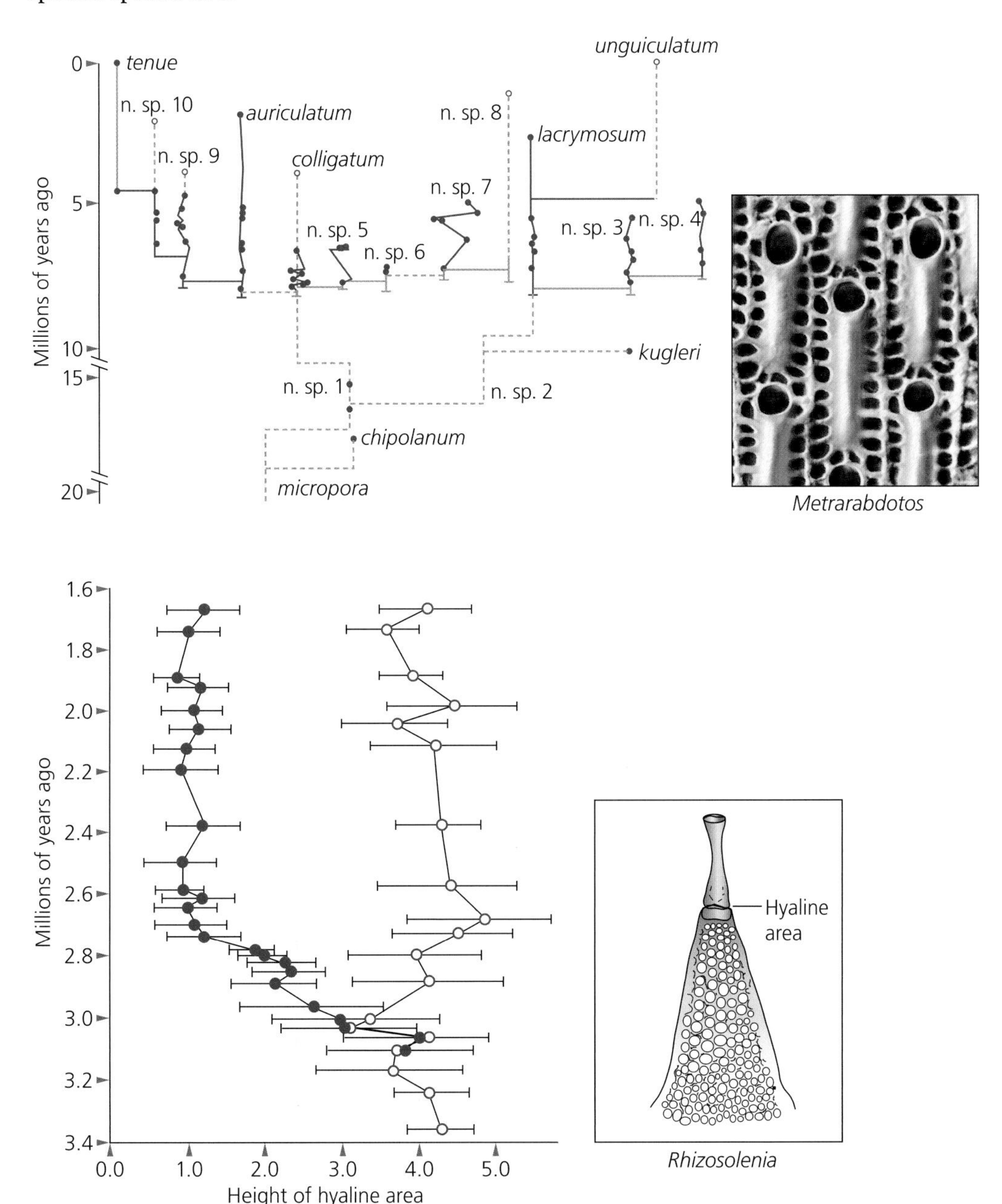

Figure 10.6 Paleontologists have documented cases of punctuated change and gradual change in the fossil record. Top: A lineage of bryozoans (*Metrarabdotos*) evolved rapidly into new species, but changed little once those species were established. Bottom: A shell-building organism called *Rhizosolenia* changed over the course of millions of years. This graph charts the size of a structure called the hyaline area. (Adapted from Benton, 2003)

At this point, paleontologists have found few well-documented cases that match the original model of punctuated equilibra, with rapid change happening only during speciation. But Eldredge and Gould's ideas have led to some significant changes in how paleontologists look at the fossil record. For example, Gene Hunt, a paleontologist at the Smithsonian Institution, recently developed a method for statistically analyzing patterns of change and used it to study 53 evolutionary lineages ranging from mollusks to fishes and primates. In 2007, Hunt concluded that only 5% of the fossil sequences showed signs of directional change. The other 95% was about evenly split between random walks and stasis. Hunt did not look for evidence of directional change during speciation, so he could not directly address the original model of punctuated equilibria. But Hunt's 2007 study does support the idea that stasis is a major feature of the history of life.

The Lifetime of a Species

Paleontologists estimate that 99% of all species that ever existed have vanished from the planet. To understand the process of extinction, paleontologists have measured the lifetime of species—especially species that leave lots of fossils behind. Mollusks (a group of invertebrates that includes snails and clams) leave some of the most complete fossil records of any animal.

Michael Foote, an evolutionary biologist at the University of Chicago, and his colleagues inventoried fossils of mollusks that lived in the ocean around New Zealand over the past 43 million years. They cataloged every individual fossil from each species, noting where and when it lived. Foote and his colleagues found that a typical mollusk species expanded its range over the course of a few million years and then dwindled away. **Figure 10.7** shows a selection of the species they cataloged. Some species lasted only 3 million years, while others lasted 25 million years.

Left: The dodo became extinct in the late 1600s, probably due to hunting and rats that ate their eggs. Right: The Carolina parakeet became extinct in the early 1900s, due in part to logging, which removed the hollow logs in which it built its nests.

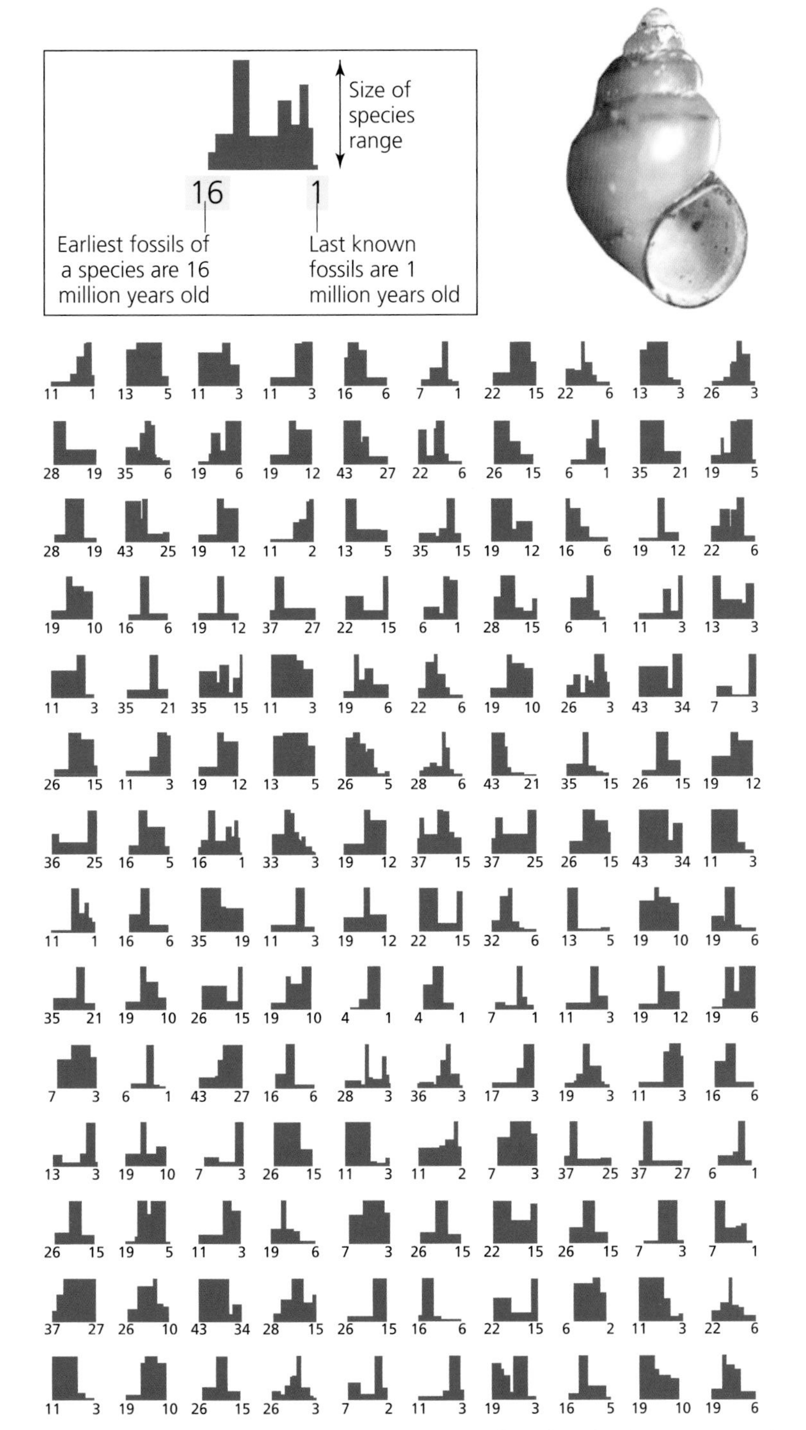

Figure 10.7 These graphs chart the rise and fall of mollusk species over the past 43 million years around New Zealand. The left number on each graph is the age of the earliest fossil in a species (in millions of years), and the right number is the age of the youngest fossil. The height of each graph represents the range over which fossils at each interval have been found. As these graphs demonstrate, some species survive longer than others, but in general they endured for a few million years. (Adapted from Foote, 2008)

To understand how species became extinct millions of years ago, biologists can get clues from extinctions that have taken place over the past few centuries. When Dutch explorers arrived on Mauritius in the 1600s, for example, they killed dodos for food or sport. They also inadvertently introduced the first rats to Mauritius, which then proceeded to eat the eggs of the dodos. As adult and young dodos alike were killed, the population shrank until only a single dodo was left. When it died, the species was gone forever.

Simply killing off individuals is not the only way to drive a species towards extinction. Habitat loss—the destruction of a particular kind of environment where a species can thrive—can also put a species at risk. The Carolina parakeet once lived in huge numbers in the southeastern United States. Loggers probably hastened its demise in the early 1900s by cutting down the old-growth forests where the parakeets made their nests in hollow logs.

Habitat loss can turn a species into a few isolated populations. Their isolation makes the species even more vulnerable to extinction. In small populations, genetic drift can spread harmful mutations and slow down the spread of beneficial ones. If the animals in an isolated population are wiped out by a hurricane, their numbers cannot be replenished by immigrants. As isolated populations wink out, one by one, the species as a whole faces the threat of extinction.

Cradles of Diversity

Understanding the long-term patterns of speciation and extinction may help scientists answer some of the biggest questions about today's patterns of biodiversity—such as why the tropics are so diverse. David Jablonski, a paleontologist at the University of Chicago, has tackled the question by analyzing the fossil record of bivalves, noting where they were located, how large their ranges became, and how long they endured.

Jablonski's analysis of 3,599 species from the past 11 million years revealed a striking pattern. Twice as many new genera of bivalves had emerged in the tropical oceans than had emerged in cooler waters. Jablonski found that once new bivalve genera evolved in the tropics, they expanded towards the poles. In time, however, the bivalves near the poles became extinct while their cousins near the equator survived. From these results, Jablonski argued that the tropics are both a cradle and a museum. New species can evolve rapidly in the tropics, and they can accumulate to greater numbers because the extinction rate is lower there as well. Together these factors lead to the high biodiversity of the tropics.

A similar pattern emerged when Bradford Hawkins, a biologist at the University of California, Irvine, studied the evolution of 7,520 species of birds. The birds that live closer to the poles belong to younger lineages than the ones that live in the tropics.

It's possible that the tropics have low extinction rates because they offer a more stable climate than regions closer to the poles. Ice ages, advancing and retreating glaciers, swings between wet and dry climates—all of these may have raised the risk of extinction in the cooler regions of the Earth. The changes that occurred in the tropics were gentler, which made it easier for species to survive. But the tropics also foster a higher rate of emergence of new species. Why the tropics can sustain more species than other regions is not clear, however; it's possible that the extra energy the tropics receive somehow creates extra ecological room for more species to live side by side.

Radiations

When biologists examine the history of a particular lineage, they discover a mix of diversification and extinctions. **Figure 10.8** shows the history of one such lineage, that of a group of mammal species called mountain beavers. About 30 species of mountain beavers have evolved over the past 35 million years in the

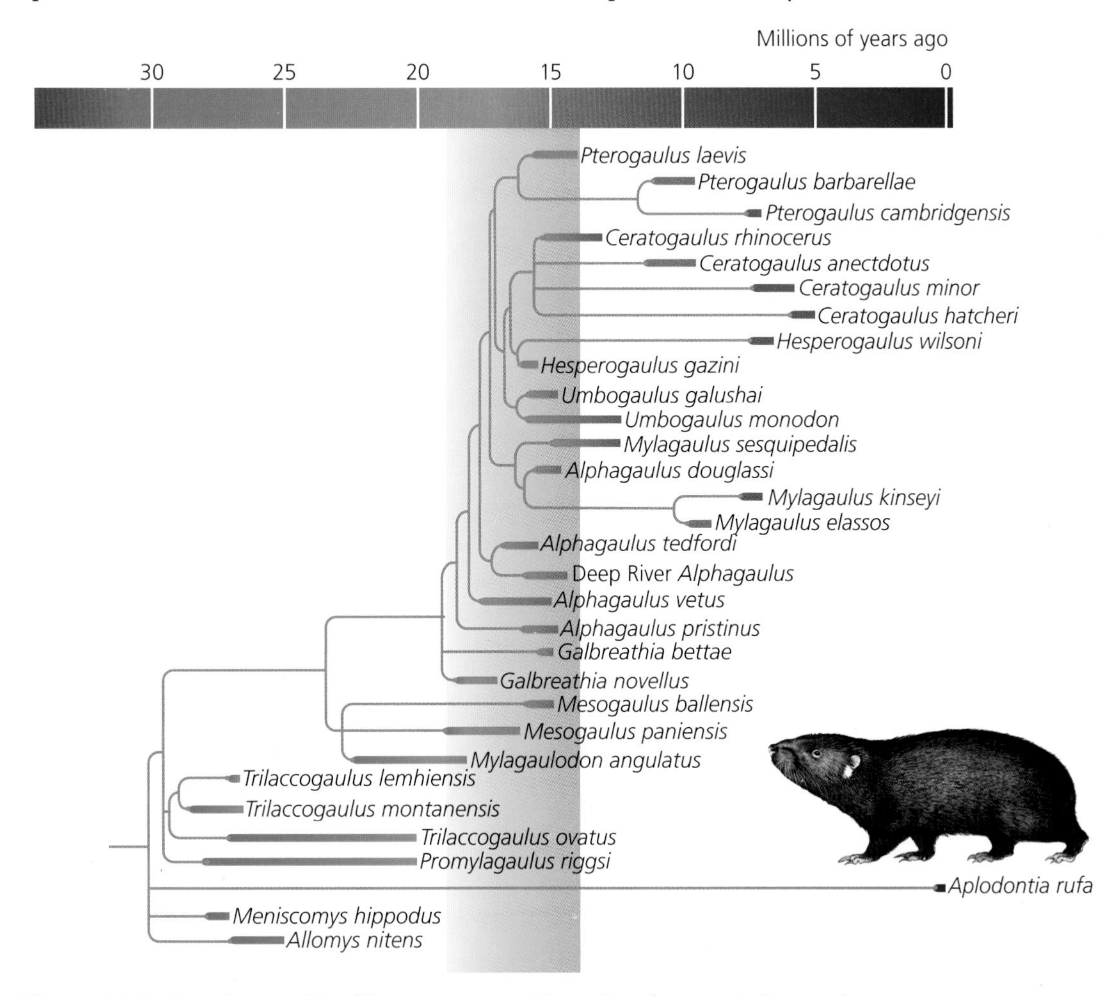

Figure 10.8 Over the past 35 million years, some 30 species of mountain beavers have existed in western North America. A burst of new lineages evolved around 15 million years ago. (Adapted from Barnovsky, 2008)

western United States, but today only a single species survives. Anthony Barnosky, a paleontologist at the University of California at Berkeley, and his colleagues have gathered fossils of mountain beavers, and they have found that new species of mountain beavers did not emerge at a regular pace. Instead, there was a period of rapid speciation around 15 million years ago. The number of mountain beaver species then gradually shrank as one species after another became extinct without new ones evolving to make up for their loss.

Sometimes a burst of diversification is accompanied by dramatic morphological evolution—an event known as an adaptive radiation. When the ancestors of Darwin's finches arrived on the Galápagos Islands a few million years ago, they did not simply evolve into 14 barely distinguishable species. They evolved distinctive beaks and behaviors that allowed them to feed on cactuses, crack hard nuts, and even drink the blood of other birds. The Great Lakes of East Africa also saw an adaptive radiation of cichlid fishes. These enormous lakes are geologically very young, in many cases having formed in just the past few hundred thousand years. Once they formed, cichlid fishes moved into them from nearby rivers. The fishes then exploded into thousands of new species. Along the way, the cichlids also adapted to making a living in a staggering range of ways—from crushing mollusks, to scraping algae and eating other cichlids.

Biologists don't yet know exactly what triggers adaptive radiations. One thing the African cichlids and Darwin's finches have in common is that they were able to move into a new ecosystem that was not already filled with well-adapted species. Without any established residents offering competition, the colonizers may have been able to evolve into a wide range of forms. Yet ecological opportunity cannot be the only factor behind adaptive radiations. Among the close rela-

Cichlid fishes that live in East Africa are a striking example of an adaptive radiation. Small founding populations entered each lake and then rapidly evolved into a wide range of forms and thousands of species.

tives of Darwin's finches are a lineage of birds that settled on the islands of the Caribbean. But they have only evolved into a narrow range of new sizes and shapes. It's possible that some lineages are somehow "preadapted" to take full advantage of ecological opportunity, while others are not.

A Gift for Diversity

Mountain beavers enjoyed a burst of new species 15 million years ago, but it's been pretty much downhill ever since. The story of insects is very different: they've enjoyed a durable success. They first evolved about 400 million years ago, and they've diversified fairly steadily ever since (**Figure 10.9**). The rise of insect diversity is all the more striking when you compare them to their closest relatives, a group of arthropods called entognathans that includes springtails.

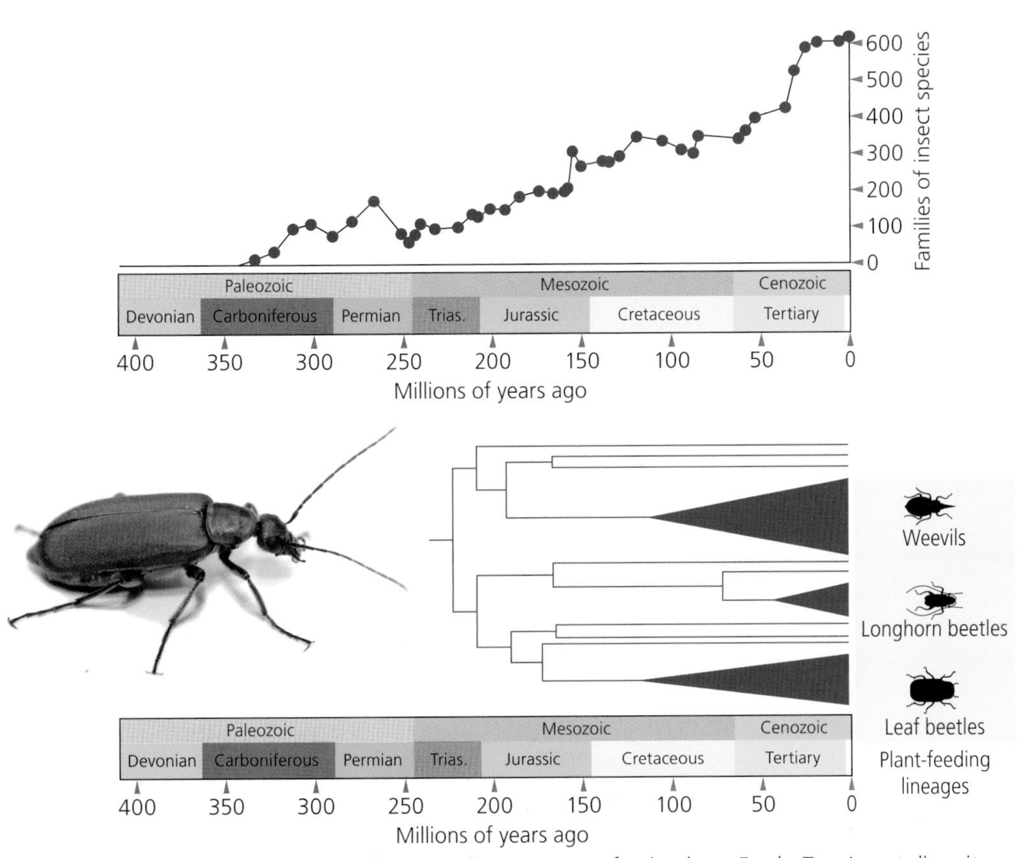

Figure 10.9 Insects are the most diverse group of animals on Earth. Top: Insect diversity has gradually grown over the past 400 million years. Bottom: Plant-feeding may have helped spur insect diversity, judging from the fact that many insect lineages that evolved the ability to eat plants became more diverse than their closest relatives. (Adapted from Mayhew, 2008, and Mayhew, 2009)

The entognathan lineage is just as old as the insect lineage. And while there are a million known insect species, there are only 10,600 entognathan species.

A number of biologists have probed the history of insects to determine what factors account for their huge diversity. Peter Mayhew, a biologist at the University of York, has tested the leading hypotheses. Insects don't seem to have a particularly high rate of speciation, he has found, but they do seem good at withstanding extinctions. Fifty percent of all families of insect species alive today existed 250 million years ago. None of the families of tetrapod species alive 250 million years ago exists today; all have been replaced by newer groups.

So what gives insects their sticking power? Mayhew argues that a few key factors are at work. The ability to eat plants provides insects with a huge amount of food; plant-eating has evolved several times among insects, and the plant-eating lineages tend to accumulate more species than closely related lineages of insects that don't eat plants. The small bodies of insects may lower the amount of food they need to survive, and shortens the time they need to develop from eggs. Wings also allowed insects to disperse much farther than arthropods that can only crawl or jump. Mayhew argues that all these advantages gave insects a massive edge, allowing them to colonize new habitats quickly and to survive catastrophes.

Lighting the Cambrian Fuse

By studying Darwin's finches and East African cichlids, scientists can get clues that help them understand much older, much bigger adaptive radiations. One of the biggest was the early rise of animals.

This period of animal evolution is sometimes nicknamed "the Cambrian explosion." Unfortunately, that name gives the impression that all the modern groups of animals popped into existence 540 million years ago at the dawn of the Cambrian period. Animals evolved from protozoans, which left fossils over a billion years before the Cambrian. Some 630 million years ago, one group of living animals—sponges—was already leaving behind biomarkers. By 555 million years ago, fossils belonging to some living groups began to appear—12 million years before the Cambrian Period.

The phylogeny of early animals is also showing how the body plans of living animals emerged not in a single leap, but in a series of steps. Arthropods, for example, have a body plan with a combination of traits (such as segments and an exoskeleton) seen in no other group of living animals. But some Cambrian fossils had some of those traits and not others. In **Figure 10.10**, we can see how these fossils help document the evolution of the arthropod body plan.

Clearly, then, animals did not drop to Earth in the Cambrian Period. They evolved. Nevertheless, the fossil record of the Cambrian chronicles a remarkable pulse of rapid evolution. When paleontologists look at 530-million-year-old rocks, they mainly find small, shell-like fossils. When they look at rocks just 20

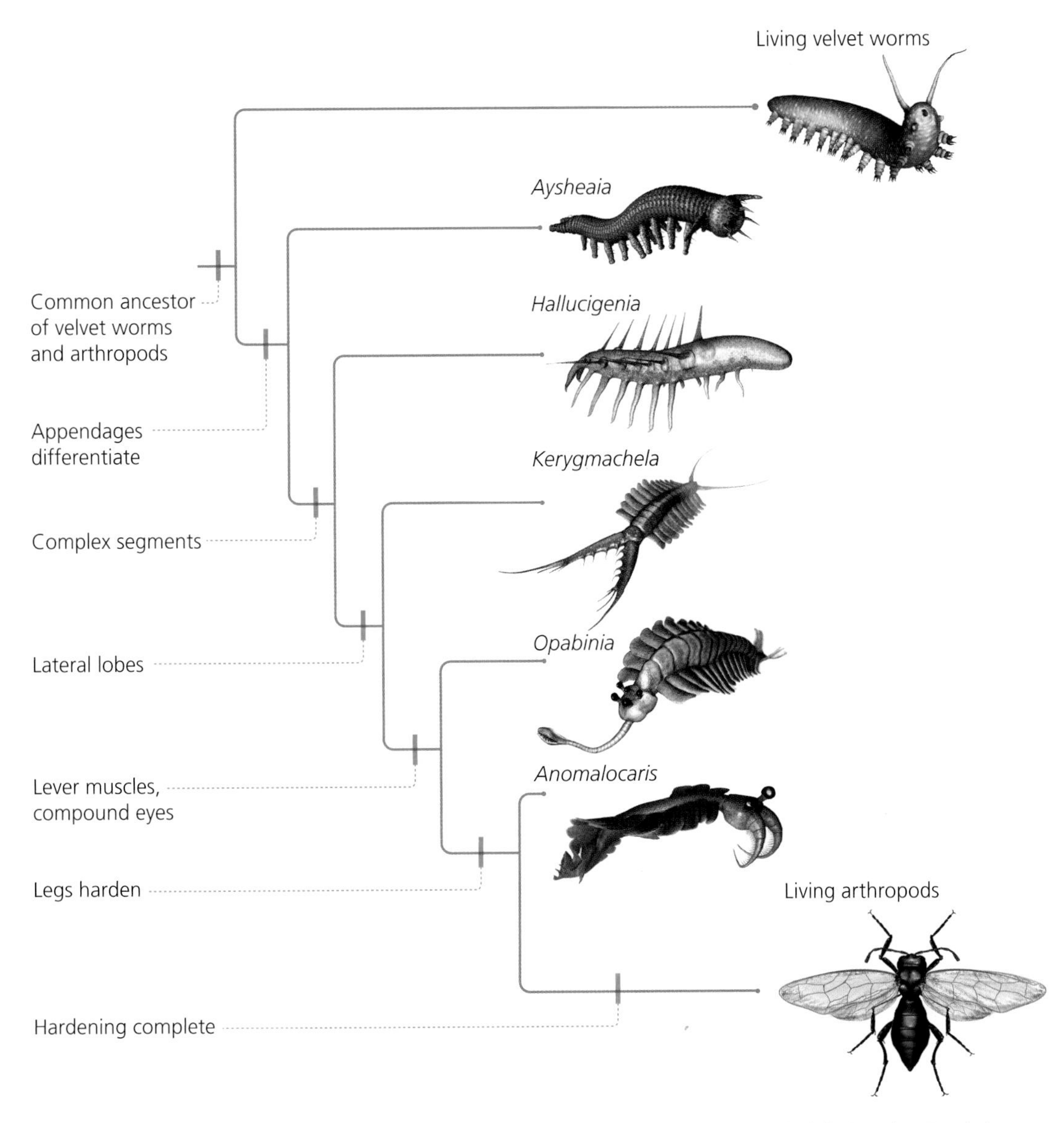

Figure 10.10 The fossil record documents how major groups of animals emerged during the Cambrian Period. Arthropods—a group that includes insects, spiders, and crustaceans—share a number of traits in common, such as jointed exoskeletons. Some Cambrian fossils belong to relatives of today's arthropods that lacked some of these traits. (Adapted from Budd, 2003)

million years younger, they find fossils that are recognizable relatives of living arthropods, vertebrates, and many other major groups of animals.

As an adaptive radiation, the early evolution of animals was unsurpassed. **Figure 10.11** is a diagram that marks the changes during the Ediacaran and Cambrian periods. One way to gauge these changes is to measure the diversity over time. **Figure 10.11** tracks diversity by tallying animal genera. Over the course of

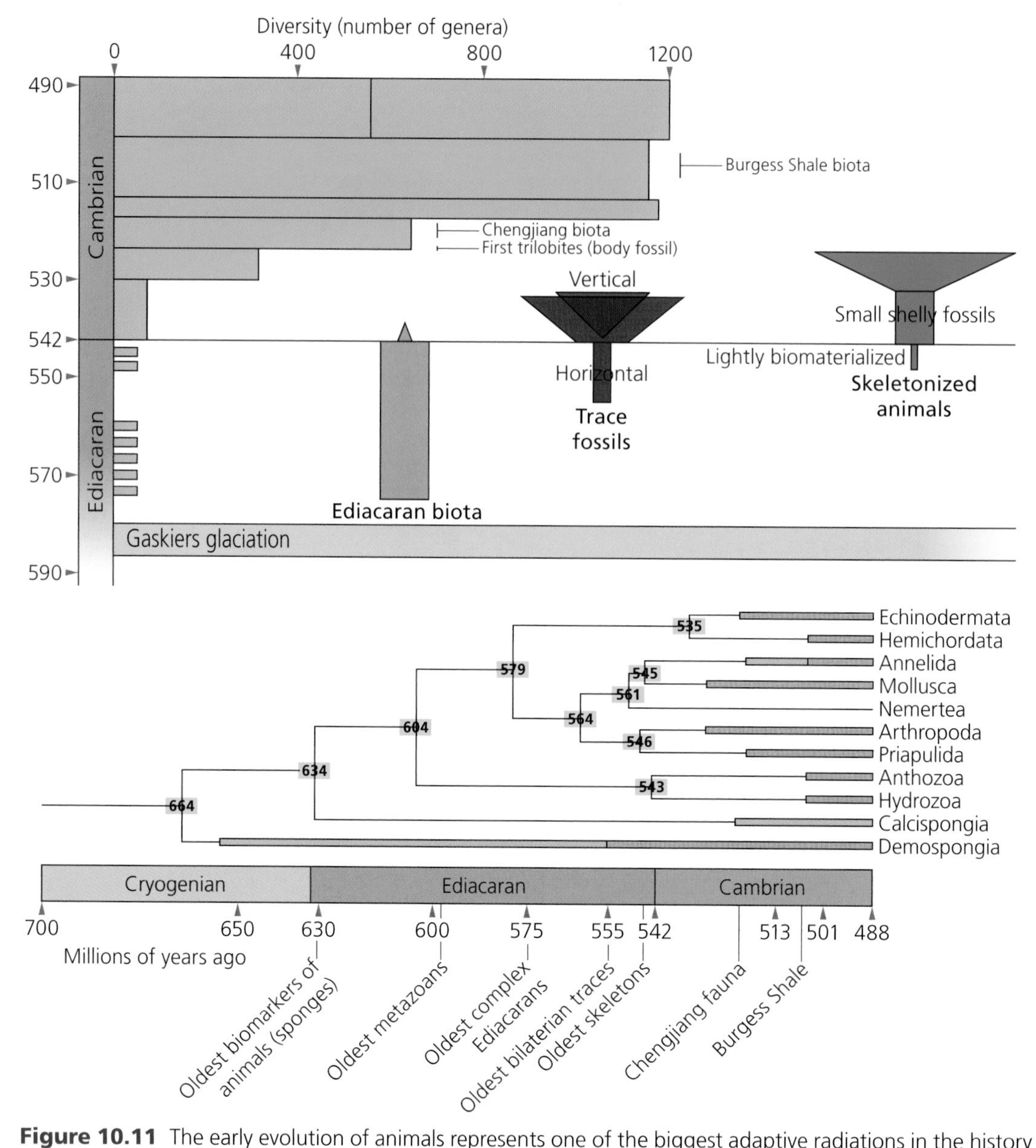

Figure 10.11 The early evolution of animals represents one of the biggest adaptive radiations in the history of life. Top: Fossils document the growing diversity of animals. Bottom: The major groups of animals evolved from common ancestors. Tan bars show the known fossil record of different groups. Pink bars show the range of fossils that have been proposed to belong to some groups. The numbers are ages based on DNA studies. (Top: adapted from Marshall, 2006. Bottom: adapted from Peterson, 2005)

about 40 million years, animal genera multiplied 100 times. But counting genera is not the only way to measure diversity. After all, there is much less variation among 100 genera of beetles than between a single species of squid and a single species of hummingbird. This morphological variation is known as disparity. The disparity of animals expanded rapidly during the Cambrian Period, before tapering off.

Paleontologists, developmental biologists, geochemists, and many other scientists are testing hypotheses that may explain this remarkable radiation. Some researchers observe that the radiation of animals came at a time when the Earth was going through some dramatic physical changes. It was emerging from a climate so cold that the entire planet was covered in glaciers. The ocean's chemistry was also changing drastically. For most of Earth's history, the oceans were almost devoid of free oxygen. While the atmosphere contained some free oxygen, the molecule could not survive for long in the ocean before bonding with other molecules. But Paul Hoffman of Harvard University and his colleagues have found evidence that oxygen levels began to rise in some parts of the ocean about 630 million years ago. By 550 million years ago, the change had spread across all of Earth's oceans.

Both the retreat of the glaciers and the rise of oxygen in the ocean may have spurred the rise of the animals. All animals need oxygen to fuel their metabolism and to build their tissues. The low levels of oxygen in the oceans may have made it impossible for the ancestors of animals to evolve into multicellular creatures.

If a rise in oxygen opened the door for animal evolution, what pushed the animals through? Part of that answer may lie within the animals themselves—in particular, in the set of genes that control their development. Most animal species alive today use the same "genetic tool kit" to build very different kinds of bodies (page 165). It was during the Ediacaran Period that this tool kit itself first evolved.

Douglas Erwin of the Smithsonian Institution argues that the animal tool kit allowed animals to evolve from a relatively limited number of Ediacaran forms to the frenzy of diversity that marked the Cambrian Period. As genetic circuits were rewired, new body parts evolved, along with new appendages, organs, and senses. It's possible that the genetic tool kit gave early animals the sort of flexibility adaptive radiations require.

The new body plans allowed animals to organize themselves into new ecosystems the Earth had never seen before. The earliest animals appear to have lived like sponges do today—trapping microbes or organic matter from the water as they remained anchored to the seafloor. But then animals evolved with guts and nervous systems, able to swim through the water or burrow into the muck. With their guts, they could swallow larger microbes, and, eventually, could even start to attack other animals. **Figure 10.12** shows a 550-million-year-old fossil *Cloudina* bearing the oldest known wounds from the attack of a predator.

Charles Marshall, a biologist at Harvard University, has proposed that the evolution of these new predators changed the fitness landscape for early animals. The old soft-bodied creatures anchored to the seafloor became easy targets for new predators. Now natural selection favored new defenses, such as hard shells, exoskeletons, and toxins. Predators in turn benefited from more sophisticated equipment for finding their prey, such as eyes. Their prey benefited from improved vision as well.

Natural selection did not converge on a single strategy for predators or for prey, Marshall argues. The new ecosystem created many different selection

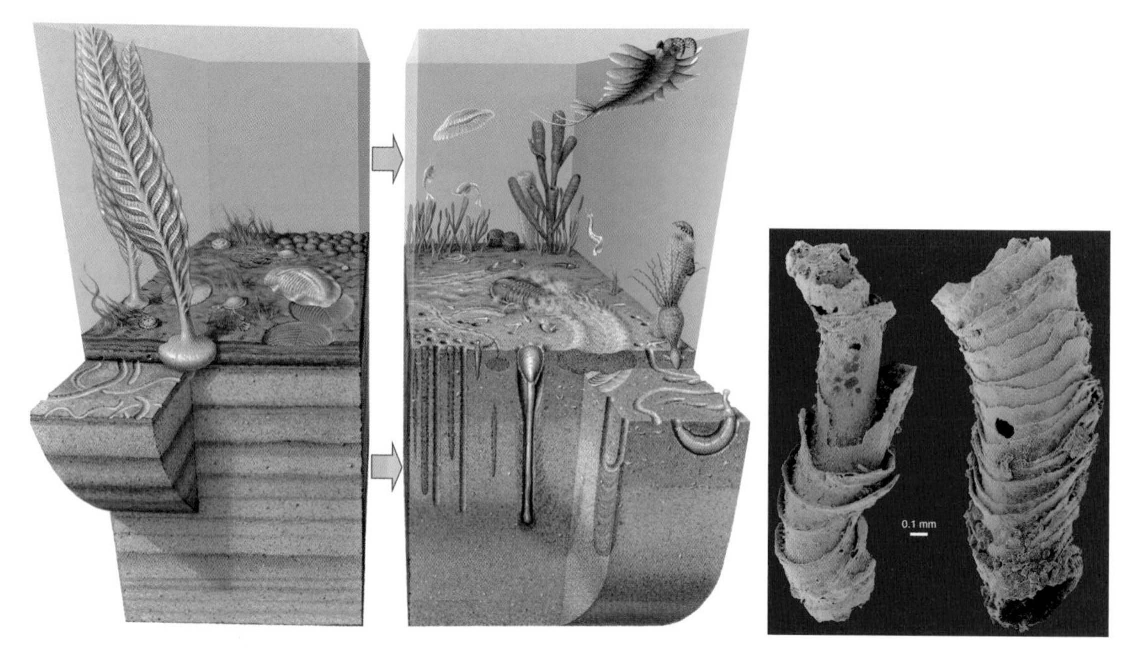

Figure 10.12 Left: During the Cambrian Period, the ecology of the ocean changed dramatically. New animals began burrowing, crawling on the ocean floor, and swimming rapidly after prey. Right: 550-million-year-old fossils bear holes bored by a predator—one of the earliest signs of predation in the fossil record.

pressures, each creating trade-offs with the other pressures. Instead of a single adaptive peak, Marshall proposes, animals now evolved on a fitness landscape erupting with a rugged expanse of hills. The complex landscape led to the extraordinary diversity and disparity of the animal kingdom.

Driven to Extinction

As new species emerged and evolved into disparate new forms, other species became extinct. And just as the origin of species and disparity form large-scale patterns, extinctions have formed patterns of their own. One of those patterns is illustrated in **Figure 10.13**. It shows how the extinction rate has gone up and down over the past 540 million years. A few pulses of extinctions stand out above the others. These mass extinctions were truly tremendous cataclysms. The biggest of all, which occurred 250 million years ago, claimed 55% of all genera. When scientists estimate the destruction in terms of species rather than genera, the event is even more catastrophic: perhaps 90% of all species disappeared. The fossil record also leaves ecological clues from that time that suggest that it was a period of global devastation. Forests and reefs drop out of the fossil record, and they do not reappear for 20 million years.

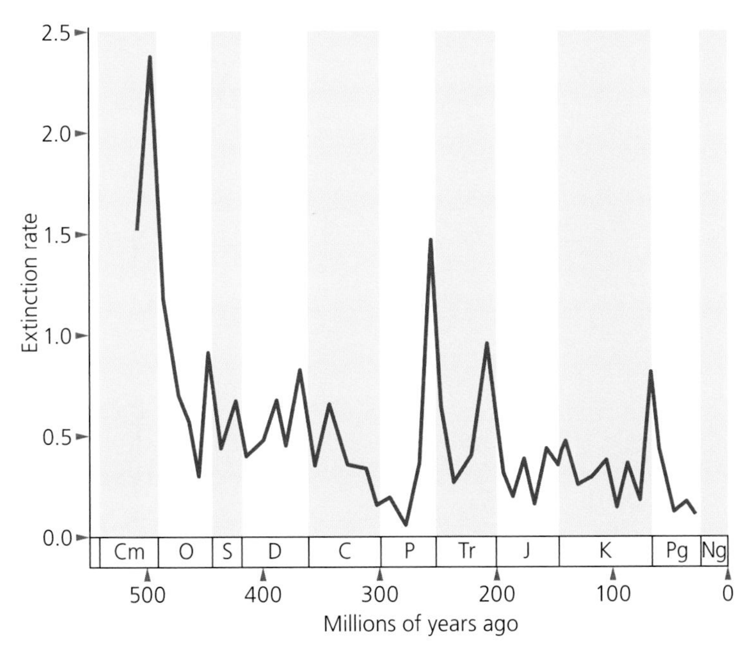

Figure 10.13 The rates at which species have become extinct has changed over time. The history of life has been marked by a few pulses of mass extinctions, in which vast numbers of species became extinct within a few million years. (Adapted from Alroy et al., 2008)

Only about 20% of all extinctions occurred during mass extinctions. The other 80% are known as background extinctions. For most individual species, scientists don't know the precise cause of extinction. But scientists can gather clues in the broad patterns of extinctions formed by thousands of species. In 2008, Jonathan Payne and Seth Finnegan, two paleontologists at Stanford University, surveyed 227,229 fossils of marine invertebrates from about 520 million years ago to 20 million years ago. They found that the bigger the geographical range of a genus, the longer it tended to survive. Small ranges raised the odds that a genus would become extinct.

One potential explanation for Payne and Finnegan's result is that a small range may make a genus more vulnerable to small-scale catastrophes, such as volcanic eruptions or an invasion by a dangerous predator. A small range may also be the mark of a genus that can survive only under very special conditions, such as a particular range of temperatures or a particular amount of rainfall. If the climate should change, the genus may not be able to adapt.

Another striking pattern in the history of life is the way in which major lineages gradually suffer extinctions as other lineages become more diverse. From 540 to 250 million years ago, the seafloor was dominated by invertebrates like trilobites and lamp shells (known as brachiopods), many of which fed by trapping bits of food suspended in the water. But today only a few hundred species of lamp shells survive, and trilobites disappeared entirely 252 million years ago. Now the seafloor is dominated by other animals, such as clams and other bivalves that bury themselves in the sediment.

Shannan Peters, a paleontologist at the University of Wisconsin, studies the shift from the old species (known as the Paleozoic Fauna) to the new ones (the Modern Fauna). Peters found that most fossils of the Paleozoic Fauna are found in sedimentary rocks known as carbonates, which formed from the bodies of microscopic organisms that settled to the seafloor. Most of the Modern Fauna fossils are found in rocks known as silicoclastics, which formed from the sediments carried to the ocean by rivers. Over the past 540 million years, carbonate rocks became rare, while silicoclastic rocks became common, possibly as rivers delivered more sediments to the oceans. Peters proposes that as the seafloor changed, the Modern Fauna could expand across a greater area, while the Paleozoic Fauna retreated to a shrinking habitat where it suffered almost complete extinction.

Another force influencing the extinction rate is the planet's changing climate. Geologists can estimate the average temperature in the distant past by measuring oxygen isotopes in rocks. High levels of oxygen-18 are a sign of a warm climate, because warm water can hold more of it than cold water. Peter Mayhew and his colleagues compared these climate records to the diversity of species over the past 300 million years. Diversity has gone down when the climate has been warm, Mayhew found, and it has been higher when the climate has been cool. The researchers found that most of the change was due to extinction rates going up rather than the speciation rate going down. As we'll see later, Mayhew's results are an ominous warning about our future.

When Life Nearly Died

Paleontologists have long debated whether mass extinctions shared the same causes as background extinctions, or whether some fundamentally different process was responsible. To test the alternative hypotheses, they have searched for rocks that formed during those mass extinctions that may chronicle those exceptional times.

In recent years researchers have discovered some new formations in China that record the mass extinctions at the end of the Permian in exquisite detail. These rocks are loaded with fossils from before, during, and after the mass extinctions, and they're also laced with uranium and other elements that geologists can use to make good estimates of their ages. The rocks indicate that these mass extinctions actually came in two pulses. The first pulse, a small one, came about 260 million years ago. Eight million years passed before the next one hit. The second strike was geologically swift—less than 300,000 years. How much less is a subject of debate; a few scientists have even proposed that it took just a few thousand years.

In Siberia, 252-million-year-old rocks reveal a potential culprit for the mass extinctions. They contain huge amounts of lava, spewed about by volcanoes. All

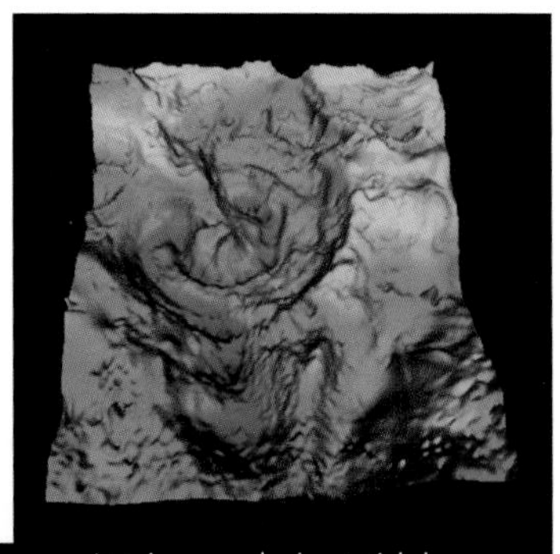
An asteroid struck the coast of Mexico 66 million years ago. It triggered giant tidal waves, vast forest fires, and a global environmental crisis. Right: Remnants of the crater have been found deep underground. Many researchers argue that the impact was at least partially responsible for mass extinctions at about the same time.

told, those eruptions covered a region as big as the United States. They released a harsh cocktail of gases into the atmosphere that would have disrupted the climate.

Atmospheric scientists have built computer simulations of the eruptions that suggest they could have devastated life in several ways. Heat-trapping gases, such as carbon dioxide and methane, could have driven up the temperature of the atmosphere. A warmer atmosphere could have warmed the oceans, driving out much of the free oxygen in the surface waters. Bacteria that thrive in low-oxygen water may have undergone a population explosion, releasing toxic gases such as hydrogen sulfide. Meanwhile, other gases released by the Siberian eruptions may have risen up to the stratosphere, where they could have destroyed the protective ozone layer. High-energy particles from space may have penetrated the lower atmosphere, creating damaging mutations. This chain of events could explain the puzzling appearance of deformed pollen grains during the Permian–Triassic extinctions.

Giant volcanic eruptions may not be the only things that can affect life across the planet. In the late 1970s, the University of California geologist Walter Alvarez was searching for a way to estimate precisely the ages of rocks. His father, the physicist Luis Alvarez, suggested that Walter measure levels of a rare element called iridium. Iridium falls to Earth from space at a relatively steady rate, and so it might act like a geological clock.

However, when Walter Alvarez collected rocks in Italy from the end of the Cretaceous Period 66 million years ago, he discovered concentrations of iridium far higher than average. The Alvarezes and their colleagues proposed that an asteroid or comet, rich in iridium, struck the Earth at the end of the Cretaceous

Period. In 1991, geologists in Mexico discovered a 110-mile-wide crater along the coast of the Yucatan Peninsula of precisely that age.

What made Alvarez's discovery electrifying for many paleontologists was the fact that the end of the Cretaceous also saw one of the biggest pulses of extinctions ever recorded. Through the Cretaceous, the Earth was home to giants. *Tyrannosaurus rex* and other carnivorous dinosaurs attacked huge prey such as *Triceratops*. Overhead, pterosaurs as big as small airplanes glided, and the oceans were dominated by whale-sized marine reptiles. By the end of the Cretaceous Period, these giants were entirely gone. The pterosaurs became extinct, leaving the sky to birds, which were the only surviving dinosaurs. Marine reptiles vanished as well. Along with the giants went millions of other species, from shelled relatives of squid called ammonites to single-celled protozoans.

The impact on the Yucatan may have had enough energy to trigger wildfires thousands of kilometers away and to kick up tidal waves that roared across the southern coasts of North America. It may have lofted dust into the atmosphere that lingered for months, blocking out the sunlight. Some compounds from the underlying rock in the Gulf of Mexico mixed with clouds to produce acid rain, while others absorbed heat from the sun to raise temperatures.

Many researchers argue that this impact was in large part responsible for the mass extinctions at the end of the Cretaceous Period. But some geologists point out that, not long before the impact, India began to experience tremendous volcanic activity that probably disrupted the atmosphere and the climate as well. Meanwhile, some paleontologists question how much effect the impact or the volcanoes had on biodiversity at the end of the Cretaceous. The diversity of dinosaurs and other lineages was already dropping millions of years earlier. Moreover, if a sudden environmental cataclysm wiped out the dinosaurs and millions of other species, it's strange that snakes, lizards, turtles, and amphibians did not also suffer mass extinctions. Those are the animals that today are proving to be exquisitely vulnerable to environmental damage.

Whatever the exact causes of mass extinctions turn out to be, it is clear that they left great wakes of destruction. After the Permian–Triassic extinctions 252 million years ago, for example, forests were wiped out, and weedy, fast-growing plants called lycopsids formed vast carpets that thrived for a few million years before giving way to other plants. And when ecosystems finally recovered from the mass extinctions, they were fundamentally different than before. On land, for example, ancient reptile-like relatives of mammals were dominant before the extinctions. They took a serious blow, however, and did not recover. Instead, reptiles became more diverse and dominant—including dinosaurs, which would thrive for 200 million years.

A similar pattern unfolded 66 million years ago, with the Cretaceous extinctions. After large dinosaurs became extinct, mammals came to occupy many of their niches, evolving into large carnivores and herbivores. In the oceans, mammals evolved into whales, taking the place of marine reptiles (page 8). Even as mass extinctions wipe out old biodiversity, they may open the way for the evolution of new radiations, either by wiping out predators or by clearing out ecological niches.

The New Die-off

The dodo was not alone as it headed for oblivion. Other species were also being driven extinct by humans at the same time. There are written accounts of a few hundred species that have become extinct, but scientists suspect that many others have also quietly vanished. Some scientists have tried to estimate the current rate of extinctions by focusing on groups of species that have enjoyed a lot of scientific scrutiny. Birds are one such group, because they're relatively big, bright, and adored by bird-watchers around the world.

In 2006, Stuart Pimm of Duke University, an expert on bird extinctions, tallied up the total number of bird extinctions known to have been caused by humans. He looked at historical records of birds such as the dodo, but he also included extinct birds discovered by archaeologists on islands in the Pacific Ocean. Pimm and his colleagues calculated how quickly birds were becoming extinct and compared this to background rates of extinction documented in the fossil record. They concluded that birds are disappearing 100 times faster. And Pimm warns that this rate will only accelerate in the coming decades. Many bird species that aren't extinct are already endangered, their populations vanishing thanks to hunting and lost habitat. Given the growing human population and the

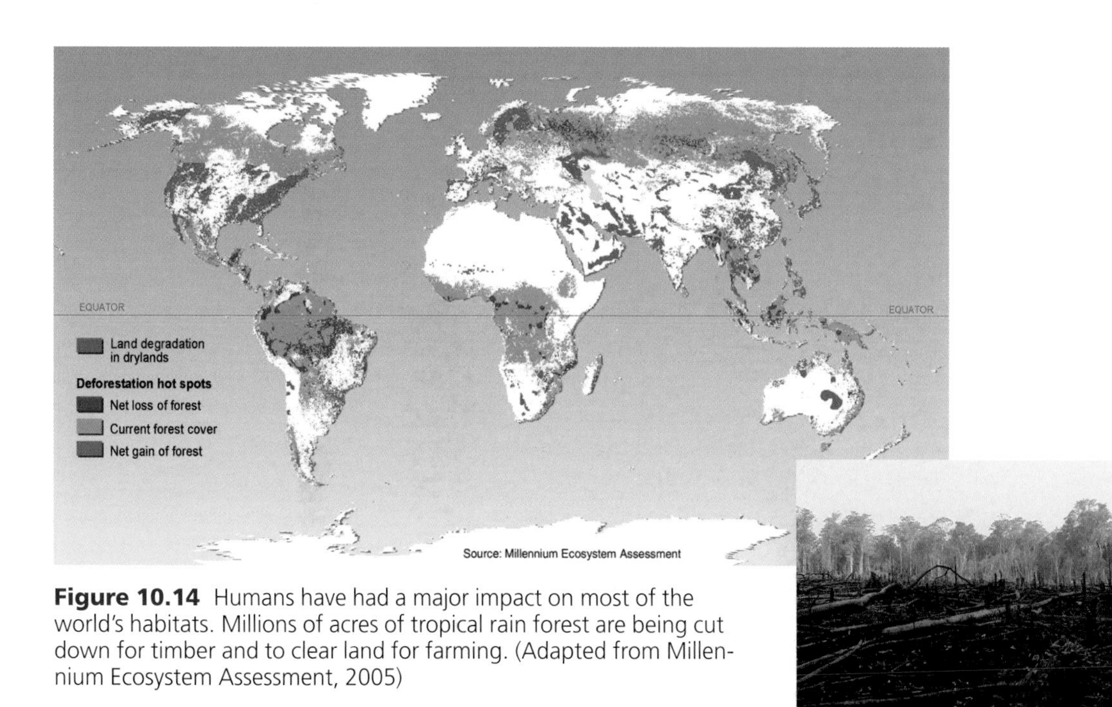

Figure 10.14 Humans have had a major impact on most of the world's habitats. Millions of acres of tropical rain forest are being cut down for timber and to clear land for farming. (Adapted from Millennium Ecosystem Assessment, 2005)

continuing deforestation (**Figure 10.14**) in many bird habitats, Pimm fears that these endangered species will become extinct as well. He predicts that, in a few decades, birds will become extinct 1,000 times faster than the background rate.

Other scientists have come up with equally grim predictions for other groups of animals and plants (**Figure 10.15**). These studies indicate that we are entering a new phase of mass extinctions on a scale not seen for 66 million years. And these studies were carried out before researchers began to grapple with another major threat to biodiversity: the release of carbon dioxide from fossil fuels.

Every year, humans release more than seven billion metric tons of carbon dioxide into the atmosphere. Over the past two centuries, humans have raised the concentration of carbon dioxide in the air from 280 parts per million in 1800 to 383 parts per million (**Figure 10.16B**). Depending on how much coal, gas, and oil we burn in the future, levels of carbon dioxide could reach 1,000 parts per million in a few decades.

This extra carbon may have two kinds of devastating effects. Carbon dioxide entering the oceans is making the water more acidic. James Zachos of the University of California, Santa Cruz, and his colleagues have studied the effects of acidic ocean water on animal life. They find that it interferes with the growth of coral reefs and shell-bearing mollusks, such as snails and clams. These animals may simply die and the reefs may disintegrate. The collapse of coral reefs could lead to more extinctions, because they serve as shelters for a quarter of all marine animal species.

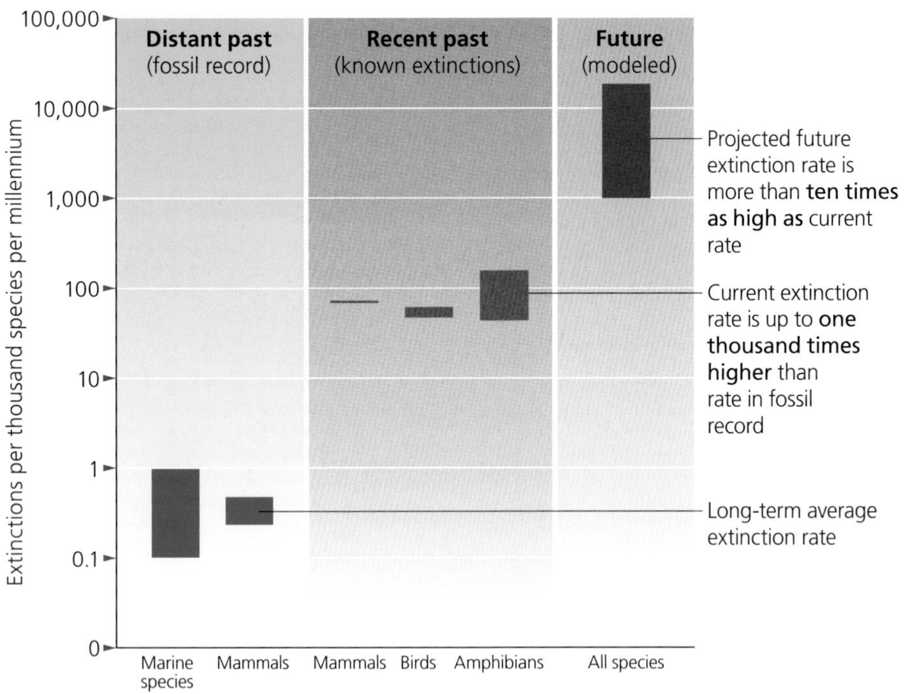

Figure 10.15 The rate of extinction is now much higher than the historical background rate. If it increases, as many scientists now predict, we are entering a new pulse of mass extinctions. (Adapted from Millennium Ecosystem Assessment, 2005)

Carbon dioxide has another effect on life: it warms the atmosphere by trapping the heat from the Sun (**Figure 10.16A**). The average global temperature has already risen 0.74 degrees Celsius (1.33 degrees Fahrenheit) over the past century (**Figure 10.16C**).

Over the next century, computer models project, the planet will warm several more degrees unless we can slow down the rise of greenhouse gases in the atmosphere. Animals and plants have already responded to the change. Thousands of species have shifted their ranges. Some species now live beyond their historical ranges, tracking the climate to which they've adapted. Other species that live on mountainsides have shifted to higher elevations.

The effects of climate change on biodiversity in the future are far from clear, but many scientists warn that they could be devastating. Among the first victims

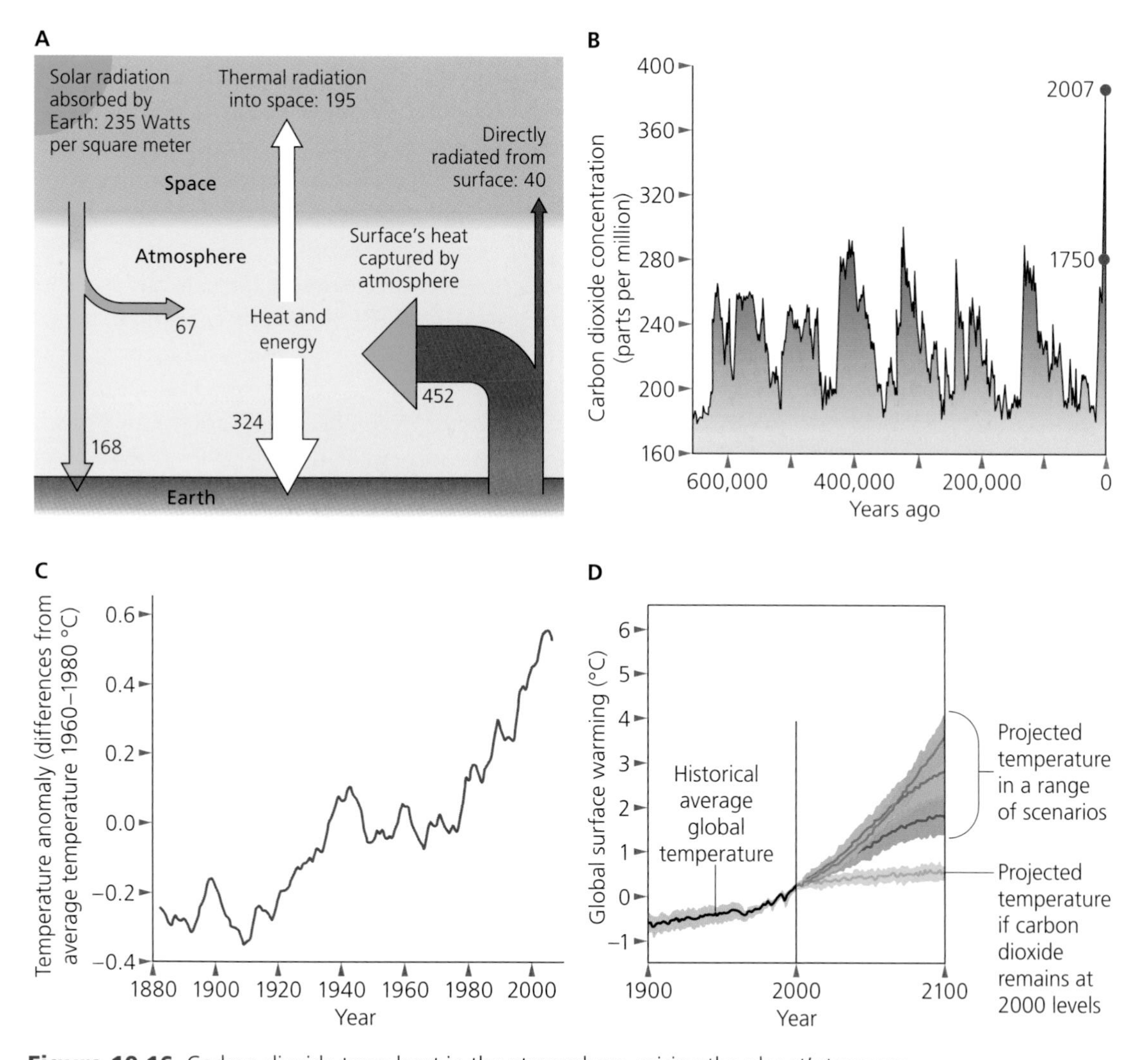

Figure 10.16 Carbon dioxide traps heat in the atmosphere, raising the planet's temperature. Humans are dramatically increasing the concentration of carbon dioxide. The rising temperature may lead to the extinction of many species.

of climate change may be mountain-dwelling species. As they move to higher elevations, they will eventually run out of refuge. In northern Australia, for example, the rare white lemuroid possum has only been found living on mountainsides at elevations higher than 1000 meters. In 2008 Australian biologists could not find a single possum and fear that it has become the first mammal to be driven extinct by manmade global warming. Polar bears and other animals adapted to life near the poles may also see their habitats simply melt away. In other cases, the climate envelope will shift far away from its current location. Some species may be able to shift as well, but many slow-dispersing species will not. Conservation biologists are now debating whether they should plan on moving species to preserve them.

It is reasonable to ask why we should care about these coming mass extinctions. After all, extinction is a fact of life, and life on Earth has endured through big pulses of extinctions in the past, only to rebound to even higher levels of diversity.

Mass extinctions are a serious matter, even on purely selfish grounds. People who depend on fish for food or income will be harmed by the collapse of coral reefs, which provide shelter for fish larvae. Bees and other insects pollinate billions of dollars of crops, and now, as introduced diseases are threatening to wipe them out in the United States, farmers will suffer as well.

Biodiversity also sustains the ecosystems that support human life, whether they are wetlands that purify water or soil in which plants grow. A single species can disappear without much harm to an ecosystem, but the fossil record shows that extinctions can lead to the complete collapse of ecosystems for millions of years. Paul Ehrlich, an evolutionary biologist at Stanford University, likens the process to someone popping rivets from a plane in flight. One or two rivets may go missing without causing trouble, but, eventually, taking away yet another rivet leads to an abrupt crash.

Coral reefs, which support much of the biodiversity of the ocean, are suffering from pollution and overfishing. As levels of carbon dioxide rise, they face further damage from the acidification of sea water.

Evolution has also generated many molecules that have enormous economic value. Antibiotics are produced by fungi and bacteria, for example. Snake venom has been adapted to treat blood pressure. The enzymes made by archaea in hot springs are now used to rapidly make copies of fragments of DNA in laboratories. Many potentially valuable molecules are waiting to be discovered. In 2007, Scott Strobel, a molecular biologist at Yale University, took 15 undergraduates on an expedition to a rain forest on the border of Bolivia and Peru. There they collected fungi and bacteria from plants and took them back to Yale to analyze. On that single field trip, the students gathered 135 species, many of which were only distantly related to any known lineage of bacteria or fungi. Strobel's students tested 88 of the new species and found that 65 of them contained molecules that could stop the growth of disease-causing microbes. If the plants on which they live become extinct, these fungi and bacteria may become extinct along with them. With every species that becomes extinct, we lose more of the raw material for essential industries ranging from biotechnology to chemical engineering to medicine.

Moreover, as species become extinct, we lose the opportunity to learn about our own evolutionary history. All of the great apes—our closest living relatives—are endangered, as they are hunted for food and driven out of forests that are being logged or cleared to make way for plantations. As we'll see in Chapter 14, some of the most important discoveries about how our own species evolved language, culture, reasoning, and even consciousness came from comparisons with those cousins now on the edge of extinction.

TO SUM UP...

- **Biogeography is the study of the distribution of species around the world.**
- **Species can move from where they originated to continents and islands. This process is known as dispersal.**
- **As continents split and drift, they can carry species with them. This process is known as vicariance.**
- **Lineages can produce new species at fast or slow rates. Lineages can also experience stasis and bursts of change.**
- **Species on average last a few million years before becoming extinct.**
- **Extinction can be caused by predation, loss of habitat, or other factors that reduce a species' population.**
- **One hypothesis for the diversity in the tropics is that the region fosters the origination of new species and allows species to last longer.**
- **Adaptive radiations are rapid diversifications of lineages.**
- **Some lineages are more diverse than others. Insects appear to be more diverse because insect species are less likely to become extinct.**
- **The early evolution of animals was a major adaptive radiation, possibly triggered by worldwide environmental changes. The struggle between predators and prey accelerated the diversification of animal lineages.**
- **The extinction rate has varied over time. On at least three occasions, there have been mass extinctions.**
- **Periods of high extinction rates coincide with major environmental changes, including asteroid impacts, volcanic eruptions, and global warming.**
- **Humans are the agents of a new pulse of extinctions.**

Intimate Partnerships

11

How Species Adapt to Each Other

Anne Gaskett, a biologist at Cornell University, spends her days crouching quietly in front of orchids in Australia. It may seem like an uneventful way to pass the time, but she is actually observing one of nature's weirdest displays of sexual deception. From time to time, a wasp lands on the orchid and tries to mate with it, convinced that it has found a female wasp.

Anne Gaskett (above) studies orchids that deceive male wasps into thinking that they're female wasps. When a male wasp tries to mate with an orchid (opposite page), it becomes covered in pollen. When the wasp tries to mate with another orchid, the insect fertilizes the flower's ovules.

The wasps are fooled by the orchids. Male wasps normally seek out females by sniffing for chemicals they produce, known as pheromones. Each species of wasp makes a unique pheromone, which means that male wasps rarely end up with the wrong females. The tongue orchids produce chemicals that precisely mimic the pheromone made by females of a single species, called *Lissopimpla excelsa.* The male *L. excelsa* wasps that pick up the scent of the orchids race to the flowers, expecting to find a mate.

Once an orchid's scent lures a wasp to its flowers, the trickery does not end. The pheromone-like scents are produced by a part of the flower that is colored like a female dupe wasp. When a male wasp lands on the flower to investigate, he finds that his body fits snugly against it, much as it would against a female wasp. The dupe wasp is so profoundly fooled that it even extends pincers called genital claspers into the flower.

One day Gaskett was watching a wasp embracing an orchid when she noticed a tiny drop of fluid on the flower when the wasp finally flew away. She wondered whether the plant or the wasp had produced it. When she and her colleagues observed other encounters between wasps and orchids, they found the same drops. They gathered some drops and analyzed them in a laboratory, applying special stains designed to give off a glow if they attached to sperm cells. The drops glowed, revealing that the wasps had actually ejaculated on the orchids. The male dupe wasps had indeed been truly duped.

Orchids do not go to such great lengths simply as a practical joke. They use the wasps to reproduce. The orchids produce a packet of pollen in just the right spot so that it gets stuck to a male dupe wasp trying to mate with the flower. When the male wasp finally flies away, it carries a packet of pollen. When it tries to mate with another orchid the wasp deposits the pollen from the original orchid flower. Without this deception, the orchid species would become extinct.

The tongue orchid's trickery is new to science, but it is an example of an evolutionary process that Charles Darwin first recognized more than 150 years ago. A species doesn't just adapt to its physical environment. It adapts to its biological environment—the species with which it shares intimate ecological relationships. A gazelle, for example, is hunted by lions and other predators; it eats grasses; it may be host for mites, ticks, and viruses. Each of the species the gazelle interacts with can raise or lower its fitness. That means that natural selection can potentially shape its relationships with all of them. The threat of lions may favor faster gazelles, for example, while the threat of a virus may favor the evolution of a particular response from the gazelle's immune system.

What makes the biological environment different from the physical environment is that it can also evolve. As a species adapts to its ecological partners, its partners can adapt to it as well. In some cases, two species may become intimately linked by their evolutionary influence on each other. Biologists refer to this reciprocal evolutionary change in interacting species as coevolution.

In this chapter, we'll consider how coevolution has shaped the natural world, from the race between parasites and their hosts to the interdependent web of life that supports the world's ecosystems.

Friends and Enemies

The relationships that species have with one another take a dizzying range of forms, from shrimp that clean fish to cuckoos that trick other birds into raising their chicks. One way to organize all this diversity is to measure the effect that one species has on another species' fitness. That effect can be positive, neutral, or negative.

Positive–Negative (Predators, Parasites, Deceivers, and Their Victims)

Predators and prey have a positive–negative relationship. Predators depend on their prey as food, while a predator can reduce its prey's fitness to zero. Parasites can also have a positive–negative relationship with their hosts. They may kill or sicken their hosts as they replicate (**Figure 11.1**).

An organism does not have to die in order to have its fitness lowered by another species, though. Fitness is not merely a matter of survival; it is ultimately about reproductive success. A caterpillar munching on the leaves of a tree may not seem to be affecting the plant's fitness. After all, the tree is still

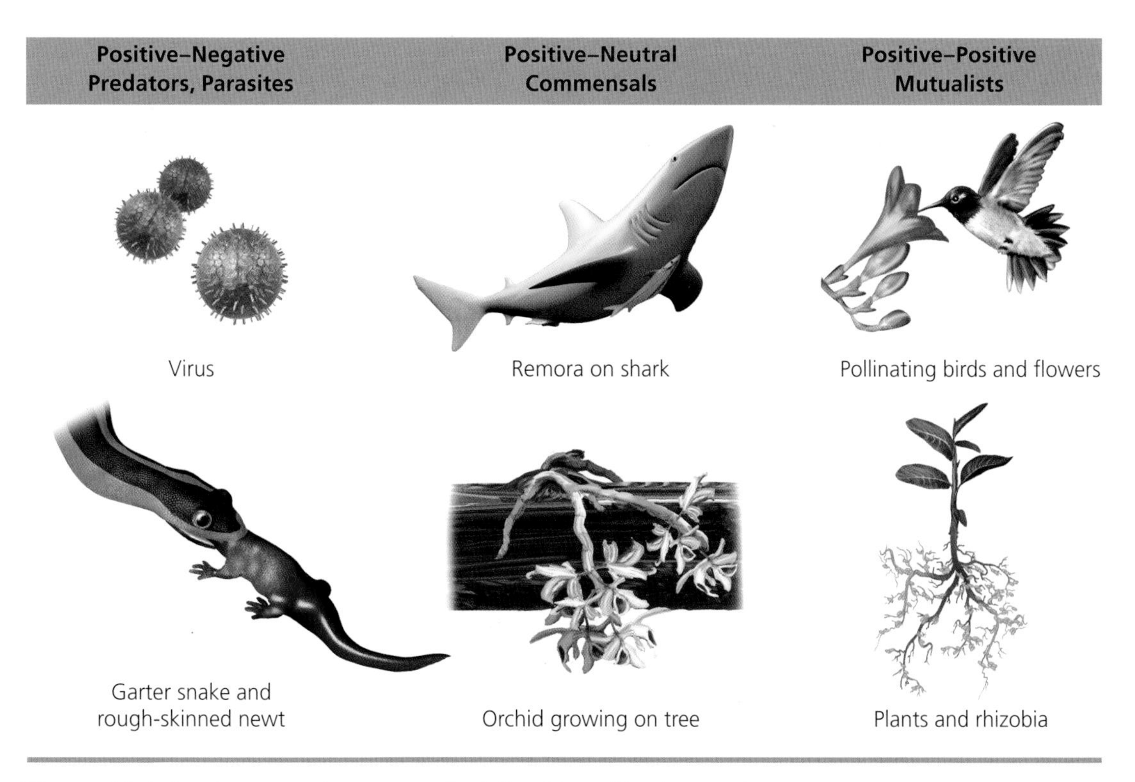

Figure 11.1 Species can evolve a range of relationships with other species.

alive when the caterpillar stops eating and develops into a moth. But insects that feed on trees drastically reduce the amount of energy the trees can capture from sunlight. By eating a tree's leaves, the insects destroy its solar panels. With less energy, the trees produce fewer seeds, which can translate into fewer descendants.

A parasitic barnacle called *Sacculina* has an equally subtle but devastating effect on the crabs it infects. The barnacle burrows into the crab's body and grows tendrils that extend through its host's tissues. The crab does not show any outward sign of being sick; it goes on searching for food and eating as it did before. But *Sacculina* destroys the crab's sexual organs, so that it can no longer reproduce. The parasite benefits, because its host no longer diverts any energy from food to eggs or searching for mates. The crab, on the other hand, is at an evolutionary dead end, its fitness having been reduced to zero.

Anne Gaskett's research suggests that the tongue orchid may also lower the fitness of the wasps that visit its flowers. If a male wasp leaves its sperm on an orchid, it may not be able to fertilize a female wasp it meets soon afterwards. Over the long term, Gaskett's research suggests, the tongue orchid lowers the reproductive success of male dupe wasps. On the whole, male dupe wasps would fare better if they weren't in a relationship with the tongue orchids. For the orchid, however, the relationship is very beneficial. Many plants can pollinate themselves, but tongue orchids can only pollinate other tongue orchids—and the only way they can pollinate those orchids is with the help of tricked wasps. So it is clear that their relationship with wasps raises their fitness rather than lowering it.

Because the interactions between the wasps and the orchids alter their fitness, they can drive natural selection in both species. Mutations that let male wasps tell the difference between orchids and female wasps should be favored by natural selection. As the wasps improve their sense of smell, natural selection will favor orchids that match the odor of wasp pheromones even more precisely.

Positive–Neutral (Commensals)

Some species depend on other species for their survival, but they have no negative or positive effect on the fitness of the partner. Remoras, for example, clamp onto shark and other fish just to catch a ride. When a host fish finds prey, its remora lets go. After the kill, the remora feeds on the scraps left behind.

Commensals can be exquisitely adapted to their hosts, just like parasites or mutualists. Pitcher plants are carnivorous plants that feed on insects that fall into their cone-shaped leaves and plunge into an enzyme-rich fluid at the bottom. But there are a number of insects, such as mosquito larvae, that feed on the decaying bodies of the dead animals inside the pitcher. These commensals have evolved adaptations that let them survive in this botanical stomach. They are so precisely adapted, in fact, that they live nowhere else.

Positive–Positive (Mutualists)

When two species interact in a way that raises the fitness of both partners, the relationship is called a mutualism. In other words, the relationship is *mutually* beneficial.

Tongue orchids deceive wasps into pollinating them, but most plants are mutualists with their pollinators. In exchange for spreading their pollen, they make nectar for birds or insects to drink. Some plants also depend on animals to spread their seeds. Birds, bats, and other animals feed on fruits. They digest the fruit pulp, and the seed passes out with their feces. Because the animals can travel long distances before releasing the seeds, they can spread plants over a wide area.

Many plants also depend on mutualists to help them get nutrients out of the soil. The roots of most plants are enmeshed in a fine web of fungal threads. These so-called mycorrhizal fungi can break down nutrients that plants cannot, pumping them into the roots to help the plants grow. In exchange, the plants pump some of the organic carbon they make through photosynthesis out of their roots to the fungi. Some plants such as beans also develop nodules in their roots that are loaded with bacteria, known as rhizobia. The bacteria can convert nitrogen from the atmosphere into a form that plants can use (ammonium, NH_4^+, the main plant nutrient in many chemical fertilizers). The plants, in turn, provide the bacteria with a steady supply of nutrients.

Many animals also form mutualisms with organisms that can break down food that they cannot. We humans have thousands of species of microbes in our guts, totaling several trillion individual organisms. Some of those species produce vitamins, amino acids, and other nutrients for us. We, in turn, provide the microbes with a warm, stable home and a steady supply of food.

The relationship between mutualists is not always purely positive. In the western United States, yucca plants can pollinate each other only with the help of visiting yucca moths. A female yucca moth will gather pollen from a yucca flower, then fly away to another yucca plant, where she will comb the gathered pollen into other yucca flowers. But then she extracts a reward for her services: she backs down into the flower and lays eggs in some of the seeds. When the eggs hatch, the larvae feed on them.

Yucca plants have evolved strategies that keep the yucca moths from taking advantage of this arrangement. Sometimes yucca moths lay too many eggs in a yucca flower. If all of the moth larvae hatch out, they will devour all the seeds, leaving none for the flower's reproduction. If a yucca moth steps over this line, a yucca plant will often abort its own seeds, killing the larvae inside them. Yucca plants also abort their seeds if the moths lay eggs without pollinating their flowers.

In both cases, the yuccas police the moths, eliminating any cheaters. They select for moths that use only an intermediate number of seeds, which is the best solution for the yucca plants. Without this enforcement, there would be no harmony between these mutualists.

Locks and Keys

Coevolution can often be diffuse, meaning that a species evolves in response to its relationship with a large number of species. Some plant species can be pollinated by several bird species, for example. Coevolution can also be specific, when two species depend on each other completely—such as yucca moths and yucca plants.

Specific coevolution can sculpt the morphology of two species like a lock and its key. Some acacia trees, for example, are protected from herbivorous insects by ants. The ants prowl up and down the trees, chasing off invaders that would harm the trees. The trees, in turn, grow hollow swellings on their branches where the ants can live. Inside these shelters, the acacias produce a sugary fluid that the ants feed on. Carine Brouat at the University of Montpellier and her colleagues studied the mutualism of three closely related species of trees in the African rain forest (*Leonardoxa* species) and their protective ants. Brouat discovered that each tree is guarded by its own species of ant, and the entrances to the shelters on each tree precisely match the size of the ants that guard it.

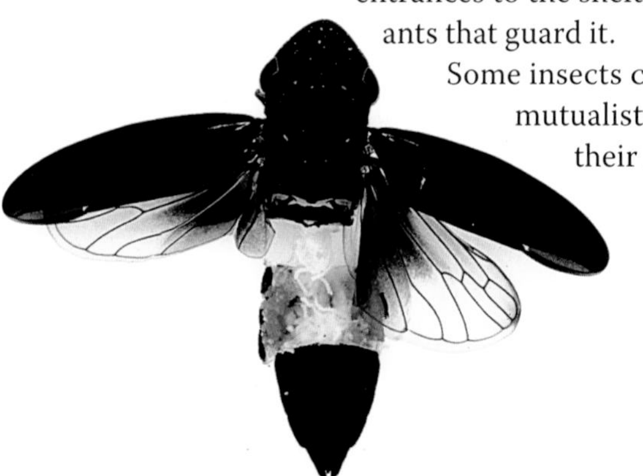

Sharpshooters have coevolved with bacteria that supply the insects with essential nutrients. The bacteria benefit by getting food and shelter from the insects. The sharpshooters even develop special organs called bacteriomes (the orange spots on the sides of the insect).

Some insects can turn their own bodies into shelters for their mutualists. Sharpshooters are large insects that push their needle-like mouthparts into plants to drink the fluid in woody fibers called xylem. Unlike fruit or seeds, this fluid is a poor source of nutrition, and so the sharpshooters have to drink huge amounts of the stuff, excreting much of it in little droplets that shower down like rain from trees. To extract what little nutrition there is in the fluid, the sharpshooters depend on several species of bacteria. The bacteria live only in a pair of special organs called bacteriomes, which look like a pair of orange spots on the back end of the insect. The cells in the bacteriomes can house huge numbers of the bacteria, which the insects then pass down to their offspring.

Mirror Trees

The history of interacting species can be recorded in their evolutionary trees. In Chapter 7, we saw how the closest relatives of HIV are viruses that infect certain populations of chimpanzees and monkeys. This pattern was the result of these viruses changing hosts over the past century. But other partners have remained linked together for millions of years. Evolutionary trees can reveal these ancient associations. Species of gophers, for example, carry particular species of lice. When biologists compare the evolutionary trees of the gophers and their lice, they see a mirror-like symmetry (**Figure 11.2**). That suggests that when a population of gophers became isolated from the rest of its species and evolved into a

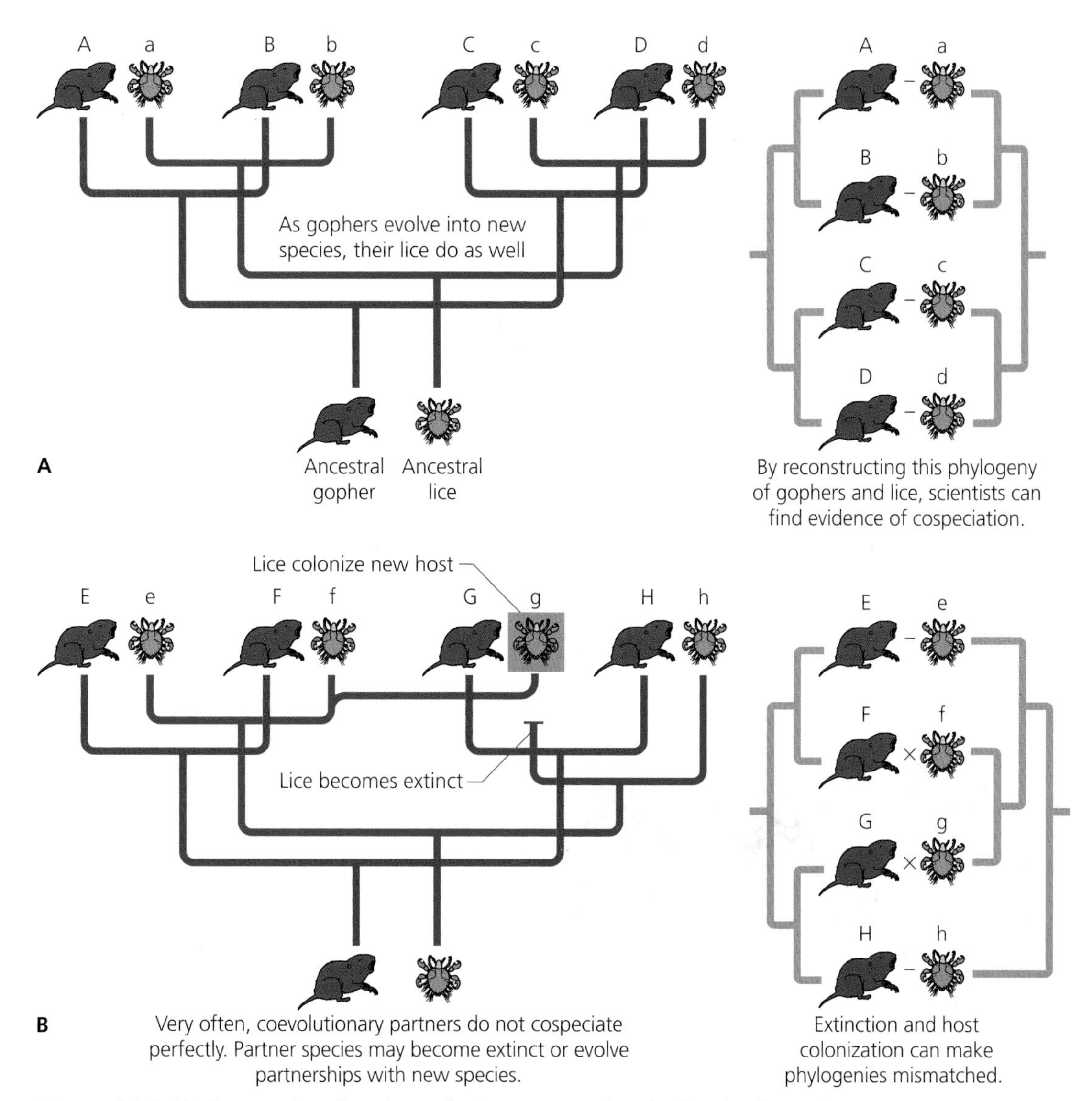

Figure 11.2 (A) As a species of gopher splits into new species, the lice that live on it diverge into new species as well. This cospeciation is recorded in the evolutionary trees of the parasite and its host. (B) The lice on gopher species G become extinct, and some lice from species f colonize them. This extinction and species-hopping breaks the mirror symmetry of the evolutionary trees.

new species, its lice became a new species as well. This process is known as cospeciation.

These mirror trees can also shed light on the steps by which intricate partnerships have evolved. Consider, for instance, the bacteria that live inside the sharpshooter and allow it to live on xylem fluid. One of these bacteria is known as *Sulcia*. Nancy Moran, an evolutionary biologist at the University of Arizona, and her colleagues have found *Sulcia* in sharpshooters, as well as their closest living relatives, a group that includes other sap-feeding insects with bacteriomes. But the relationship is much more intimate, as Moran found when she compared the evolutionary tree of *Sulcia* strains with that of the insect species in which they live. The two trees match almost perfectly (**Figure 11.3**). The common ancestor of these insects, Moran concluded, acquired *Sulcia*. Judging from the earliest fossils of the insect hosts, this mutualism must have arisen 270 million years ago at the latest. As new lineages of insects evolved, their strains of *Sulcia* also evolved.

The lineage that produced today's thousands of sharpshooter species evolved roughly 50 million years ago, judging by the oldest sharpshooter fossils. The early sharpshooters shifted from other plant fluids to xylem sap. They also

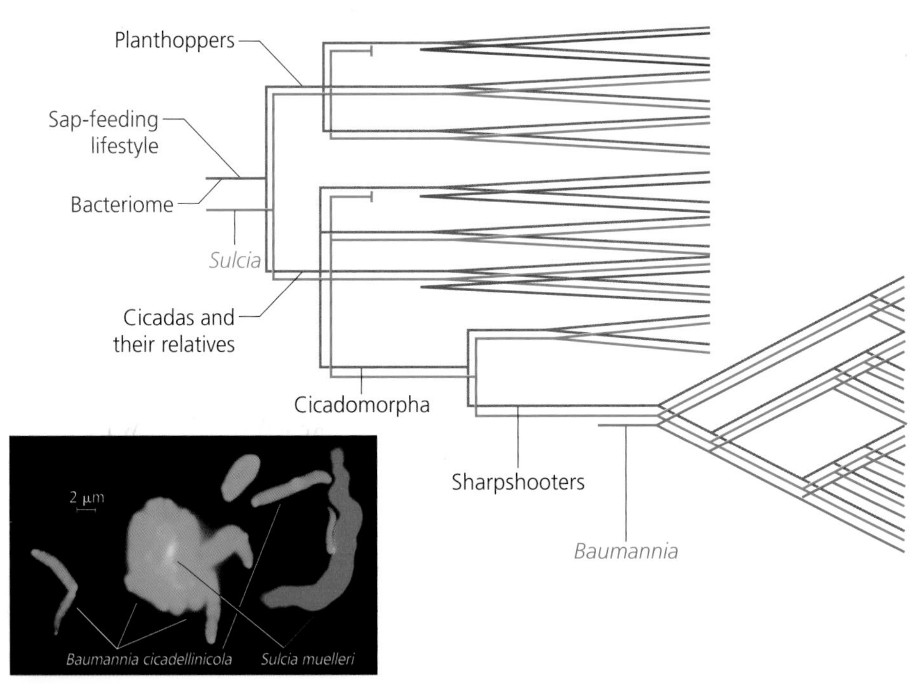

Figure 11.3 Sharpshooters rely for their survival on a collection of bacteria, including *Sulcia* and *Baumannia* (shown here in a micrograph). Studies on the DNA of the insects and their bacteria have revealed that *Sulcia* was already living inside the sap-feeding ancestor of sharpshooters and related insects some 270 million years ago. In some lineages, *Sulcia* was lost and new bacteria came to take its place (blue, purple, and green branches). In the more recent ancestors of sharpshooters, *Baumannia* joined *Sulcia* inside the insects, allowing their hosts to feed on xylem, a very poor source of nutrition. (Adpated from Moran, 2007)

became hosts to a new mutualist, a bacterium called *Baumannia cicadellinicola*. The evolution of *Baumannia* tightly tracked sharpshooter evolution as well.

Over the course of this long, intimate coevolution, *Sulcia* and *Baumannia* have changed in some significant ways. They combined their efforts inside the insect. *Sulcia* provides its insect host with amino acids, while *Baumannia* takes care of vitamins and other nutrients. Each microbe has lost many of the genes that it does not use for its particular tasks. For example, *Baumannia* has lost all of the genes for producing amino acids, except for the gene encoding histidine. *Sulcia*, however, has retained all of the genes for producing amino acids, except for one: the one encoding histidine. In other words, these two microbes have not only coevolved with their insect host, they have coevolved with each other.

Natural Arms Races

In the 1950s, a story circulated in Oregon about three hunters who died mysteriously at their campsite. The only peculiar thing about the campsite was a rough-skinned newt that had crawled into the coffee pot. A biologist named Edmund Brodie decided to investigate. He discovered that the newt produced a powerful toxin in its skin. The toxin could slip through channels on the surface of neurons and paralyze its victims. Brodie and his colleagues calculated that a single newt made enough toxin to kill 100 people. The Oregon hunters had probably died after drinking the poisoned coffee.

These toxins are obviously a powerful deterrent against predators, and yet rough-skinned newts still face a dangerous enemy. Garter snakes can eat them without dying. Brodie and his colleagues discovered that the snakes have evolved neuron channels with a distinctive shape that makes it more difficult for the toxins to enter.

The extreme adaptations of the rough-skinned newt and the garter snake are an example of what scientists call a coevolutionary arms race: One coevolutionary partner—a prey species, in this case—evolves a new defense. Now natural selection favors variations in the predator that allow it to overcome the defense. Any prey that can boost its defenses further will now, in turn, be favored by natural selection. Back and forth the arms race goes, until both partners reach the sorts of extremes seen in the rough-skinned newt and the garter snake.

The idea that coevolution could escalate this way was first proposed in 1964 by two biologists named Peter Raven and Paul Ehrlich. They were inspired by their studies of the ways milkweed plants defend themselves from their insect enemies. They grow hairs that make it difficult for the insects to reach down to their tissues. If a caterpillar does manage to bite into a milkweed, sticky white fluid bursts out of the plant (the "milk" that gives milkweed its name). And even if the insect should still manage to keep eating, the plant produces a cocktail of toxic molecules that can cause the caterpillar serious harm.

Yet the caterpillars have many of their own defenses, Raven and Ehrlich recognized. They can disarm the toxins in the milkweed. They can sabotage the milk defense by cutting holes in the vessels through which it flows. Raven and Ehrlich proposed that each time the milkweed evolved a new defense, natural selection favored a new counterdefense from the insects, which led to the evolution of more defenses in the milkweed.

The arms-race theory helps to explain many of the puzzling patterns of coevolution. Take, for example, the tongue orchids and dupe wasps. Dupe wasps do not always fall for the orchid's trick. In fact, male wasps can learn to avoid the flowers. Avoiding orchids means not wasting sperm, which may translate into a higher reproductive fitness for the smart male wasps. They, in turn, will pass on more of their genes to their offspring. The selection for wasps that are less easily fooled, Gaskett suggests, may drive the evolution of more sophisticated deception in the orchids. If so, only those orchids that do a better job of smelling, looking, and feeling like wasps will succeed in getting pollinated.

Arms races may last for millions of years, but they can also come to an end—or, rather, the two evolutionary enemies may shift their strategies. Recently, Anurag Agrawal of Cornell University and Mark Fishbein of Portland State University took a new look at the coevolution of milkweeds and insects. They constructed the evolutionary tree of 38 species of milkweed and traced the evolution of their defensive traits. In 2008, they reported a trend in the evolution of the milkweeds: the plants have evolved to grow back more vigorously after being attacked by insects. Meanwhile, the defenses of the milkweeds—the toxins, latex, and sticky hairs—have been getting weaker. Agrawal and Fishbein argue that milkweed plants have slowly been shifting towards a new strategy. Instead of resisting the insects, they are evolving to grow back faster after being feasted on.

Cheaters in the Commons

In the Middle Ages, cattle herders could graze their cows in public meadows known as commons. But this arrangement created the opportunity for tragedy. In the long term, all the cattle herders would be better off if each one only grazed the commons lightly. That way, the grass would be able to stay healthy. For each individual herder, though, the best short-term strategy was to graze his cows as much as possible to produce the most milk and meat to sell. If everyone followed the short-term strategy, they would overgraze the commons and destroy it.

This so-called tragedy of the commons applies not just to humans but to coevolutionary partners as well. Cheating can be a very successful strategy. One of the most striking examples comes from rhizobia, nodules of bacteria that live in plant roots. The best long-term strategy for all of the bacteria is to provide the plant with nitrogen so that it can grow and return more nutrients to the bacteria.

But fixing nitrogen demands energy. Some of the bacteria may mutate so that they don't supply the plant with nitrogen. In other words, they cheat by taking nutrients from the plant without returning any service in exchange. Natural selection ought to favor these cheaters, because they can use their extra energy to grow faster and outcompete the other bacteria.

Why don't the cheaters win? Mathematical models suggest that cheating can be squelched if the plant imposes sanctions in response. R. Ford Denison, a University of Minnesota biologist, and his colleagues tested the sanction hypothesis in 2003 with an elegant experiment. They grew soybean plants with rhizobia in special chambers, replacing the nitrogen in the chamber air with argon, an inert gas. The plants could still survive, but now the bacteria could not provide them with any nitrogen. Denison had, in effect, turned all the rhizobia into cheaters.

The plants responded with a brutal punishment. The cheating rhizobia could form only about two-thirds as many nodules on each plant as true mutualists. Denison's team found that the nodules contain less oxygen than normal ones. The scientists suggest that plants can shut down the supply of oxygen to cheating bacteria to slow their spread, giving an evolutionary advantage to true mutualists.

A Geographic Mosaic

Coevolution can produce some exquisitely precise adaptations, like the chemicals made by tongue orchids that attract the males of a single species of wasp. But coevolution does not produce these adaptations in a simple assembly-line process, like a factory churning out locks and matching keys. Coevolution is messy. Its raw ingredients are mutations, recombinations, and other sources of genetic variation. A particular mutation may boost the fitness of a particular parasite or mutualist, and natural selection may then spread it through a population. But genetic drift can also spread mutations that are neutral or even harmful. Natural selection may adapt one population to one set of local conditions, but act differently on another population. Organisms can then migrate from one population to another, mixing genes together and blurring precise adaptations between coevolutionary partners.

In the 1990s, John Thompson of the University of California, Santa Cruz, and his colleagues developed a theory to account for this complexity, which they dubbed the geographic mosaic theory of coevolution. According to their theory, a pair of coevolutionary partners do not have the same relationship everywhere across their range. Thompson proposed that there are coevolutionary "hotspots" where they are coevolving rapidly, as well as coevolutionary "coldspots" where little coevolution is occurring. In some parts of their range, the species may be precisely adapted to each other, but mismatched elsewhere. At any moment, Thompson argued, a map of two coevolutionary partners is a mosaic, made up of many populations in different relationships. Over time, the mosaic changes its pattern, as genes flow between populations and alter their coevolutionary relationships.

Testing Thompson's theory proved to be a challenge. Scientists had to study coevolutionary partners across much bigger ranges than they had in the past. But those experiments are now finally yielding results, and they generally support the geographic mosaic theory of coevolution. In the Rocky Mountains, for example, birds known as Clark's nutcrackers eat seeds from pine trees. The pine trees, in turn, depend on the birds to spread their seeds in their droppings. A single Clark's nutcracker can spread 98,000 seeds as far as 22 kilometers (14 miles). Craig Benkman, an evolutionary ecologist at the University of Wyoming in Laramie, and his colleagues have found that some populations of pine trees have adapted to the birds. They grow pine cones with thin scales and abundant seeds, an arrangement that allows the birds to feed quickly.

But Benkman has found other populations of pine trees that are maladapted to Clark's nutcrackers. Their cones produce fewer seeds, which are lodged between thick scales. The cause of this mismatch, Benkman discovered, was the presence of another animal with a taste for the seeds: red squirrels. While Clark's nutcrackers spread pine seeds, red squirrels hoard them in caches, where few of the seeds manage to germinate and grow into trees. Pine trees that made their seeds easier for birds to eat would lose many of their seeds to the squirrels.

Meanwhile, in South Africa, biologists Steven Johnson and Bruce Anderson have documented another geographic mosaic in the coevolution of long-tongued flies (*Prosoeca ganglbaueri*) and the flowers they pollinate. The tongues of the flies are actually tubes that they usually keep coiled under their head. When they find a flower from which they want to drink nectar, they inflate their tongue, stretching it out like a balloon until it is several times longer than their entire body.

These flies use their extraordinary tongue to feed on the nectar of long-necked flowers, especially the flowers of a species called *Zaluzianskya microsiphon*. As they try to push their tongues to the bottom of the deep nectaries, the flies rub their heads on the flowers, picking up pollen. When the flies visit another flower, the pollen from the first plant can fertilize the second one's ovules.

Anderson and Johnson traveled to 16 sites in South Africa, where they caught hundreds of long-tongued flies. They measured the length of each fly's tongue, along with many other traits of the flies. At the same 16 sites, they also gathered *Zaluzianskya* flowers, measuring them as well. The scientists found a striking pattern. At some sites, the flies have tongues as long as 5 centimeters (2 inches), closely matching the depth of the *Zaluzianskya* flower tubes at those sites. At other sites, the flies have tongues only half that length. Their shorter tongues are matched by shorter flower tubes.

Anderson and Johnson argue that natural selection must be creating the local tongue lengths. They suspect that *Zaluzianskya* flowers at some sites dominate the sources of nectar for the flies. Natural selection favors deeper tubes for nectar in these places, because it ensures that the flies will pick up pollen as they struggle to reach their tongues into the tubes. As the flower tubes get deeper, natural selection favors longer tongues. But in other populations, the flowers may have to compete with other species with smaller flowers. If a fly can't reach to the bottom of a *Zaluzianskya* flower, it can always drink from a flower of another species. At these sites, natural selection may favor shorter tubes on *Zaluzianskya* flowers.

Mutualists are not the only organisms to form these mosaics. So do rough-skinned newts and the garter snakes that eat them. The maps in **Figure 11.4** chart the toxicity of the newts and the resistance of the snakes across western North America. There's a tremendous range in these two traits. In some places the newts produce very deadly toxins, and the snakes are highly resistant. In other places, the snakes have no resistance to speak of, and the newts produce only barely detectable levels of toxins. But the newts and snakes are also mismatched in about a third of their territory. At many sites, the snakes are resistant, but the newts have weak toxins. It's possible that this mismatch evolves because the snakes can become resistant very quickly. With the change of a single amino acid, the neuron channels of the snakes can block the newt toxin very

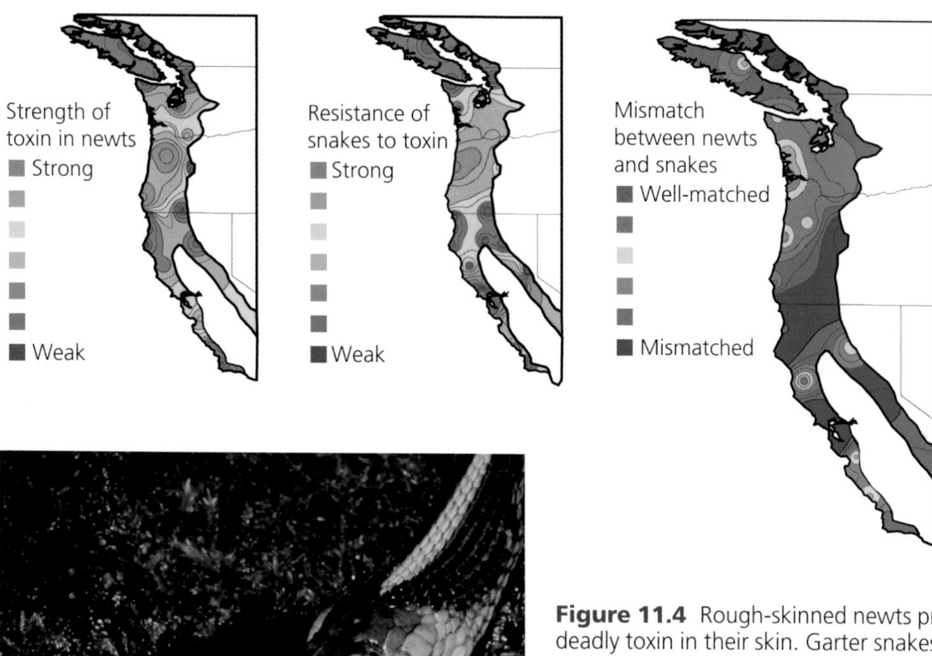

Figure 11.4 Rough-skinned newts produce a deadly toxin in their skin. Garter snakes have evolved so much resistance to the toxin that they can devour the newts. In some places, the newts have evolved even stronger toxins as a result. These maps show the strength of newt toxins and the strength of snake resistance in western North America. Across much of their range, the newts and snakes are poorly matched. The strength of toxins and resistance are well matched only in the orange and red regions. (Map adapted from Brodie et al., 2008)

well. Newts, on the other hand, have to acquire a series of mutations to make their toxins more deadly. Brodie and his colleagues argue that when a population of snakes becomes highly resistant, the arms race is essentially over. No small increase in the strength of a newt's toxins will make it any less likely that the newt will be eaten.

Engines of Diversity

Coevolution helps boost the world's biodiversity. Biodiversity exists at many scales—in the number of major lineages of life, in the number of species, and in the diversity of populations within species. Biodiversity emerges as mutations arise and spread in populations. These mutations may spread through genetic drift or through natural selection in response to the climate or some other physical aspect of the environment. But coevolution also spreads new mutations. Even if a species is well adapted to its physical environment, it may still change because it is coevolving with its parasites, mutualists, and predators.

Coevolution can also help drive some populations to evolve into distinct species. One of the few tree species to survive in the Mojave Desert is the Joshua tree (*Yucca brevifolia*). In the western part of its range, it is pollinated by a species of yucca moth called *Tegeticula synthetica*; in the east, it is pollinated by another species called *Tegeticula antithetica*. Olle Pellmyr, a biologist at the University of Idaho, and his colleagues have reconstructed the evolution of the tree and the moths on which it depends. They found that the ancestor of both moth species was already pollinating Joshua trees. Only later did the two species divide, each specializing on one population of the trees. In response, the scientists have found, Joshua trees have evolved canals in their flowers to match the length of the tongues of their own species of yucca moth. If the trees and moths continue to diverge, there will be less and less opportunity for pollen to move from one population of Joshua trees to another. It could set the trees on the path to speciation.

Over millions of years, this kind of coevolution can have a profound impact on biodiversity. Szabolcs Lengyel, a biologist at North Carolina State University, and his colleagues recently took a look at the diversity that arises from the coevolution of flowering plants and the ants that spread their seeds. About 11,000 known plant species around the world grow fleshy handles on their seeds called elaiosomes that serve as food for the ants. After ants bring the seeds to their nests and eat the elaiosomes, they discard the seeds in a special room in their colony. There the seeds can sprout, protected from being eaten by other animals. Elaiosomes have evolved independently at least 101 times, as Lengyel and his colleagues reported in 2009. They also found that ant-dispersing lineages contain over twice as many species as the most closely related lineages of plants. Ants may foster plant diversity by protecting seeds and by keeping them growing in a small range around their colonies, causing them to become reproductively isolated.

The world's biodiversity is now experiencing mass extinctions on a scale rarely seen in the history of life. Some scientists are trying to understand how coevolution will affect which species survive and which disappear. If one species depends on another one for its survival, then it will not be able to endure after the other species becomes extinct. If it can shift to a new partner, however, it may be able to survive.

Mass extinctions in the past offer some clues to how coevolution makes species vulnerable. Some species of corals live mutualistically with algae, while some do not. In the last major mass extinctions, 66 million years ago, the mutualist corals suffered about four times more extinctions than the nonmutualists. The mass extinctions coincided with a huge asteroid impact that blocked out the light of the Sun for months. It's possible that this black-out killed off photosynthetic algae, as well as the corals that depended on them for survival.

Today we can see a growing number of species that have lost their coevolutionary partners. In Central America and South America, a number of trees grow giant fruits with massive seeds at their core. Daniel Janzen, a biologist at the University of Pennsylvania, has argued that these giant fruits are the result of coevolution with giant mammals, such as ground sloths. The mammals ate the fruits and spread the seeds in their droppings. Around 12,000 years ago, giant sloths and other big mammals disappeared, probably hunted by newly arrived humans. The trees survived, still producing their giant fruits. They can be dispersed today by cows and rats, but not as successfully as they were in the past.

In coming years, global warming may also put strains on many partnerships. Rising temperatures are changing the ranges in which species can survive, in some cases causing their ranges to shrink to dangerously small sizes. Animals and plants are also responding to the changing climate by becoming active earlier each spring. As the climate continues to change, animals and plants will continue this shift. But some species are shifting faster than others.

In 2007, Jane Memmott of the University of Bristol and her colleagues looked at the plants and their pollinators in western Illinois in order to project the effects that climate change will have on their mutualisms. They predicted that between 17% and 50% of all pollinators will face a disrupted food supply. In some cases, plants will flower too early for the insects, running out of nectar while the insects still need to eat. In other cases, the insects will face gaps in their food supply as some flowers run out of nectar while others have not yet bloomed. Memmott and her colleagues predicted that some populations of pollinating insects will become extinct, while the populations of some plants will decline because they can no longer spread their pollen. Some plants will withstand this change because they can be pollinated by several insect species, while some insects will be able to shift to new flowers. But scientists don't yet know exactly how this complicated interplay of coevolution will play out in the years to come.

Two Species Become One

A typical human cell is packed with thousands of sausage-shaped structures known as mitochondria. They are essential to our survival, using oxygen, sugar, and other molecules to produce energy for the cell. Mitochondria also carry out other important jobs for our cells, such as building clusters of iron and sulfur atoms that are then attached to certain proteins. What makes mitochondria puzzling is the fact that they seem very much like little cells within our cells. They are surrounded by two membranes and carry their own DNA, which replicates as they divide.

At the dawn of the twentieth century, Russian biologists proposed that mitochondria were once free-living, oxygen-consuming bacteria that entered a single-celled host. That host was an early eukaryote, and its descendants today include animals, plants, fungi, and protozoans. Their proposal was generally forgotten, but, in the 1960s, Lynn Margulis, a biologist at the University of Massachusetts, revived it. In the 1970s, scientists were able to test her hypothesis by examining bits of mitochondrial DNA. This DNA did not closely resemble any human genes, nor any genes of any animals, or even any eukaryotes. The closest matches came from bacteria.

Today, scientists know much more about mitochondria, and about their closest living relatives among bacteria (**Figure 11.5**). It appears that all mitochondria in eukaryotes—from shiitake mushrooms to poplar trees to humans—all evolved from a single bacterial ancestor. The evidence also indicates that mitochondria had already evolved in the common ancestor of all living eukaryotes.

For many years, scientists thought mitochondria had evolved much later. They based this hypothesis on the fact that some single-celled protozoans lack mitochondria. Their ranks include *Giardia lamblia*, a protozoan that can cause painful diarrhea. It appeared that only after *Giardia* and other mitochondria-free eukaryotes branched off did bacteria become mitochondria.

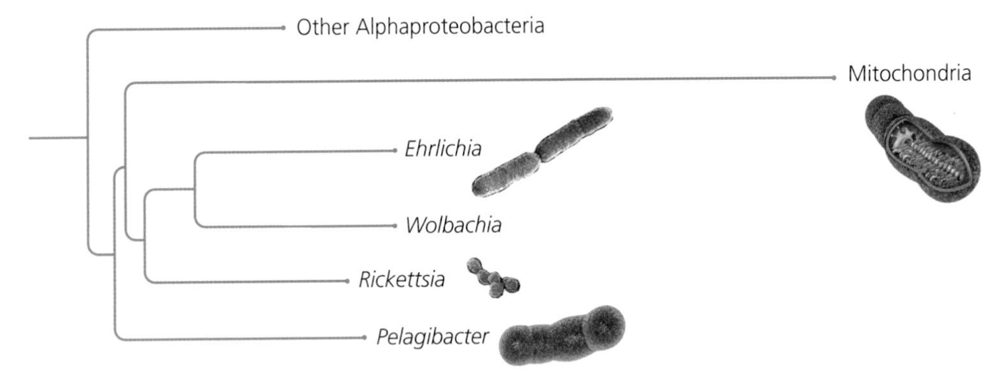

Figure 11.5 Mitochondria generate energy for our cells. These sausage-shaped structures were originally free-living bacteria that were later engulfed in our single-celled ancestors. They are now present in the cells of most eukaryotes. This evolutionary trees shows the relationship of mitochondria to their closest bacterial relatives, based on a study of their DNA. (Tree adapted from Williams, 2007)

But most experts now reject this hypothesis based on recent research on eukaryotes, including studies on *Giardia*. In 2003, Jorge Tovar, of Royal Holloway College in England, and his colleagues discovered proteins in *Giardia* that were very similar to the proteins in mitochondria that build iron and sulfur compounds. The scientists manipulated the proteins so that they would light up inside *Giardia*. It turned out that the proteins all clumped together in a tiny sac that had, until then, gone unnoticed.

Tover and his colleagues proposed that these sacs, which they dubbed mitosomes, are vestiges of full-blown mitochondria. As *Giardia* adapted to an oxygen-free life in the intestines of animals, it lost its ability to use oxygen and its mitochondria became mitosomes. Similar results have emerged from other supposedly mitochondria-free eukaryotes. They have genes, proteins, and compartments that all show signs of being remnants of full-blown mitochondria.

The most compelling hypothesis that accounts for these results is that oxygen-consuming bacteria took up residence inside some of the first eukaryotes some two billion years ago. Judging from the fact that single-celled eukaryotes today are loaded with bacteria, such an infection would have been routine. Early eukaryotes exploited the energy provided by their new resident, and they gradu-

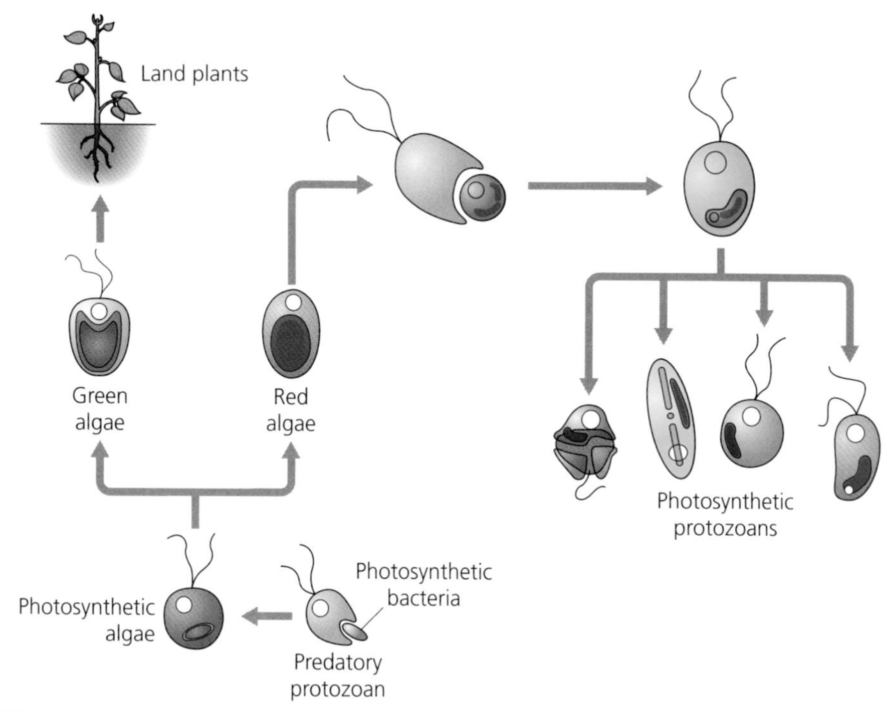

Figure 11.6 Plants can use sunlight to grow with the help of bacterial partners. Lower left: A single-celled ancestor of plants engulfed photosynthetic bacteria. The descendants of these photosynthetic cells evolved into several living lineages, including red algae and green algae (upper left). One lineage of green algae evolved into land plants. Upper right: one lineage of red algae were engulfed by another host, which evolved into many lineages of living photosynthetic protozoans. (Adapted from Bhattacharya, 2007)

ally abandoned their own energy-generating proteins. From those early hosts, all of the lineages of living eukaryotes have evolved.

Mitochondria today are not typical bacteria by any measure. They would not survive outside cells for a second. That's because they've been evolving inside eukaryote cells for billions of years. As they replicated inside their host cells, their DNA mutated. Some of those mutations deleted genes that were no longer necessary to their survival, now that they were protected inside another cell. Other mutations caused genes to be transferred from the mitochondria to the DNA in the nucleus of the cell.

Another coevolutionary merger took place hundreds of millions of years later, in the ancestors of plants (**Figure 11.6**). Plants descended from single-celled eukaryotes that originally were most likely microscopic predators, feasting on bacteria. At some point, however, they became hosts to bacteria that could carry out photosynthesis. At first they may have gotten energy both from eating other organisms and from capturing sunlight. Some photosynthetic eukaryotes live this way today in the ocean. But the ancestors of plants gradually shifted over to rely entirely on photosynthesis. Along with their mitochondria, they also carry remnants of the photosynthetic bacteria, called plastids. Every patch of living green you see, from a blade of grass to a forest of redwoods, got its start with a merger of two species.

Invasion of the Genomic Parasites

In 2006, Thierry Heidmann, a researcher at the Gustave Roussy Institute in Villejuif, France, resurrected a virus that had been dead for millions of years. Heidmann and his colleagues did not discover the virus buried in ice or hidden in a cave. They found it in the human genome. All human beings on Earth carry remnants of the virus's genetic sequence in their own DNA. This spectacular revival has helped scientists to understand one of the strangest yet most important forms of coevolution: the coevolution that takes place between different parts of our genome.

The virus that Heidmann revived belongs to a group known as retroviruses. They infect their hosts by creating an RNA copy of their genome, which is then turned into DNA that is inserted into the genome of a host cell. Typically, these embedded viruses hijack the biochemistry of their host, using it to produce new viruses that then burst out of the cell. But scientists have also discovered retrovirus-like DNA that is a permanent part of the human genome, passed down from one generation to the next.

It's likely that these viruslike stretches of DNA descend from retroviruses that infected sperm or egg cells. An organism produced from one of those infected sex cells carried the virus in all the cells of its body, including its own sex cells. Over the course of many generations, mutations to this viral DNA robbed the viruses of their ability to make new viruses that could escape their host. The best they could manage was to make copies of themselves, which could be inserted back into the same cell's DNA. Eventually, even that trait was lost, and the virus's DNA became inert.

To test this idea, Heidmann tried to revive one of these retroviruses into its once-active form. He and his colleagues selected a viruslike segment of DNA found only in humans. They found slightly different versions of the segment in different people. These differences presumably arose as the original retrovirus mutated in different lineages of humans. Heidmann and his colleagues compared the variants to determine what the original sequence had been. They built a piece of DNA that matched the original sequence and inserted it into a colony of human cells reared in a petri dish. Some of the cells produced new viruses that could infect other cells. Heidmann named the virus Phoenix, for the mythical bird that rose from its own ashes.

Phoenix has been found only in humans, which indicates that the virus infected our ancestors after they had branched off from the apes some seven million years ago. We share other endogenous retroviruses with other apes, as well as with monkeys in Africa and Asia. These viruses must be much older, because the common ancestor of all these primates lived about 30 million years ago. After the ancestors of those viruses infected early primates, they continued to replicate and to insert new copies back into the genomes of their hosts. There are almost 100,000 fragments of endogenous retrovirus DNA in your own genome, making up about 8% of your DNA. They take up about four times more of the human genome than the 20,000 genes that encode proteins.

Endogenous retroviruses were not the first "jumping genes" scientists have discovered in the genome. In the 1950s, biologist Barbara McClintock was studying the genes that control the color of corn kernels. She discovered that the genes could move within the corn genome from one generation to the next. In 1983, she was awarded the Nobel Prize for her work.

Later generations of scientists discovered a vast menagerie of DNA elements that can move through the genome. About half of the human genome is made up of these mobile elements, which number in the millions. While most mobile elements are "dead"—that is, they cannot replicate themselves—a few of them still do. One out of every 20 to 100 human babies acquires a new insertion of a mobile element.

It appears that at least some of these mobile elements got their start as retroviruses. Mutations deleted much of their original DNA, leaving behind just the bare minimum instructions for making new copies of themselves that could be reinserted into the host genome. Other mobile elements may have entered the genomes of our ancestors by hitchhiking with viruses from other species.

Mobile elements and endogenous retroviruses often behave—and evolve—like genomic parasites. As they spread within a host genome and hop to new hosts, they can harm their hosts. Mobile elements that insert themselves into new places in the genome can disrupt a cell's normal rhythms of growth and division. A cell may begin to multiply out of control, giving rise to cancer. If a mobile element inserts itself in the middle of an essential gene, a cell may no longer be able to produce a vital protein.

Experiments on mobile elements in *Drosophilia melanogaster* have allowed scientists to measure the effect of mobile elements on fitness. Flies with two copies of a mobile element called *mariner* lived for 57.6 days on average, while *mariner*-free flies lived for 61.4 days. In other words, the *mariner* mobile element cuts short the life of its host.

By lowering the fitness of their hosts, mobile elements and endogenous retroviruses lower their own fitness, because there will be fewer hosts to carry them. But many of these genomic parasites can replicate themselves so quickly that they can still spread through a population despite the harm they cause. Mobile elements demonstrate how selection can take place at several levels at once. A mutation can raise the fitness of a mobile element, even as it lowers the fitness of the organism that carries it.

The coevolution of genomes and genomic parasites is as complex as the coevolution of two free-living species. The genomic parasites evolve ways to spread themselves, and host genes evolve to halt them. Sometimes DNA that originally behaved as a parasite takes on a new function that benefits the host. These "domesticated" parasites blur the line between coevolutionary partners even more than mitochondria and plastids.

Ironically, one of these domesticated parasites helps us to fight off other parasites. The immune systems of sharks, bony fishes, and all land vertebrates are able to recognize a vast number of pathogens. They do so thanks to a set of genes that can produce receptors and antibodies with an equally vast number of different shapes. To generate these molecules, our immune cells must cut apart the corresponding genes and then join them together. Depending on what gets cut out, the genes produce molecules of different shapes.

The genes that encode the cutting proteins are called *Rag1* and *Rag2*. In 2005, Vladimir V. Kapitonov and Jerzy Jurka, two geneticists at the Genetic Information Research Institute in Mountain View, California, discovered that the two genes were significantly similar to a family of mobile elements called Transib. Transib mobile elements don't just resemble *Rag1* and *Rag2* in their DNA sequence. They also cut and paste DNA in the same way. Kapitonov and Jurka argue that in some ancient fish that lived 500 million years ago, Transib mobile elements mutated so that they began to cut and paste immune-system genes. Now, instead of being a burden to their hosts, they helped their hosts fight off disease.

When Darwin first recognized coevolution, he saw its effects on separate partners—on flowers and bees, for example, or on predators and their prey. Today, however, scientists can see coevolution's marks within us. Our genome has emerged out of a coevolutionary history of cooperation and conflict.

TO SUM UP...

- Species in ecological relationships adapt to one another in a process known as coevolution.
- Species can have a range of relationships with each other, ranging from parasitism to mutualism.
- During cospeciation, coevolving partners branch into new lineages together.
- Parasites and their hosts can evolve in an arms race. Predators and their prey can do so as well.
- Mutualisms are sometimes maintained by punishment.
- Coevolution takes place in a geographic mosaic.
- Coevolution can promote biodiversity. The extinction of a species can endanger its coevolutionary partners.
- Some organisms have become permanent residents inside other organisms. Mitochondria, for example, began as free-living bacteria.
- Some viruses can become established permanently in the genome of their host. The human genome contains 100,000 segments of DNA from viruses.

Sex and Family

12

Being a scientist can mean learning to do some pretty strange things. For Patricia Brennan, an evolutionary biologist at Yale University, those strange things include measuring the length of a duck's phallus—the duck equivalent of a human penis.

Sexual selection has produced many extravagant traits, such as the long tails of widowbirds (left). Above: Patricia Brennan has discovered that an evolutionary arms race between male and female birds can produce gigantic sexual organs.

A male duck normally keeps his phallus retracted inside his body, only extending it during mating. To measure it is thus a two-person job. Brennan has a colleague grab a bird and hold it upside down, its legs sticking out in the air. If the maneuver is done with care, the duck does not quack or struggle; it just gazes off into the distance. Brennan gently presses around a small dome of muscle below the bird's tail, and after a little coaxing the phallus emerges. Brennan grabs a ruler.

The measurement she makes depends on the time of year. After the breeding season is over, a duck's phallus shrinks to a fraction of its former size. When the next breeding season approaches, the phallus grows back again. Shaped like a spiraling tentacle, it can grow to astonishing lengths. In some species of ducks, it grows as long as the bird's entire body.

The length of a duck's phallus is all the more remarkable when you consider the fact that only 3% of all bird species have phalluses at all. (In the other species, the male has only a simple opening that he positions against a similar opening in the female.) Brennan wants to understand why duck phalluses are so elaborate, especially when other bird species have none. To discover why, she has embarked on a study of the forces driving the evolution of duck phalluses.

To find a diversity of ducks and other birds with phalluses, Brennan traveled to Alaska. She caught 16 species of ducks and other waterfowl that migrate there for the summer. Brennan measured the phalluses on the male birds, and then she turned her attention to the females. The male duck phallus enters a long tube in the female known as the oviduct. After the male deposits his sperm, they swim up the oviduct to reach the female's eggs in her ovaries.

In some species, Brennan observed, the oviduct had pouches branching off along its length. The females of some species had more pouches than others. Brennan also noticed that the oviduct is twisted in a clockwise spiral, twisting more in some species than in others. The clockwise twist is very striking when you consider that the phalluses of the male birds are twisted counterclockwise. It's as if the females were trying to hamper the male birds, not help them.

But most remarkable of all was the discovery that Brennan made when she compared her measurements of male and female genitalia. In species where males have longer phalluses, the females have oviducts with more pouches and more coils.

From this correlation, Brennan concluded that the sex organs of male and female ducks have been coevolving, much like long-tongued flies and the flowers they pollinate. In the case of the flies and flowers, two species coevolve. In the case of birds, however, it's the males and females of a single species that are coevolving.

Brennan's ducks demonstrate that evolution does more than produce adaptations for finding food or withstanding the rigors of the environment. To pass its genes down to future generations, a duck cannot simply find food to eat and avoid dying young. It must also find a mate and produce healthy ducklings, and those ducklings must survive until they can have ducklings of their own. At every stage, there are opportunities for the selection of some individuals over others. Some males have more success in attracting mates than others. Some have sperm cells that are more successful at reaching eggs. Some females manage to rear more offspring than others—sometimes with the help of males, and sometimes without. In this chapter, we'll take a look at the remarkable diversity of ways in which organisms mate and rear their young, as well as the underlying patterns of evolution that produced it.

Why Sex?

Some species can reproduce without sex. A microbe can make a copy of its genome and then divide itself in two. An aspen tree may send roots through the ground that sprout new saplings that grow into trees themselves. Over time, a new tree may become physically separated from the original one, but the two plants still share an identical genome. While most species of animals must have sex to reproduce, it's not a universal rule. Some species of whiptail lizards in the southwestern United States are made up entirely of females. In order to reproduce, a female first goes through a mock courtship, with another female mounting her like a male might. This experience triggers one of her eggs to begin to divide into an embryo without any DNA from a male lizard.

Many species have a choice about how they can reproduce: they can do so with sex or without it. Aspens can send up new shoots, but they can also produce pollen and ovules, which join with the gametes of other aspens to produce seeds. Many species of flatworms develop both male and female sexual organs; they can fertilize their own eggs, or they can play the part of a male and try to mate with another flatworm. Even bacteria can engage in a microbial version of sex through horizontal gene transfer.

All of these forms of reproduction are made possible by genes, and those genes are subject to natural selection. And therein lies a puzzle: As an evolutionary strategy, sexual reproduction imposes huge costs on organisms—costs that could well make sex a recipe for extinction.

To appreciate the toll sexual organisms pay, imagine a population of lizards in which some need to mate and some don't. Each asexual female can produce many daughters, each of which can produce daughters of her own. Meanwhile, the sexually reproducing females have to mate with males before they can reproduce. Half of their offspring will be males, which cannot produce young themselves. Sex, in other words, effectively cuts the reproducing population of these lizards in half. The population of asexual lizards should expand much faster than the sexual one. The late British evolutionary biologist John Maynard Smith called this disadvantage the twofold cost of sex.

Evolutionary biologists have developed a number of competing hypotheses to explain why sex is so common, despite its steep cost. They've tested these explanations on animals, plants, and microbes. They've made their models more complex and tested them yet again. After decades of research, a few hypotheses have emerged as particularly promising.

One hypothesis explains the benefit of sex as a way for populations to adapt quickly to their environment. If an asexual organism picks up a beneficial mutation, it can pass the mutation down only to its direct offspring (**Figure 12.1**). If two organisms each acquire a different beneficial mutation, there's no way for the two mutations to be combined. Asexual organisms may also acquire mutations that have harmful effects that offset earlier, beneficial mutations. There's no way for them to separate the bad mutations from the good.

Sexual reproduction, on the other hand, splits up genotypes and shuffles them into new combinations. Beneficial mutations can be combined with other beneficial mutations, and they can be separated from the harmful ones. Over many generations, sexual reproduction can raise the fitness of a population faster than asexual reproduction.

Tim Cooper, an evolutionary biologist at the University of Auckland in New Zealand, ran an experiment to test this fast-adaptation hypothesis. He selected the gut microbe *Escherichia coli* to study, because some strains of *E. coli* can readily exchange genes while others cannot. The "sexual" strain carries an extra ring of DNA, called a plasmid, which encodes proteins for tubes that can attach to other *E. coli*. The microbes then pump both the plasmid DNA and some of their own DNA through the tube. Cooper was thus able to watch how bacteria that can or cannot have sex evolve under identical conditions.

Cooper bred a line of genetically identical *E. coli* and then inserted the sex plasmid into some of them. He predicted that the bacteria that could have sex (thanks to the plasmid) would adapt faster than the bacteria that could not. The differ-

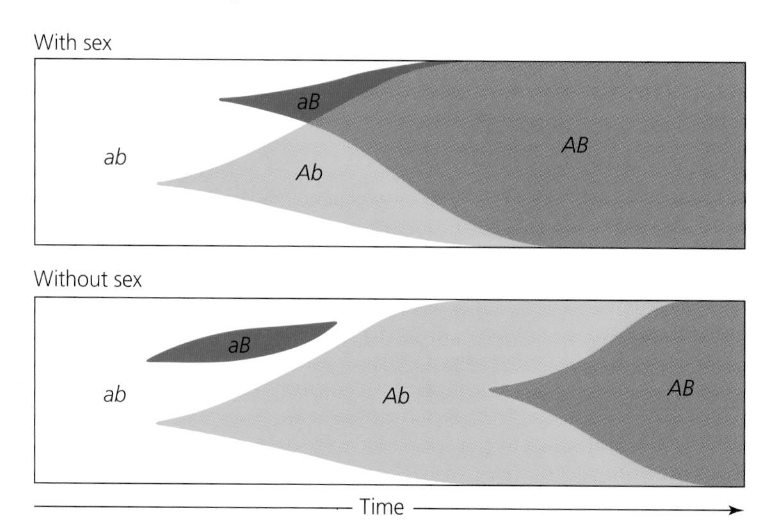

Figure 12.1 One hypothesis for why sex persists is based on its ability to produce new genotypes. This figure shows a sexual population (top) and an asexual one (bottom). Each population starts out with a single allele, *a*, for one gene, and a single allele, *b*, for another gene. Over time, a mutation changes *a* to a new allele, *A*, in a single individual. This mutation is beneficial, and so descendants with *A* become more common (yellow). Meanwhile, another individual acquires a beneficial mutation to *b*, producing the allele *B* (dark green). In a sexual population, *aB* and *Ab* individuals can mate and produce offspring with the *AB* genotype (light green), which has even higher fitness. Meanwhile, in an asexual population, it takes longer for the *AB* genotype to evolve, because each gene has to mutate in the same lineage.

ence between the two strains would depend on the rate at which new mutations arose. The more mutations that became available, the faster sexual bacteria would evolve when compared with asexual bacteria.

Another advantage of studying *E. coli* is that scientists can carefully control the mutation rate. *E. coli* carries a gene called *mutS* that makes an enzyme involved in repairing damaged DNA. By altering the *mutS* gene in some of the bacteria, Cooper made them sloppier. As a result, their mutation rate went up. So now Cooper had four populations of bacteria: they were either sexual or asexual, and either slowly mutating or rapidly mutating.

Cooper then allowed the bacteria to evolve, using a method pioneered by evolutionary biologist Richard Lenski (page 118). He filled flasks with nutrient broth and placed bacteria in each of them. Each flask had bacteria with one of the four combinations of traits. Each morning, Cooper transferred a drop of the broth (and the bacteria living in the drop) to a fresh flask. He repeated this procedure for 130 days, allowing the bacteria to adapt to the identical flasks for 1,000 generations. Along the way, he froze some of the bacteria so that he could track their evolution.

The results matched Cooper's expectation. At low mutation rates, the sexual bacteria had only a slight edge over the asexual ones: their fitness increased 32%, versus 29% in the asexual ones. At high mutation rates, however, the sexual bacteria did much better. Their fitness rose by 43%, while the fitness of the asexual ones rose by only 32%. The difference in fitness increased by about a factor of four (**Figure 12.2A**).

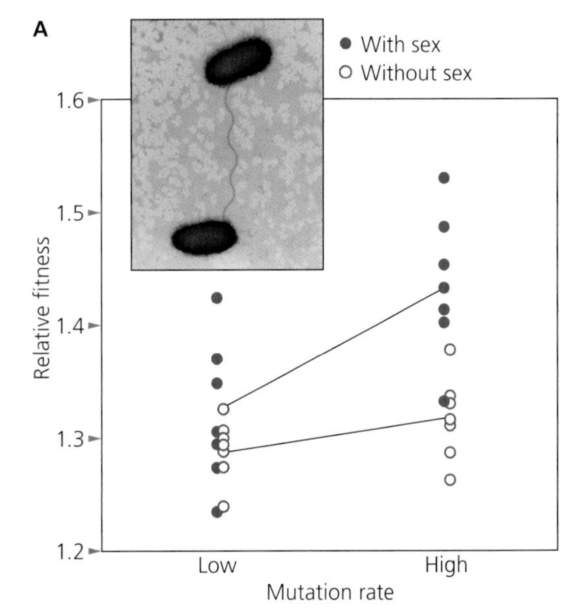

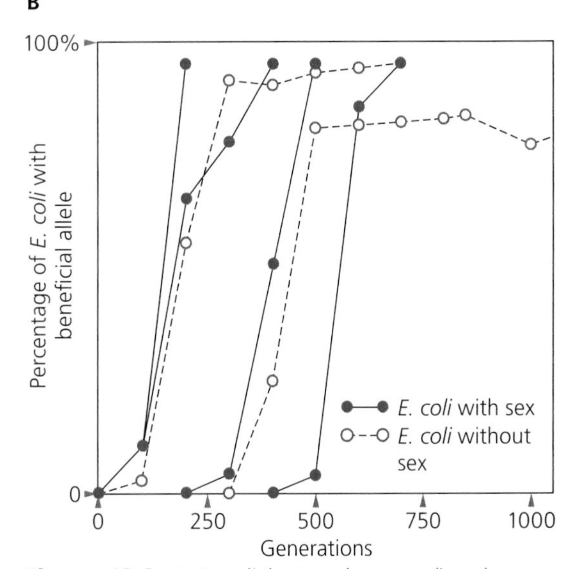

Figure 12.2 A: *E. coli* that can have sex (inset) can evolve greater fitness when mutation rates are high. B: A beneficial allele can spread faster through a population by natural selection if sex is possible. (Adapted from Cooper, 2008)

Cooper found that a beneficial mutation arose in all of his populations of *E. coli*. Looking back over the experiment, he determined how long it took for natural selection to fix the mutation in each population (**Figure 12.2B**). He discovered that it took 900 generations for the mutation to become fixed in the asexual bacteria, but only 300 generations for it to become fixed in the sexual ones. All of these results supported the hypothesis that sex speeds up the evolution of adaptations.

Running in Place

Parasites may also help explain the benefits of sex. Their relentless assaults on their hosts pose a special kind of threat. A population of hosts may evolve strong defenses against parasites, only to have the parasites evolve countermeasures to get around those defenses.

In the 1970s, several evolutionary biologists suggested that this host–parasite coevolution proceeded in cycles. Parasites adapted to the most common genotypes in the population of their hosts. Those hosts died off, while other host genotypes became more common. The most common parasite was now no longer well adapted to the most common host genotype, but over time new parasite genotypes evolved that could exploit them. But eventually those new victors also killed off their hosts, and the cycle began again. This model came to be known as the "Red Queen hypothesis." The name comes from the Red Queen in Lewis Carroll's book *Through the Looking Glass*, who takes Alice on a run that never seems to take them anywhere. "Now here, you see, it takes all the running you can do to keep in the same place," the Red Queen explained.

The Red Queen hypothesis makes sense mathematically, and that's prompted evolutionary biologists to design experiments to test it. Ellen Decaestecker of Catholic University in Leuven, Belgium, and her colleagues published a particularly clever experiment in 2007. They realized that they could actually replay 40 years of Red Queen coevolution. The host they studied was a water flea that lived in ponds in Belgium. Each year the water fleas laid vast numbers of eggs, some of which were buried in the muck at the bottom of the pond. The water fleas have many parasites, but the one Decaestecker chose to study was a disease-causing strain of bacteria called *Pasteuria*. These bacteria float around in pond water, and some of them get trapped in the muck as well.

Decaestecker drilled into the bottom of a Belgian pond and hauled up a core of sediment. The annual layers were so distinct that she could identify which year they formed. From each layer, Decaestecker isolated eggs of water fleas and found that even the oldest ones could still hatch into healthy water fleas. She also isolated *Pasteuria* from the same layers, and found that they could grow healthy colonies in her lab.

To test the Red Queen hypothesis, Decaestecker infected water fleas with bacteria. She infected some of them with bacteria that had been trapped in mud in the same year. She infected others with bacteria from older layers, which had been buried in the mud a year or two before the water fleas. And she infected still others with bacteria that had been trapped a year or two later. In other words, the water fleas were infected from the past, the present, and the future.

Decaestecker and her colleagues found that the bacteria from the same year as the water fleas did the best job of infecting them. On average, 65% of the water fleas got sick when the scientists exposed them to bacteria from the same year. The parasites from earlier years could only infect 55% of the water fleas, presumably because resistant water fleas became more common. And when the scientists used bacteria from future years, they could infect only 57% of the water fleas. If the Red Queen were driving the evolution of the water fleas, then you'd expect the bacteria from future years to be better at infecting water fleas from their own time, not water fleas from earlier cycles of coevolution.

According to one hypothesis, parasites drive hosts through a cycle of booms and busts. This explanation is sometimes called the Red Queen hypothesis, named after the character in *Through the Looking-Glass* who runs very fast just to stay in place. Some experiments indicate that sex helps organisms reproduce faster in spite of the Red Queen.

Thanks to their Red Queen effect, parasites may drive the evolution of sex in their hosts. Parasites are continually evolving new variations that adapt them to the genotypes of their hosts. It may be easier for them to do so if their hosts are asexual, because the only new variation that evolves through asexual reproduction comes from rare mutations.

Sexual reproduction, on the other hand, creates lots of genetic variation in every generation by shuffling alleles into new combinations. Among those new variations will be some that are very resistant to the dominant parasite genotype. They will be favored by natural selection and their descendants will thrive.

Over many generations, this adaptive edge may cause asexual individuals in a population to become more and more rare, while sexual individuals become more and more common. That's what mathematical models of the Red Queen effect suggest, and a number of studies on real organisms offer support as well. Jeremiah Busch, a biologist at Indiana University, and his colleagues tested the Red Queen hypothesis by looking at plants. In 2007, Busch and his colleagues reported a striking pattern: plants attacked by more disease-causing fungi reproduce sexually more often than species attacked by few species of fungi. That's the sort of result you'd expect if parasites raised the benefits of sex.

Cheap Sperm and Costly Eggs

Sexual reproduction can only take place if organisms make male and female gametes. In humans, the males make sperm, and the females make eggs. Male holly bushes make pollen, and female hollies make ovules. The body of a tapeworm is made of hundreds of segments, each of which contains both eggs and organs for making sperm. But in every case, the male gametes are small and fast, and the female gametes are big and slow. And as a rule, organisms make lots of male gametes and relatively few female ones. Over a woman's entire lifetime only about 400 eggs will reach maturity and become ready to be fertilized. Men, by contrast, make tens of millions of sperm every day.

The difference between eggs and sperm has far-reaching effects on the evolution of males and females. It means that the strategies for boosting reproductive success are different for the sexes. Each faces a different limit on how many offspring it can have. A single man makes so many sperm that he could fertilize every egg in every woman on the planet. So the number of offspring a female can have is usually not limited by a scarce supply of sperm. Instead, her reproductive success is limited by how many eggs she can nurture and rear into adults that can reproduce themselves. The reproductive success of males, on the other hand, is limited by the relative scarcity of eggs. A mutation that helps an individual male fertilize more females than other males may be strongly favored by selection.

This imbalance explains why males in most animal species compete with each other for the opportunity to mate with females. These male struggles can take many forms, from mountain sheep slamming their horns against each other to male fiddler crabs flipping each other over with their outsized claws. Despite these battles, males do not actually kill each other very often. Their contests usually end peacefully, after two males have a chance to size each other up. Rather than fight, male fiddler crabs often wave at other male crabs that challenge them for their mating burrows. The male with the smaller claw then skulks away. He may lose a chance to reproduce, but at least he is still alive.

A sperm fertilizes an egg. Male gametes are small and mobile, while female gametes are large and move relatively little.

As males compete for females, some end up having more offspring than others. The difference between the winners and the losers is especially stark among the southern elephant seals that breed on Sea Lion Island, one of the Falkland Islands in the South Atlantic Ocean. A. Rus Hoelzel, a biologist at Durham University in England, and his colleagues surveyed all the seals on the island, year in and year out. They snipped small pieces from skin from all of the adult elephant seals, as well as from all 192 baby seals that were born in 1996 and 1997. They could identify the father of almost every seal pup, because the females were mating only with males that were also lying around on Sea Lion Island. Of all the males, 72% failed to have any offspring at all. The remaining 28% did not have an equal share of reproductive success. Many had only one or two pups, while a few managed to have many offspring. One particularly successful male seal fathered 32 pups.

The reproductive success of male southern elephant seals is not random. Male elephant seals fight to mate with large numbers of females, which gather together in groups known as harems. They rear up on their flippers and throw their tremendous bodies against each other, sometimes drawing blood with their teeth. The losers slink away, to lurk at the edges of the colony. Hoelzel found that 90% of the seals that fathered pups were the heads of harems. The other 10% of the seals that fathered pups were lurkers that managed to sneak off with a female.

Male fiddler crabs grow one oversized claw. They use it to compete with other males for territory and to attract females.

Among elephant seals only a few males father pups. This steep variation in mating success strongly favors mutations in males that help them compete for mates.

The bigger a male elephant seal, the better his chances of defending a harem and fathering a lot of pups. In other words, large size is selected in the male seals, in much the same way large beaks are selected in medium ground finches on the Galápagos Islands when there are a lot of hard seeds to eat. But the selection for size that takes place among elephant seals is not based purely on how long the seals survive. A trait that boosts the opportunities the seals have to mate is being selected. This special type of selection is known as sexual selection.

Sexual selection may account for why male elephant seals are several times larger than female ones. Bigger males tend to win fights with smaller ones, and so bigger males tend to hold onto harems. The big males thus have more offspring. Female elephant seals, on the other hand, don't fight with each other, and so large females don't have a reproductive advantage over small ones. In other words, sexual selection for body size is much stronger in male elephant seals than in females.

Songs and Dances

With males competing with each other to mate with females, the females have the opportunity to choose which male they'll accept. In many species, scientists have found strong preferences among the females for certain traits in males. In many bird species, for example, females prefer males who sing complex songs

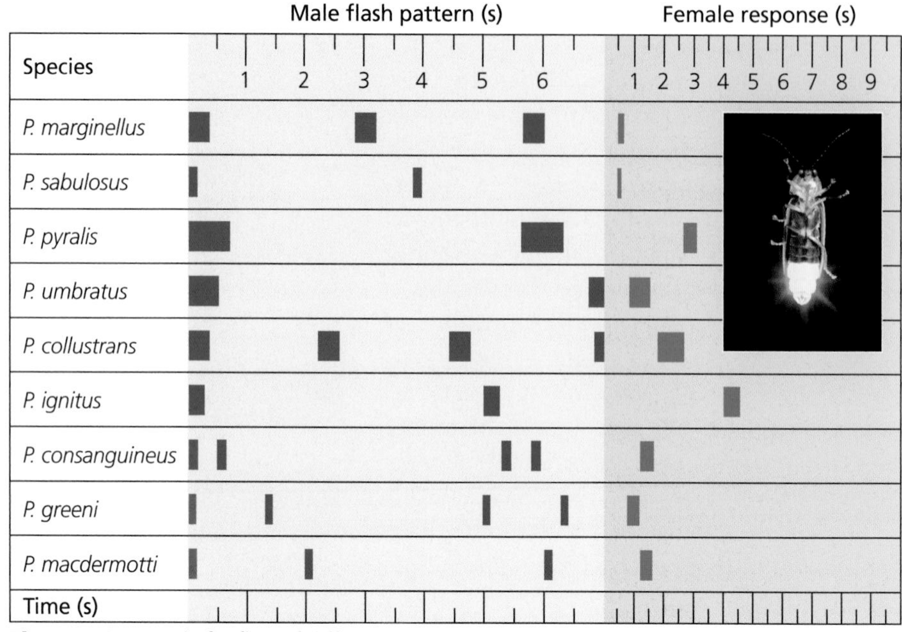

Figure 12.3 Male fireflies of different species produce flashes in different patterns. Some flash many short pulses, while others produce a few widely spaced flashes. Females only respond to signals from males of their own species. (Adapted from Lewis, 2008)

over males that produce simpler ones. Female frogs prefer males that croak more loudly over ones that croak quietly at night.

These preferences can be remarkably precise. Male fireflies draw the attention of females with flashes. Each species makes its own distinctive pattern of light (**Figure 12.3**). Females show their interest by flashing in response, allowing males to approach them and continue the courtship. But the females of each species have their preferred kind of flash. In some species, the females like a shorter pulse over a longer one; many female fireflies also prefer shorter pauses between pulses.

Stalk-eyed flies get their names from the long stalks on which their eyes develop. The distance between a fly's eyes can be longer than its entire body. Female flies prefer to mate with male flies with longer stalks rather than shorter ones. Their preference selects for genes for long stalks.

Females judge males not only on what they do, but also on how they look. Male jungle fowl with big combs on their heads have more reproductive success than males with smaller combs. Male swordtail fish develop a long extension on their lower tail fin, and the males with longer swords are more likely to attract females than short-sworded males. One of the most surreal effects of female choice can be found in stalk-eyed flies. There are several hundred species of these flies, and in most of them the males have eyes on sideways-pointing stalks that can be longer than their entire body. The distance between their eyes has a strong effect on the preference of the female flies: the longer the stalks, the more likely they are to mate.

But where do those female preferences get their start? Some studies suggest that the females have ancient preferences for certain shapes or colors, and that bias triggers sexual selection in males. On the Caribbean island of Trinidad, for example, the male guppies sport bright orange patches, and the females prefer to mate with the males with the biggest, brightest patches. But the guppies also eat bright orange fruits that plop into their streams. Helen Rodd, a biologist at the University of Toronto, and her colleagues tossed little orange disks into fish tanks and found that guppies pecked at them more than they did at disks of other colors. Males were as enthusiastic as females, indicating that the female guppies were not confusing orange disks with attractive male fish. What's more, Rodd and her colleagues found that some populations of guppies responded more strongly to orange disks than other populations. The populations that responded strongly to orange disks were also the ones in which females had the strongest preference for orange patches on males. Rodd proposes that the attraction guppies have for orange fruit drove the evolution of the orange spots on the male guppies.

In 1915, the British biologist Ronald Fisher developed a model to explain this type of sexual selection. Once females developed a preference for certain males, a runaway process of sexual selection was triggered. In Fisher's model, a population contained choosy females that preferred males with some ornament as well as unchoosy females that were equally likely to mate with any male. In such a population, all the choosy females mate with the showy males, along with some of the unchoosy females. As a result, having an attractive ornament would raise the odds that a male would reproduce. At the same time, females that chose males with ornaments had more reproductive success as well, because they gave birth

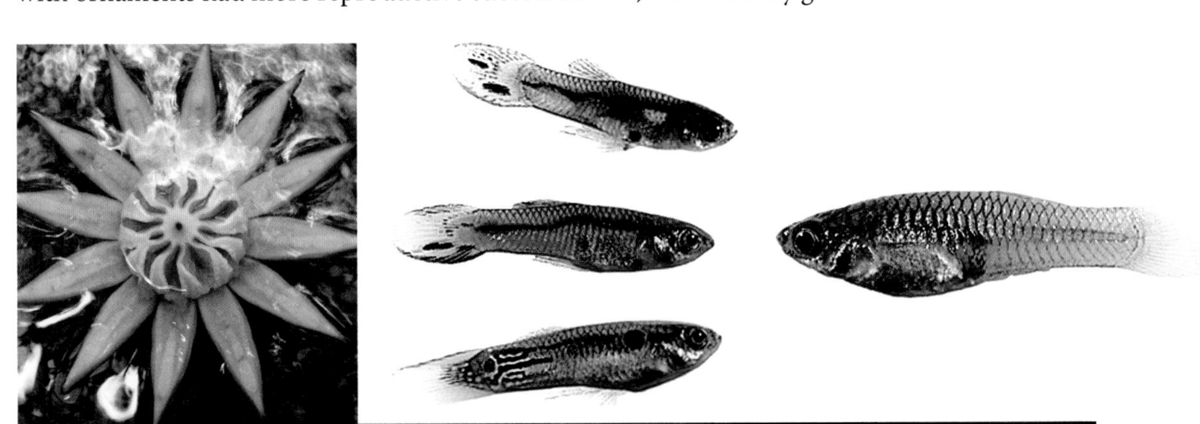

Female guppies prefer to mate with males with bright orange spots. Their preference may have originated with an attraction to orange-colored fruits that sometimes land in their streams.

to showy sons that could also attract females. A positive feedback cycle emerges in Fisher's model, as female preferences and male ornaments continue to coevolve.

Fisher called this feedback a "runaway" mechanism. It can only be stopped when a sexually selected trait starts to become a serious threat to a male's survival. Orange patches don't merely attract female guppies, for example; they also attract predatory fish. And studies on guppies suggest that their evolution is limited just as Fisher would have predicted. In streams with predators, the orange patches on male guppies are drab compared with the vibrant guppies that live in predator-free waters.

Another hypothesis holds that sexual selection takes place because displays communicate information from one sex to the other. Animals can use them like billboards to advertise their value. Scientists who favor this explanation point out that the choice of a mate can have a huge effect on an animal's reproductive success. Depending on the quality of a mate's genes, the fitness of an animal's offspring may go up or down. If animals can improve their odds of picking mates with good genes over bad, they can potentially increase their fitness in the long term. But animals can't order genetic tests on their mates. Instead, they can only have a preference for mates with some reliable sign of good genes—be it a song or a spot or a feather.

There is a risk, however, in relying on such signals in choosing a mate. Imagine, for example, that female birds are attracted to blue-headed males, because blue-headed males happen to father fitter chicks. Other male birds may gradually evolve blue heads, attracting the females even though the blue-headed imposters don't have high-quality genes. Sexual signals open the door to deception, just as coevolution offers the opportunity for cheating (page 250).

Mathematical models suggest that one way a sexual display can become a reliable signal is for an animal to have to pay a cost to make it. That cost might be the energy it takes to build antlers or to croak all night long. Strong animals can afford that extra energy, while weaker ones cannot. Such costly signals ought to be particularly difficult for weaker animals to make when they're under other kinds of stress, such as sickness or starvation.

These models have inspired a number of biologists to search for evidence of honest signals in real animals. Sarah Pryke, of the University of New South Wales in Australia, has tested the hypothesis by studying red-collared widowbirds in Africa. Male red-collared widowbirds have tail feathers that measure around 22 centimeters long—longer than their entire body. During the mating season, they fly around the grasslands, fanning out their feathers so that they flap in the breeze. Female red-collared widowbirds spend this time choosing where to make their nests, mating with the male whose territory they've picked.

To probe the choices made by females, Pryke trimmed the tails of 120 male red-collared widowbirds. On 60 birds, she snipped their tails down to 20 centimeters, while on the other 60 she cut the tails down to 12 centimeters. Then she compared how well the males fared in the mating game. (While 12 centimeters is short for a red-collared widowbird tail, it's still within the natural range of the birds.) Pryke found some dramatic differences. A long tail costs male red-collared widowbirds extra energy, because of the drag it creates in the air. The birds with the shortest tails were able to spend more time flying around, attracting females. Yet that extra time did not translate into more success. The birds with the longest tails ended up with three times more nesting females in their territories than the short-tailed birds.

Another way to measure the honesty of displays is to study both mating animals and their offspring. Anders Forsman of the University of Kalmar in Sweden and Mattias Hagman of the University of Sydney in Australia studied male poison dart frogs, which chirp to attract females. They found that male frogs that chirped faster and with longer chirps attracted more mates than other males. Forsman and Hagman then compared the offspring of those preferred males to those of males that didn't sing so attractively. The eggs fertilized by males with high calling rates and longer chirps were more likely to hatch. Once they developed into tadpoles, the offspring of these attractive fathers were more likely to survive to become full-grown frogs. Forsman and Hagman concluded that the speed and length of chirps was an honest advertisement for the quality of genes in the male frogs.

The evolution of honest signals may account not just for female preferences, but also for the horns, tusks, antlers, and other extravagant weapons that male animals can grow. These weapons may be not just for fights, but also for avoiding fights. Battling another male is very risky, potentially leaving both animals wounded or dead. In many species, males reduce this risk by sizing up their opponents from a distance. A big weapon may serve as a clear signal of the strength of the male that carries it, and weaker males may back away just at the sight of it.

Of course, such a signal cannot survive very long if it's easy to counterfeit. If weak males can make big weapons, then natural selection may favor males that ignore the weapons and fight opponents anyway. There's growing evidence, though, that weapons are indeed honest signals that can't easily be faked. Douglas Emlen, a biologist at the University of Montana, has studied the cost of beetle horns. In order to grow big horns, male beetle larvae have to dedicate bigger patches of cells on their heads to those organs. That means they have fewer cells to develop into neighboring organs, such as eyes or antennae, which they need to survive. Vanessa Ezenwa of Oregon State University and Anna Jolles of the University of Montana have found a similar trade-off in African buffaloes. Male buffaloes burdened with more parasites grow smaller horns. A male's body size and the amount of food he has eaten can also influence the size of his horns.

This sort of research is revealing that sexually selected traits can send different signals to different viewers. Male fiddler crabs use their oversized claws as a signal to other males that they are strong enough to win a fight. But female crabs are watching, too, and they prefer to mate with males with bigger claws. What is a warning to other males of a fearsome opponent is a sign to females of a desirable mate.

Mating Games

In the animal kingdom, there are many different kinds of mating systems. A male and a female bald eagle may develop a lifelong bond—a mating system known as monogamy. Among elephant seals, a single male mates with several females—a system known as polygyny. Central American birds known as wattled jaçanas are at the other extreme, polyandry, in which each female may mate with several males.

Until the late 1900s, naturalists could only study mating systems by observing animals. Yet looks are often deceiving. Many pair-bonding birds are not as loyal to one another as they may appear. When scientists analyzed the DNA of the chicks in nests of pair-bonding birds, they often found that a large fraction of the eggs did not carry the DNA of their mother's partner. Their mothers were mating with other males, and their partners were doing the hard work of raising the chicks.

To make sense of these intricate mating systems, evolutionary biologists consider how different strategies boost the reproductive success of males and females. The low cost of sperm, for example, can make polygamy a good strategy for males, because they can fertilize many females, which can bear many offspring. But polygamy has its downsides. Each male has to compete with lots of other males, and, even if he succeeds in mating with females, the females may end up using sperm from other males to fertilize their eggs.

Under some conditions, a male may actually be better off mating with a single female, trying to fight off other males, and helping to raise their offspring. While most male mammals are not monogamous, for example, male California mice are. If a male California mouse helps to rear his pups, more than twice as many of them will survive than if the mother rears them alone.

Polyandry can benefit females in a number of ways. By mating with a number of different males, females may be able to get the highest quality of genes possible for their offspring. Female superb fairy-wrens form bonds with males, for example, but then slip off at night to mate with other males. The males they select on these forays show signs of being in particularly good condition, suggesting that these females are "upgrading" the quality of the sperm they use to conceive their chicks.

But there may be other explanations for polyandry. Females may be benefiting not from good genes but from genes that are different from their own. If animals mate with partners that are too similar genetically, their inbred offspring may have poor immune systems. Mating with multiple males may allow females to boost the health of some of their offspring by giving them a defense against a wider range of pathogens.

Sperm Wars

Because females often mate with more than one male, they can end up with sperm from many males inside them at once. The stage is set for a new kind of competition between males: a competition between sperm.

A number of strategies have evolved that make sperm more successful in fertilizing eggs. Not surprisingly, they tend to be more common in species in which males compete more intensely for females—a pattern that you'd expect, given that strong competition leads to strong selection. Among primates, for example, the species with strong male–male competition have bigger testicles than the species in which males and females tend to mate monogamously. Bigger testicles supply more sperm, and more sperm can raise a male's odds of fertilizing a female's eggs when sperm from other male are present. Sperm competition has also driven the evolution of faster-swimming sperm, as well as sperm that can join together to propel themselves faster than they could on their own.

Males can also increase the success of their sperm by keeping other males from mating with females. In some species, males linger around females after mating, chasing off other males that might also mate with them. Male *Drosophila* flies produce chemicals in their semen that may kill off the sperm of other males and then make females unreceptive to more mating. After a male rat mates with a female, he inserts a plug of mucus into her reproductive tract. This copulatory plug may make it harder for other males to mate with the female afterwards. Steven Ramm, a biologist at the University of Liverpool in England, and his colleagues found that in species of rodents where competition between males is high, the male rodents tend to make bigger plugs.

Sexual Conflict

Faster sperm, manipulative semen, mucus plugs, and other adaptations evolve as a result of males competing with one another for reproductive success. But while they may raise the fitness of a male, some of them also lower the reproductive success of females. Such evolutionary clashes between males and females are known as sexual conflict. When male *Drosophila* flies inject chemicals into their mates to ward off other males, for example, they also harm the females. The chemicals are actually toxic to the females, cutting their lifespans short.

Male damselflies have sharp ridges on their sex organs, which they use to scrape sperm from other males out of female damselflies before they mate.

Sexual conflict can drive the evolution of counteradaptations in the opposite sex. Female *Drosophila* flies, for example, have evolved proteins that can destroy some of the proteins in male semen. As these defenses evolve, selection favors males that evolve new proteins that can overcome them. Scientists have detected strong natural selection (page 148) in the genes for both kinds of proteins, suggesting that males and females are locked in an arms race.

William Rice, who now teaches at the University of California, Santa Barbara, has even documented the evolution of this sexual conflict in his lab. He reasoned that the mating system of *Drosophila* was driving their evolution, and that, if he were to alter the mating system, he would alter the evolution of the flies. He and his colleagues reared pairs of males and females so that they had no choice but to be monogamous. They experienced no sexual selection, and thus no sperm competition. After 30 generations of monogamy, Rice tested the resistance of the females. He bred them with male flies from a polygamous population, which experienced strong sexual conflict. The monogamous females laid less than half the eggs as females from the polygamous population, and they also died earlier. These results showed that monogamy had caused the female flies to evolve less resistance to males.

Sexual conflict has also led females to evolve the ability to control the sperm that fertilize their offspring. Hens, for example, prefer to mate with high-ranking

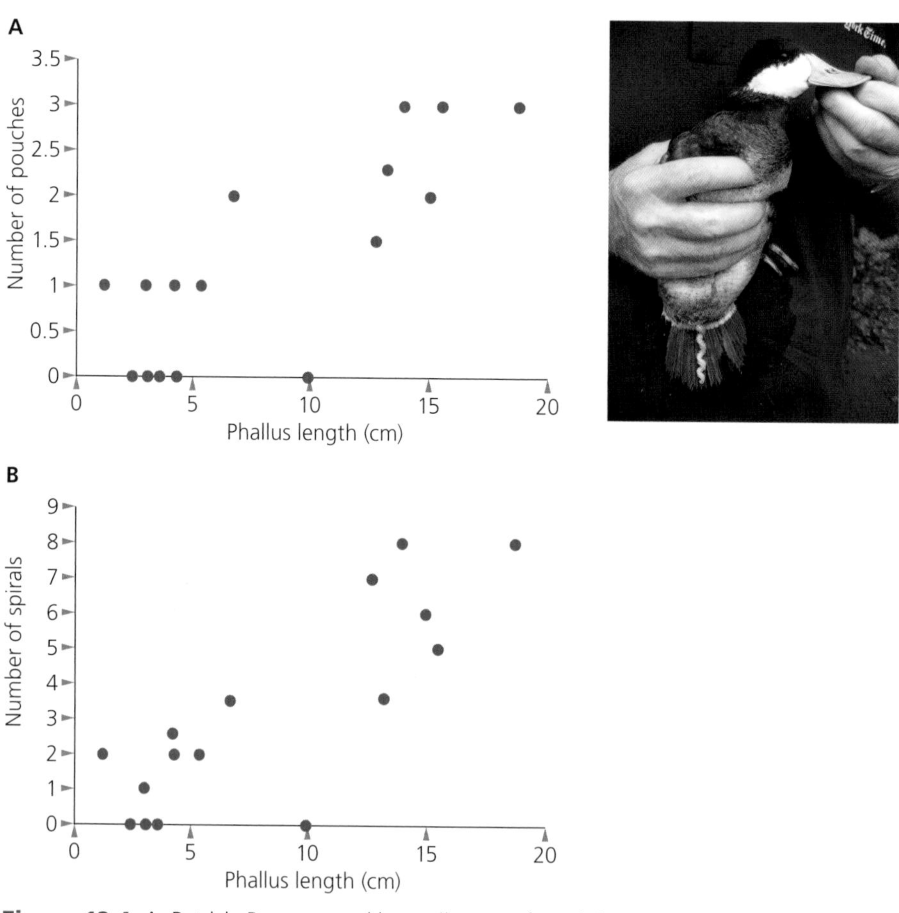

Figure 12.4 A: Patricia Brennan and her colleagues found that in bird species with long phalluses on the males, the females had more pouches in their oviducts. B: The researchers also found a similar correlation between phallus length and the number of spirals in the oviduct. Both patterns are consistent with sexual conflict between male and female birds. (Graphs adapted from Brennan et al., 2007)

roosters. But sometimes low-ranking roosters will jump on them and mate by force. The females do not have to surrender to the prospect of having chicks with inferior genes. They can squirt the sperm of low-ranking roosters out of their bodies. Some insects can store the sperm from several males and then select which male will become the father of their eggs.

Patricia Brennan's research on ducks and other waterfowl suggests that sexual conflict is driving the evolution of their bizarre sex organs. In many species of ducks and other waterfowl, the males and females bond for an entire breeding season. The males that fail to find a mate of their own at the start of a season tend to harass females and try to force them to mate. And when a bonded male's partner is busy incubating his eggs, he may search for other females and force them to mate with him as well. All told, about a third of all matings are forced in some species of ducks and other waterfowl. The harassment from male ducks can get so intense that some females die as a result.

Although forced matings are common in ducks, the unwanted males father only 3% of the ducklings each year. Brennan suspects that the female birds are controlling the sperm in their bodies, favoring the sperm from their partners over that of the intruders. They may be able to shunt sperm into the side pockets in their oviducts.

When females evolve these kinds of defenses, sexual selection favors countermeasures from the males. In the case of ducks, males may have evolved longer, more flexible phalluses; and, in response, females evolved even more twisted oviducts with more side pockets. That would explain why male and female sexual organs are so tightly matched in the birds Brennan has studied (**Figure 12.4**).

Evolving Parents

Animals mate in order to become parents. But that simple rule belies the dizzying diversity of ways to be a parent. Most species of ants, for example, are organized into colonies with a single reproducing queen, many males that do nothing but mate with her, and thousands of sterile daughters that feed the queen's offspring and tend the nest. A female chimpanzee will give birth to a single baby at a time, and she will nurse it for as long as 4.5 years before weaning and getting pregnant with a new baby. Wolf spiders carry their young on their back; some cichlid fish shelter their young in their mouth.

One of the most important advances in making sense of all this diversity came in the 1960s, thanks to a keen observation by George Williams, an evolutionary biologist at the State University of New York at Stony Brook. Williams pointed out that an animal's reproductive success depends on the total number of offspring it can produce over its lifetime that survive long enough to reproduce as well. But an animal has a limited amount of energy it can invest in rearing its young, and so it faces trade-offs. If, for example, a bird were to lay a thousand eggs at once, all the chicks would probably die because the bird wouldn't be able to supply them with enough food. On the other hand, a bird might lay a single, well-nourished egg, only to be outbred by other birds that lay more eggs. Williams argued that natural selection should favor a compromise between these demands that would allow parents to produce the greatest number of offspring possible over their lifetimes.

Scientists have confirmed that these trade-offs exist in many species. Even humans are not exempt. Beverly Strassmann, an anthropologist at the University of Michigan, has studied the Dogon, a group of farmers in the African country of Mali (**Figure 12.5**). She kept track of 55 women, recording the total number of babies to which each of them had given birth. Some women had given birth to only one baby, while others had given birth to as many as 11. Most women fell somewhere in between. In many of the families, at least one child had died. Strassmann found that women who were more fertile did not necessarily have the most reproductive success. Women with more than seven children actually had fewer children survive to the age of 10.

The exact nature of the trade-off depends on the conditions in which a species lives. If those conditions change, the best trade-off may shift. David Reznick, an evolutionary biologist at the University of California, Riverside, and his colleagues study this trade-off among the guppies that live in streams on the island of Trinidad. The guppies in some sections of the streams are hunted by bigger fish, while other sections are predator-free. Reznick and his colleagues found stark differences between the two groups of guppies. The ones that did not face predators grew up slowly, and the females produced small clutches of big eggs. In the guppy populations that were menaced by predators, however, the fishes had evolved a different strategy. The males were ready to mate much earlier than the males at predator-free sites.The females produced twice as many

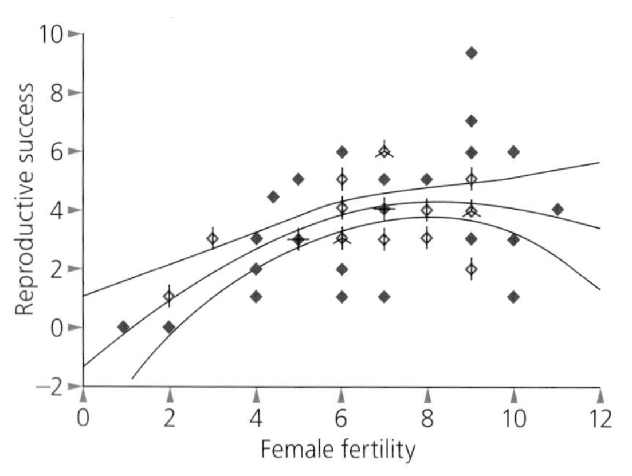

Figure 12.5 Beverly Strassmann studied the reproductive success of women who belong to the Dogon people of Mali. She compared how many babies each woman gave birth to, and how many of those babies were still alive at age 10. (Each "petal" in this graph represents an additional data point.) The central curve on the graph shows how reproductive success increases, but then drops in families of more than seven children. The curve is the result of the trade-off between large families and the resources necessary to care for each child. (Graph adapted from Koella and Stearns, 2008)

eggs, but each egg was only 60% the size of the average egg produced by the predator-free females.

Reznick and his colleagues hypothesized that the guppies had adapted to their habitats just as Williams had predicted. The guppies threatened by predators were more likely to die young. Natural selection thus favored individuals that matured quickly and laid a lot of eggs (**Figure 12.6**). In streams without predators, on the other hand, the guppies could take more time to grow to bigger sizes. They laid fewer eggs at a time, but, without the threat of predators, they lived longer and thus had more chances to reproduce. To test this hypothesis, Reznick and his colleagues moved some of the guppies that lived with predators to predator-free streams. The guppies evolved to produce fewer offspring, and those offspring grew into bigger animals—50% bigger than the guppies that still lived with predators.

George Williams also argued that animals could evolve flexible strategies for reproducing. They might evolve ways to take stock of their conditions in order to make their family decisions. And his prediction has been fulfilled in studies on some animals. Male sand gobies, for example, are unusual among fish because they help take care of their eggs. A male goby builds nests for his eggs, covers it with sand, and guards it from predators. He even cleans the algae off of

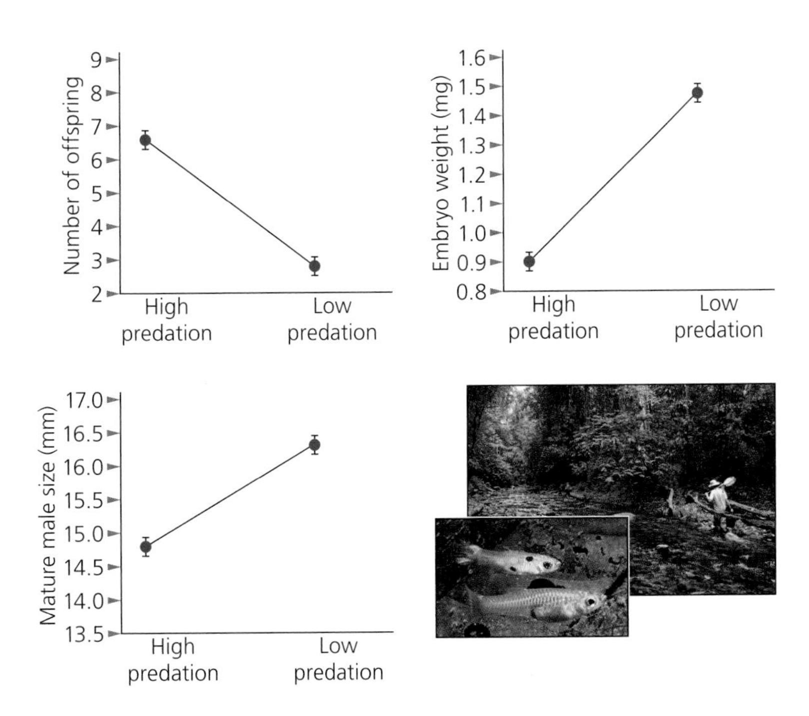

Figure 12.6 David Reznick and his colleagues compared guppies in streams with many predators to those in streams with few. As these graphs show, the guppies that faced few predators produced fewer, bigger offspring than the guppies menaced by many predators. The guppies also grew to larger sizes by the time they were sexually mature. Natural selection shaped the life history of the guppies in different ways, depending on whether they were more or less likely to be killed by predators. (Adapted from Reznick, 2008)

his nest. Sand goby eggs need to consume oxygen, and so a father will fan his nest to provide it with a fresh supply of oxygen-rich water. But sometimes male sand gobies will do something that seems unthinkable: they dig up their nests and devour their own eggs.

Hope Klug, a biologist at the University of Helsinki, found an evolutionary logic behind this cannibalism. She experimentally raised and lowered levels of oxygen in the nests. The number of eggs the father ate depended on how low the oxygen levels dropped. Klug concluded that the fathers were eating some of the eggs so that the remaining eggs would have enough oxygen to develop properly. Supporting this conclusion was the fact that sand goby fathers also adjusted the extent of their cannibalism to the density of the nest. The denser a nest, the more eggs the father was likely to eat. Rather than being a random act of self-destruction, the cannibalism appears to be a response the males make to certain changes in their environment—a response that helps raise their fitness.

Tipping Sex Ratios

On average, most animals produce sons and daughters in equal proportions. Ronald Fisher offered an elegant argument for why this balance should have evolved. Imagine that mutations lead to more female births than male. The imbalance gives males an advantage; a male is more likely to find a mate than a female. If mutations arise that make some animals produce more sons, they will be favored by natural selection. But, as the numbers of males come to equal the numbers of females in each new generation, the advantage of being male dwindles. The same process would work under the opposite conditions, with more males than females. The sex ratio of the population balances itself at one to one.

In 1973, Robert Trivers and Dan Willard, both then at Harvard, argued that natural selection could drive sex ratios away from one to one under certain conditions. Consider, for example, a polygynous species in which a few males in good condition mate with most of the females. If a female is in good condition herself, she may be able to boost her reproductive success by having more sons than daughters. Her sons, in good condition themselves, will mate with many partners, and give her more grandchildren than daughters would.

On the other hand, if a female is in poor condition, Trivers and Willard argued, she may be better off having more daughters than sons. Sons in poor condition may fail to attract any mates, and may therefore leave their mother without any grandchildren. Daughters, on the other hand, will probably have at least some offspring, even if they are in poor condition.

Trivers and Willard proposed that mothers could switch the sex ratio of their offspring to suit their condition. Some of the most compelling support for sex ratio-switching comes from the remote Seychelles Islands in the Indian Ocean, which are inhabited by a bird known as the Seychelles warbler. When a mother warbler's eggs hatch, the male and female chicks can look forward to different

Female Seychelles warblers can adjust their ratio of sons to daughters to boost their reproductive success.

lives. The male birds tend to fly away in search of other female warblers and other territories. Young female birds tend to stay behind, helping their mother incubate her eggs. A mother benefits from their help, because she is able to rear more chicks over her lifetime with the aid of her daughters.

Jan Komdeur, a biologist at the University of Groningen in the Netherlands, and his colleagues have been carefully chronicling the lives of all 2,000 or so Seychelles warblers that live on the islands. They discovered that female Seychelles warblers can adjust the balance of male and female chicks in their broods in response to their environment. An unassisted female living on a patch of land with abundant food may produce a brood that's as much as 88% daughters. But Komdeur has found that female warblers that live in places where food is scarce may produce broods in which as few as 23% of the chicks are daughters.

Komdeur hypothesized that the birds were adjusting the balance of daughters and sons to maximize their reproductive success. A female bird that lives on high-quality territory can use the help of her daughters to produce more chicks. A female that is stuck living on low-quality territory will be better off producing sons that can search for greener pastures. Komdeur tested his hypothesis by moving birds from low-quality territories to high-quality ones. As he predicted, the birds switched from mostly sons to mostly daughters.

There is, however, such a thing as too much help. When Seychelles warbler mothers have more than three female helpers when living on high-quality territory, trouble arises. The helpers eat too much food, and they may crack the mother's eggs as they clamber around the nest. In response, the Seychelles warblers adjust the sex ratio yet again. They produce more sons that will soon fly away and not be such a burden. To test his hypothesis in a new way, Komdeur took away the helpers in some of the warbler nests. The sex ratio changed again, and in exactly the way he predicted: the unassisted birds started producing more daughters again.

Inclusive Fitness

The Seychelles warblers are intriguing not just because their mothers can manipulate the sex of their chicks. The helper daughters are worth considering as well. Their altruism seems, at first, like a paradox. If natural selection is all about the relative fitness of individuals, these helper birds seem to be ignoring the prime directive of evolution. Rather than trying to reproduce themselves, they spend time helping others have offspring.

Darwin himself recognized the challenge that altruism posed to his theory. It was not until the 1960s, however, that a biologist came up with a solution for altruism that could accommodate modern genetics in a mathematically precise way. The British biologist William Hamilton observed that an animal shares 50% of its alleles with one of its offspring, but it also shares 25% of its alleles with its nephews and nieces (**Figure 12.7**). If an animal helps its sibling to reproduce, some of its own genes will be carried down to the next generation. Helping relatives has an evolutionary cost, because an animal reduces its own reproductive success; but that animal may also be able to spread some of its own alleles by helping its relatives.

Hamilton argued that the original definition of fitness—based on how many offspring that organisms with a particular genotype had—should be updated.

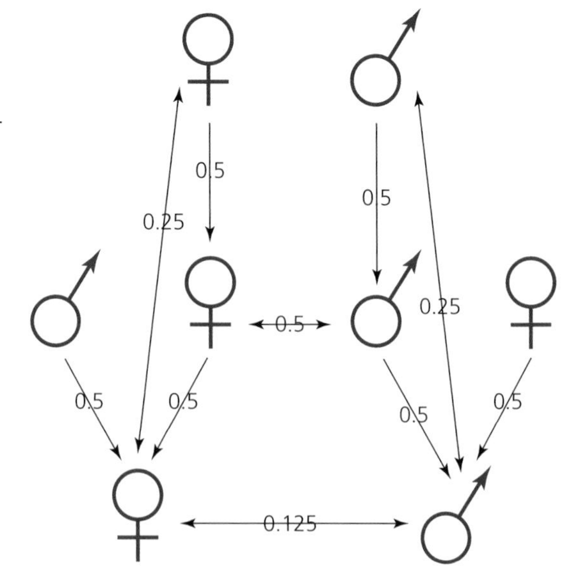

Figure 12.7 One way to think about how we are related to our relatives is to consider how much genetic material we share in common. A child inherits half of his or her genetic material from his or her mother. Biologists would say their coefficient of relatedness is 0.5. This diagram also shows the coefficients of relatedness for siblings, grandchildren, and cousins. Because families share genetic material, natural selection sometimes favors genes that cause individuals to help their relatives, rather than themselves. (Adapted from http://bjoern.brembs.net/)

The definition also needed to include the effect an individual had on the reproductive success of its relatives, which carry the genotype as well. Hamilton proposed that this new kind of fitness be called inclusive fitness.

Hamilton argued that altruism evolved if the cost of helping others was paid back in a rise in inclusive fitness. In the years since he presented this argument, evolutionary biologists have developed much more nuanced mathematical models and have run experiments to test their predictions. The concept of inclusive fitness helps scientists make sense of the parenting strategies of animals like the Seychelles warbler. A female warbler may have higher inclusive fitness by helping raise her younger siblings than by trying to start a family of her own in tough times.

Parents in Conflict

The concept of inclusive fitness helps to explain the origin of cooperative care for offspring. But families can also bring parents into evolutionary conflict. As we saw earlier, the best mating strategy for a male animal may lower the fitness of its female mate. Parents can have the same conflict of interest over raising offspring. A strategy that allows one parent to produce more offspring over its lifetime may lower the number of offspring the other parent can have.

One of the most striking examples of parental conflict is displayed by a bird called the Eurasian penduline tit (*Remiz pendulinus*). After a male and female mate, the father gets to work building a large nest. Hanging from a bough, the nest has a narrow-mouthed opening for the mother to fly into in order to lay the eggs. Either the mother or the father will take on the job of incubating the eggs and then feeding the chicks once they hatch. Despite this care, a third of breeding pairs desert their eggs, leaving them to die.

Tamas Szekely, a biologist at the University of Bath in England, and his colleagues have run experiments to discover the source of this perplexing behavior. They've concluded that sexual conflict is to blame. Each parent can boost its total number of offspring by abandoning the nest to its partner and finding a new bird to mate with. But deserting a nest also has its risks. If a father abandons a mother too soon, she may not be able to lay her eggs successfully. Another risk of an early departure is that another male bird will visit the mother and fertilize her remaining eggs. However, staying around too long also has its risks. If the father waits too long, he may fail to find any free female birds left to mate with. The mother may even abandon the nest before he does, leaving him to raise the chicks by himself.

In order to leave at the best possible moment, a male needs to keep careful track of the female's preparations for laying her eggs. The females, in turn, have evolved strategies to make it difficult for males to figure out what's going on. If a male sticks his head into the nest while she's laying eggs, she'll fight him off—in some cases, even killing him. As a female lays her eggs, she hides some of them in the bottom of her nest. If the male does manage to slip into the nest, he'll get the impression that the female needs more time to finish laying all her eggs.

Experiments confirm that females hide their eggs due to sexual conflict. If scientists uncover the eggs, for example, a female becomes far more aggressive in keeping the male away from the nest. Nevertheless, the male usually deserts her that very day. Now that he can see how far things have progressed, he moves off to find other females.

Conflict within the Genome

Sexual conflict over parenting also appears to be responsible for one of DNA's great mysteries, first discovered by mule breeders 3,000 years ago. Mules are hybrids, produced from a cross between a horse and a donkey. But if breeders use a female horse and a male donkey, they will get a different animal than if they use a male horse and a female donkey. In the latter case, the hybrid is born with a thick mane, stout back legs, and short ears. (This particular hybrid is called a hinny.)

This mismatch would seem to defy the rules of genetics. Genes should have the same effect on a mammal, no matter which parent it inherits them from. It turns out, however, that genes do have different effects, depending on which parent they come from. From place to place on mammalian DNA, groups of carbon and hydrogen atoms, called methyl groups, are clamped on. These methyl groups block the gene-reading enzymes of the cell, making it impossible for RNA molecules and proteins to be produced from the genes. Most genes have the same pattern of silencing regardless of which parent they come from. But biologists have also identified about 100 so-called imprinted genes that are peculiar. The father's copy is silenced while the mother's is not, or vice versa. They are imprinted, in other words, by the parent they come from.

Why should some genes be imprinted? David Haig, an evolutionary biologist at Harvard University, has proposed that sexual conflict drives the evolution of imprinted genes. The conflict in this case is over how much energy a female mammal should give to her embryos.

When a female mammal becomes pregnant, she must use her own body's resources to make the fetus grow. After her offspring is born, it still depends on her for milk. The best long-term strategy for a female is to produce as many offspring as possible over her whole life, and that requires that she not give all her resources to any one pregnancy. Such a sacrifice can put her own health at risk. She may die, or she may have difficulty with future pregnancies. Haig proposed that natural selection favors adaptations that let mothers rein in the nutrition they supply to their offspring.

A father, on the other hand, benefits if his mate puts lots of energy into her current pregnancy. Chances are good that her current offspring are his, but there's little guarantee that her future offspring will be his as well. So even if the mother's health is harmed, fathers benefit from more energy going to their offspring. They'll be healthier as a result, and more likely to survive until adulthood.

Haig proposed that fathers should pass down genes to their offspring that would allow them to get as much out of their mothers as possible. Mothers, meanwhile, should pass down genes that make their offspring's demand more moderate. In effect, the mother and father play tug-of-war in their offspring's DNA.

This sexual conflict, Haig predicted, should also affect the way genes are imprinted. Consider a gene that causes a fetus to draw more nutrients from its mother. One way a mother could reduce that demand would be to imprint her copy of the gene, so that it doesn't make any proteins. In a single mutation, she cuts the production of this dangerous protein in half.

Only a few imprinted genes have been carefully studied to understand how they work, but much of the evidence found so far is consistent with Haig's hypothesis. One of the most striking examples is a gene called insulin growth factor 2 (*Igf2*). It's produced by cells derived from fetuses, which invade the lining of the uterus to draw out nutrients. Normally, only the father's copy is active. To understand the gene's function, scientists disabled the father's copy in the placenta of fetal mice. Without *Igf2* to help draw nutrients from their mothers, the mice were born weighing 40% below average. It's possible that the mother's copy of *Igf2* is silent because turning it off helps to slow the growth of a fetus (**Figure 12.8**).

On the other hand, mice carry another gene called *Igf2r*, which interferes with the growth-spurring activity of *Igf2*. This may be a maternal defense gene. In the case of *Igf2r*, it is the father's gene that is silent, perhaps as a way for fathers to speed up the growth of their offspring. If the mother's copy of *Igf2r* is disabled, mouse pups are born 125% heavier than average.

Another way to test Haig's hypothesis is to look at which species imprint their genes and which do not. Among mammals, only placental mammals imprint genes, while marsupials and monotremes do not. The evolution of the placenta may have given embryos an opportunity to manipulate their mothers in ways other mammals cannot. As a result of this conflict, gene imprinting evolved.

A few other groups of vertebrates also grow placenta-like organs that nurture embryos. Even some species of guppies have placentas. David Reznick and his colleagues tested Haig's hypothesis by looking at *Igf2* in species of guppies with and without placentas. They discovered that species with placentas imprinted the gene, while species without placentas did not.

This research on the evolution of imprinted genes shows why evolution is central to modern biology. If scientists happened to shut down *Igf2* or *Igf2r* out of random curiosity, they would have had no hypothesis to explain why one gene makes mice big and another makes them small. Both genes are found not just in mice, of course, but in humans as well, and mutations to both genes are linked to

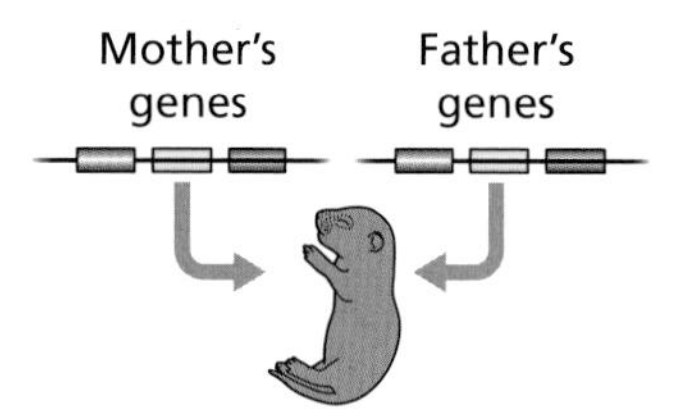

Proteins from mother and father promote growth of fetus

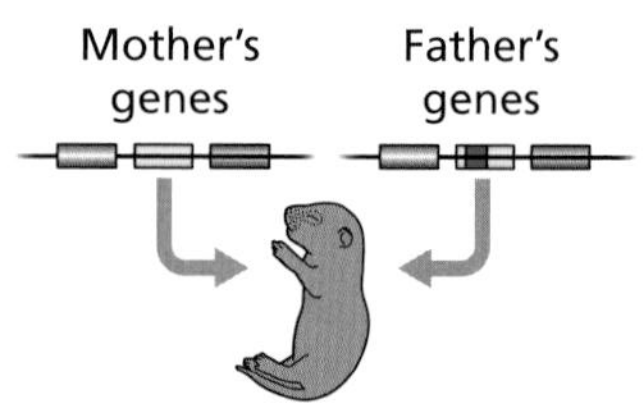

A mutation leads to faster growth. It increases the fitness of fathers because their offspring are more likely to survive. But it lowers the fitness of mothers because it lowers the total number of offspring she can have.

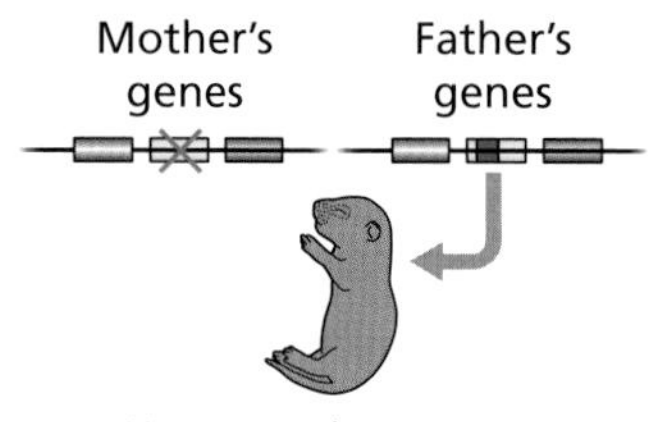

New mutations cause a mother's copy of the gene to be silenced. The supply of proteins is cut in half, and the fetus grows more slowly. Mothers who pass the silenced gene to their young have more offspring.

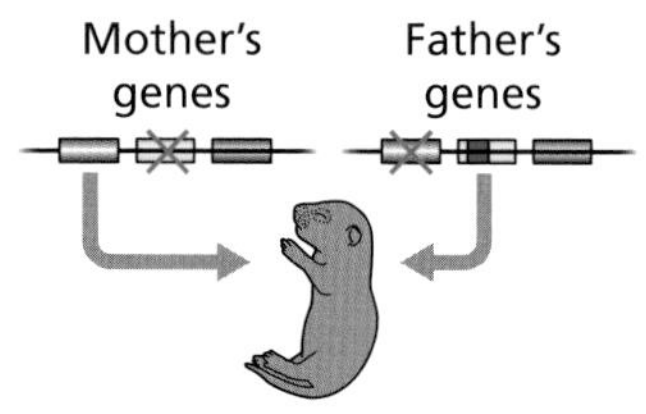

Other genes mutate so that they interfere with growth-promoting genes. These genes benefit mothers and so paternal copies are silenced.

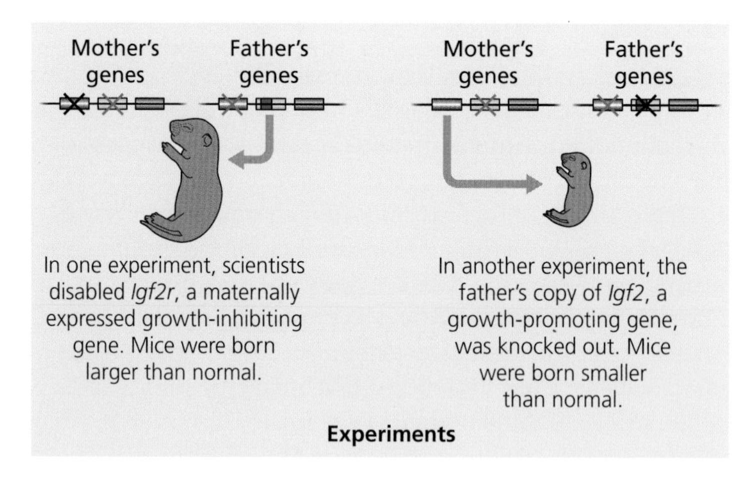

Figure 12.8 Imprinted genes behave differently depending on which parent they come from. The gene from one parent works normally, but the other parent's copy is silenced. David Haig and other researchers have theorized that imprinted genes evolve through an evolutionary conflict between parents. Experiments (summarized in the gray box) are consistent with their hypothesis. (Red X's mark genes silenced by imprinting. Black X's mark genes knocked out in experiments.)

a number of diseases. The evolution of sex, in other words, helps researchers understand human disease. This intersection of evolution and human health—known as evolutionary medicine—is the subject of the next chapter.

TO SUM UP...

- **Sex is widespread in nature, but its success is puzzling. It may allow populations to adapt faster, to keep deleterious mutations from accumulating, or to withstand parasites.**
- **Females must invest heavily in a limited number of eggs, while males can produce vast numbers of sperm. This imbalance explains some of the differences in the sexes in many species.**
- **In many species, males attract females with displays, songs, and other forms of courtship.**
- **In many species, females have been shown to prefer males with certain traits over others. Those traits, in some cases, signal good genes in the males.**
- **Females in most species mate with more than one partner.**
- **Sperm competition has evolved among males in many species.**
- **The number of offspring that females produce and the timing of their reproduction are subject to selection.**
- **Females in some species adjust the ratio of sons to daughters.**
- **Fitness can include the reproductive success of relatives.**
- **Males and females each have optimal strategies for reproductive success. Sometimes those strategies come into conflict, as when birds abandon a nest to find another mate.**
- **Conflict between parents has been invoked to explain the evolution of gene imprinting.**

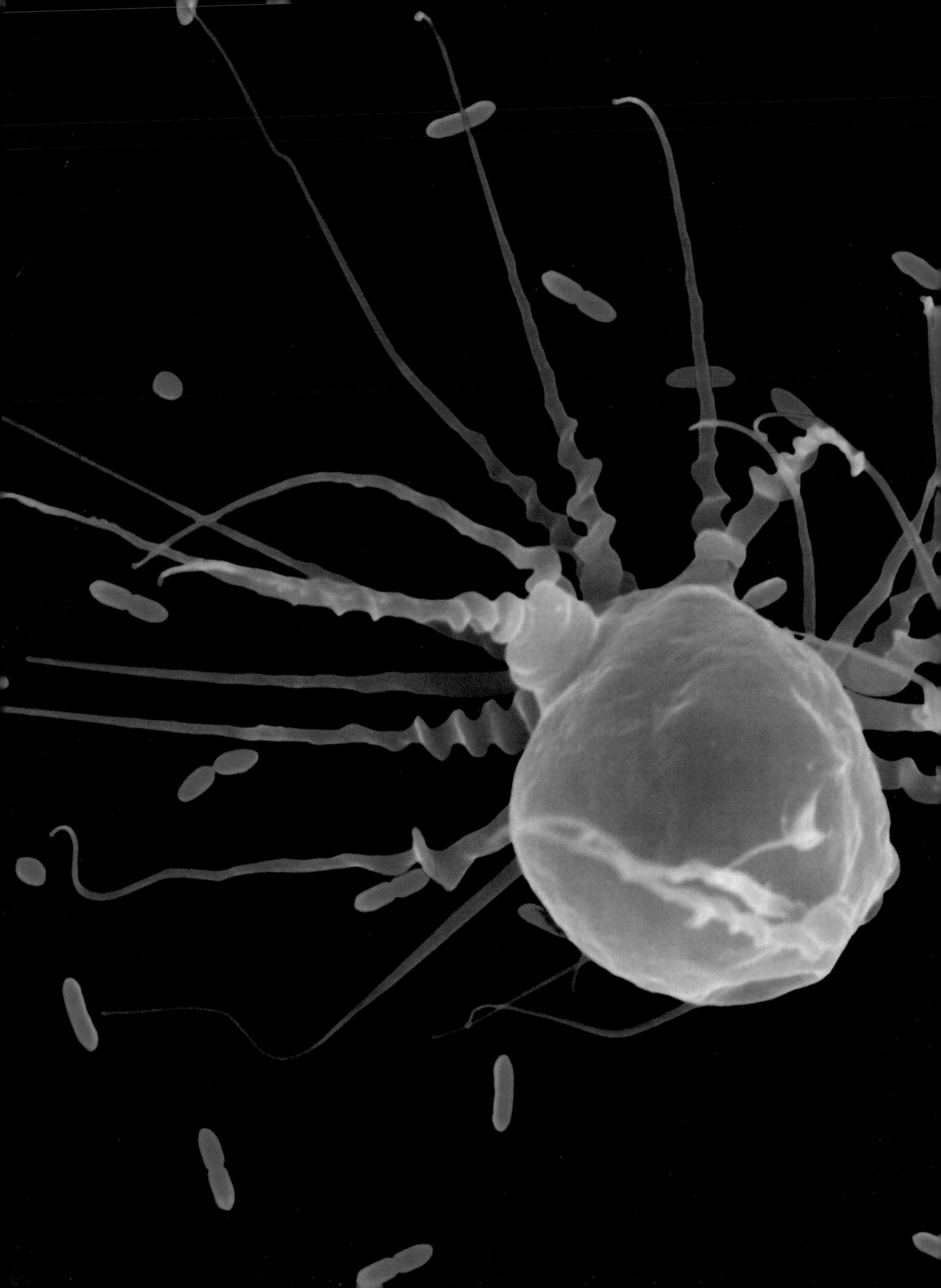

Evolutionary Medicine

13

Ananth Karumanchi is a doctor at Beth Israel Deaconess Hospital in Boston and an associate professor at Harvard Medical School. He is an expert on the risks of pregnancy, including a mysterious condition known as preeclampsia, which causes a dangerous spike in blood pressure. Preeclampsia strikes about 5% of all pregnant women. Left untreated, it can be fatal;

Left: An immune cell attacks bacteria. The evolution of pathogens plays a central role in diseases. Above: S. Ananth Karumanchi of Harvard Medical School has found evidence of an evolutionary conflict underlying pregnancy disorders.

it kills about 75,000 women worldwide each year. Thanks to the level of medical care in the United States, few American women die of preeclampsia, but they can still suffer lifelong effects if their preeclampsia leads to a stroke, a failed kidney, or a ruptured liver. When a woman develops preeclampsia, there's not a lot that doctors can do for her. The one certain way to end the condition is to deliver the baby immediately. Once the baby is out of its mother's body, her blood pressure drops. As a result, preeclampsia is among the most common causes of premature birth in the United States—which puts babies at risk as well.

Doctors would love to have a better treatment for preeclampsia, one that attacked the source of the disease without endangering the baby's health. But despite the fact that physicians have been familiar with preeclampsia for more than two thousand years, they still do not know its cause. Karumanchi wants to find it. In the year 2000, Karumanchi began to suspect that preeclampsia was triggered by the release of some kind of molecule. When he compared the blood of women with preeclampsia with that of healthy women, he discovered a potential culprit: a molecule called soluble Flt-1.

Soluble Flt-1 is only known to appear in the bodies of pregnant women; in women with preeclampsia, Karumanchi found, it appears in much higher levels than in healthy pregnant women. Experiments with soluble Flt-1 have shown that it makes other proteins in the blood stick together so that they can't nourish the walls of the blood vessels. When Karumanchi injected soluble Flt-1 into rats, he found their blood pressure went up. They also suffered damage to their kidneys, which resembled the damaged kidneys seen in women with preeclampsia.

Karumanchi and his colleagues tested his hypothesis by getting their hands on a much larger set of blood samples from pregnant women. Once more, they found more soluble Flt-1 in women with preeclampsia. This time around, however, they also discovered that levels of soluble Flt-1 actually rose five weeks *before* women showed the first signs of preeclampsia.

The discovery was an important one, but it raised a swarm of new questions. No one knew what soluble Flt-1 (sFlt-1)was for. It presumably had some function in pregnancy, because healthy pregnant women make it too. Most puzzling of all was the source of soluble Flt-1. It is not made by the mother's own cells. Instead, the placenta releases the molecule. The placenta, an organ that takes nutrients from mothers and gives them to their fetuses, develops from the same fertilized egg that produces the fetus. Genetically speaking, it's part of the child, not the mother. In other words, the fetus actually gives its mother preeclampsia.

As Karumanchi was puzzling over these mysteries, he was contacted by Harvard evolutionary biologist David Haig (page 286). Haig had predicted years before that preeclampsia would turn out to be caused by a molecule produced by the baby, not the mother. He made the prediction based on his study of evolution.

Haig argued that pregnancy in mammals brings mothers and fathers into evolutionary conflict. A father benefits from mutations that allow his offspring to get more nutrients from its mother. A larger baby is healthier at birth and more likely to survive to adulthood. Mothers get an evolutionary benefit from healthy babies, but if a mother supplies too many nutrients to a single child, her health

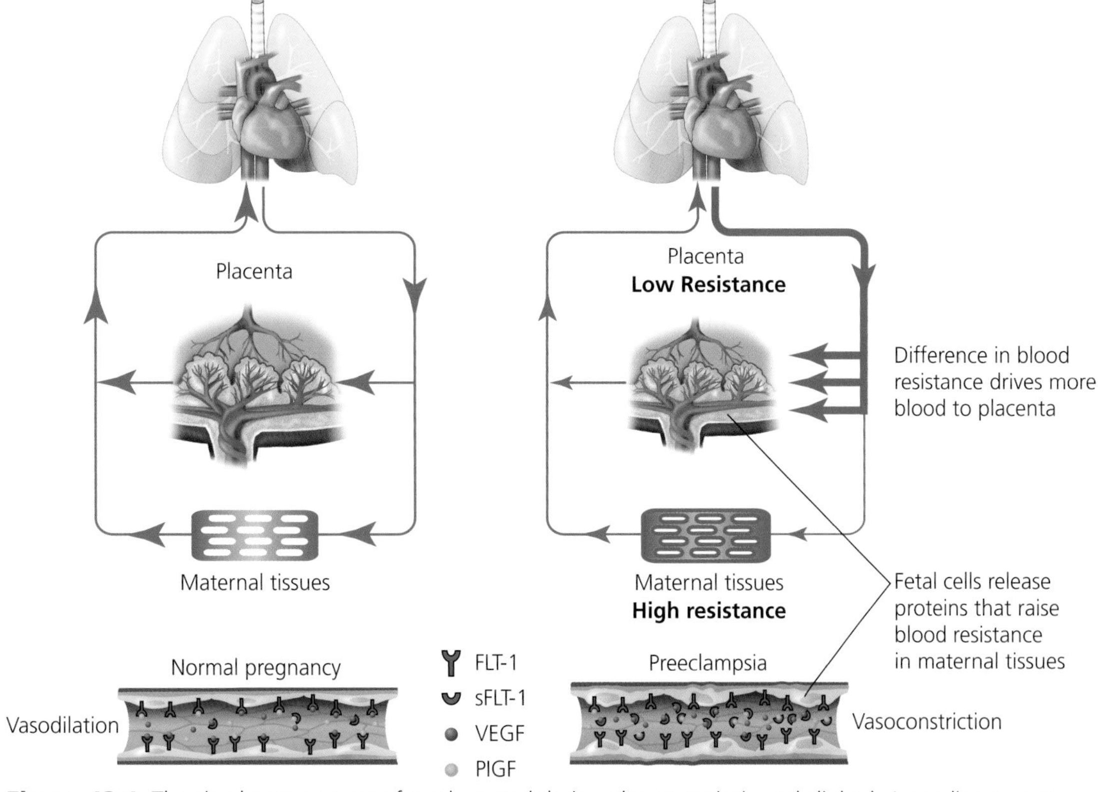

Figure 13.1 The circulatory systems of mothers and their embryos are intimately linked. According to one hypothesis, babies can increase blood flow to the placenta by releasing a molecule called soluble Flt-1 into the mother's bloodstream, raising the resistance in the mother's circulatory system. If the resistance gets too high, the mother may suffer dangerously high blood pressure. (Adapted from Koella and Stearns, 2008)

may suffer. As her health declines, she can bear fewer children. Thus, Haig argued, natural selection should favor genes in mothers that could keep the demands of a fetus in check.

In the last chapter, we saw how this conflict leads to the imprinting of genes that control the growth of fetuses. Haig also proposed that parental conflict could have other effects on pregnancy as well. For example, the amount of nutrients a fetus gets depends on how much blood flows from the mother's circulatory system into the placenta (**Figure 13.1**). If her blood pressure is high, more of her blood will get forced through it. Haig argued that mutations that allowed fetuses to raise their mothers' blood pressure would be strongly favored by natural selection. But the success fetuses had in getting more nutrients this way would favor mutations that helped the mother reduce her blood pressure. Haig argued that the opposing adaptations of mothers and fetuses are normally balanced, so that the mother and baby both survive pregnancy. But it also opened the possibility for pregnancies to go awry. If a mother could not counter the fetus's strategies for raising her blood pressure, it would shoot up to dangerous levels. Instead of simply supplying more nutrients to the fetus, the mother would suffer preeclampsia.

Haig's model of preeclampsia made a provocative prediction. If doctors ever found the factor that caused preeclampsia, they would find that it was produced in the fetal tissue, not in the mother's. Karumanchi was amazed to discover that he had unknowingly confirmed that prediction.

The two scientists began to collaborate—Haig offering insights from evolutionary biology, Karumanchi from his medical studies on pregnant women. Together, they have begun to design experiments that will allow them to test more of Haig's hypotheses about preeclampsia, so that they can better understand what triggers it in some women and not in others. For example, it's possible that genes that help produce or release soluble Flt-1 may be imprinted. In other words, the father's copies of the genes are actively boosting the mother's blood pressure, while the mother's copies are shut down. Like other insights into the basic workings of diseases, the results of their experiments may someday lead to new treatments.

Haig and Karumanchi are exploring a new kind of science—the intersection of evolutionary biology and medicine. Evolutionary medicine, as the field is known, sheds light on diseases and disorders by looking back at their history. In some cases, that history reaches back 600 million years ago, to the origins of multicellular animals. In other cases, that history reaches back over just the past few days of a virus's rapid evolution. Evolutionary medicine can lead to concrete changes in how doctors practice medicine—such as the ways in which they prescribe antibiotics to kill infectious bacteria. It also offers deeper lessons about what it means to be human. Natural selection may be able to shape complex adaptations, but it has not made our bodies perfect. We are still left vulnerable to many disorders. In some cases, evolution has actually made us more likely to get sick, not less. Evolutionary medicine doesn't just reveal our adaptations, in other words, but our maladaptations as well.

Evolving Parasites

Every year, 18.3 million people die worldwide from infectious diseases. Each of the viruses, bacteria, protozoans, fungi, and various parasitic animals that cause those diseases has its own evolutionary history. Evolutionary trees of parasites and their human hosts can reveal that history and point to new ways to treat them.

Some pathogens have been infecting our ancestors for many millions of years. Chickenpox, cold sores, and a number of other diseases are caused by a family of viruses known as herpesviruses. Herpesviruses also infect many other species of animals, including even oysters. But many strains of herpesviruses have been adapted only to humans, ever since our species evolved, and to the ancestors of humans before them. Consider human herpesvirus 5 (also known as cytomegalovirus). Most people get infected with this virus at some point in their lives. It causes fevers, fatigue, and swollen glands.

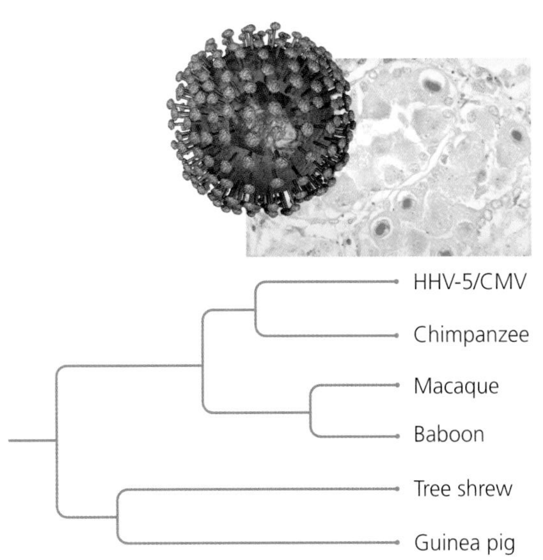

Figure 13.2 This tree shows how human herpesvirus 5 is related to other herpesviruses, with their hosts indicated on each branch. The branching pattern of the viruses mirrors that branching pattern of their hosts, suggesting that hosts and parasites have been cospeciating for tens of millions of years. (Adapted from Koella and Stearns, 2008)

There are a number of strains of human herpesvirus 5, and they are all more closely related to each other than they are to any other herpesvirus. The closest relatives of human herpesvirus 5 strains infect chimpanzees, which are the closest living relatives of humans. Studies on other strains of herpesvirus 5 reveal the same pattern: the evolutionary tree of the viruses mirrors the evolutionary tree of their hosts (**Figure 13.2**).

This mirrorlike phylogeny suggests that herpesvirus 5 has been tracking the evolution of its host for millions of years. The common ancestor of primates and other mammals was host to a herpesvirus, and as that ancestor's descendants diverged into new lineage, the virus diverged as well. It did not leap to distantly related hosts, like turtles or sharks.

One way to test this hypothesis is to estimate how long ago the common ancestor of these viruses lived. Scientists can do this with the help of a molecular clock (page 146). Mutations accumulate in a lineage of viruses at a roughly clocklike rate. The more time passes after two lineages diverge from a common ancestor, the more mutations they will have. It turns out that human herpesvirus 5 has a large number of mutations not shared by its closest relatives in chimpanzees. That's consistent with the idea that the virus has been infecting our hominid ancestors for millions of years.

Scientists get a very different result when they look at some other viruses, such as HIV (human immunodeficiency virus, pages 143 and 147). Each of the major strains of HIV is more closely related to viruses that infect chimpanzees and monkeys than it is to other HIV strains. What's more, HIV strains have acquired relatively few mutations since they diverged from viruses in other primates. The best explanation for these results is that the ancestors of HIV shifted from other primates in the early 1900s.

It took nearly 20 years for scientists to pinpoint the origins of HIV after its discovery in 1983. Thanks to advances in isolating viruses and sequencing DNA, this sort of detective work now takes much less time. In November 2002, for example, a mysterious new disease began to spread through China. At first a Chinese farmer came to a hospital suffering from a high fever and died soon afterwards. Other people from the same region of China began to develop the disease as well, but it didn't reach the world's attention until an American businessman flying back from China developed a fever on a flight to Singapore. The flight stopped in Hanoi, where the businessman died. Soon, people were falling ill in countries around the world, although most of the cases turned up in China and Hong Kong. About 10% of people who became sick died in a matter of days. The disease was not the flu, not pneumonia, nor any other known disease. It was given a new name: severe acute respiratory syndrome, or SARS.

In March 2003, the World Health Organization established a network of labs around the globe to pin down the cause of SARS. Just a month later, they had identified a virus as the culprit. But what kind of virus was it?

To answer that question, Christian Drosten of the Bernhard Nocht Institute for Tropical Medicine in Hamburg, Germany, and his colleagues drew an evolutionary tree based on the virus's RNA. In May 2003, they reported that the SARS virus was most closely related to pathogens called coronaviruses, which can cause colds and stomach flu.

Based on their experience with viruses such as HIV, scientists suspected that SARS had evolved from a virus that infects animals. They began to analyze viruses in animals with which people in China have regular contact. As they discovered new viruses, they added their branches to the SARS evolutionary tree.

In 2003, panic swept southeast Asia as a new disease, known as SARS, broke out. At first, doctors did not know what caused it.

The first big breakthrough came with the discovery of a closely related virus in cat-like mammals called palm civets. The civets are sold in animal markets in China for meat, suggesting a route the virus could have taken from one species to the other. In late 2003, a second SARS outbreak swept China, and scientists traced it to a separate invasion of palm civet viruses into human hosts. But further research revealed that civets are not the main host for the viruses that give rise to SARS in humans. Both the viruses from civets and humans belonged to tiny tufts of branches nestled in a large evolutionary tree of coronaviruses that infect Chinese bats. Viruses from bats regularly infect other animals, which can then pass them to humans (**Figure 13.3**).

This rapid response from evolutionary biologists is becoming increasingly important to the world's health. New diseases continue to emerge every year, and the only way to know what they are is to know what they evolved from. This knowledge also helps doctors and public-health officials to control the outbreaks. Knowing which animals are the sources of diseases allows public-health workers to block the transmission—in the case of civets, by banning their sale in open markets.

Once scientists identify the source of new diseases, they can compare the pathogens in their original hosts and the ones in humans to see how they have evolved to adapt to our biology. Those new adaptations may provide clues to how to fight those diseases, and understanding the origin of diseases also helps scientists to predict where new diseases will come from. Scientists now recognize that Asian bats are a reservoir for many kinds of viruses, for example, and may well produce new outbreaks. Thus many scientists are now trying to discover new viruses in bats and understand how they spread. African primates, on the other hand, have proven to be the source of multiple invasions of HIV. The leap from primates to humans was probably made possible by the large-scale deforestation of central Africa, along with a growing trade in the meat of forest animals,. Scientists have identified many other viruses in the same lineage that gave rise to HIV. There's good reason to expect that some of them will make the leap as well. How prepared we'll be for the next leap, however, is an open question.

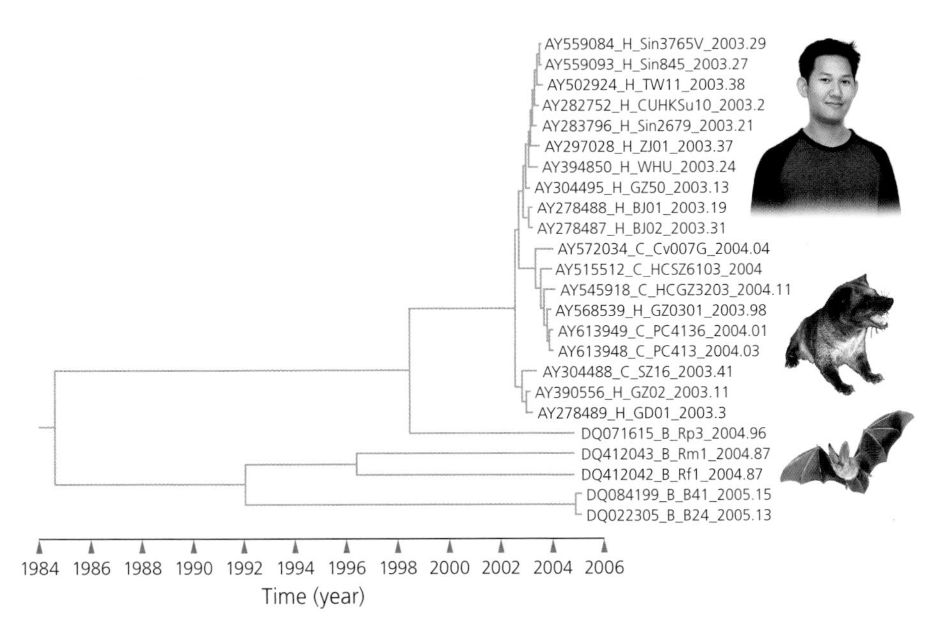

Figure 13.3 Once the SARS virus was isolated in humans, researchers searched for its closest relatives in animals. As this evolutionary tree shows, the SARS virus evolved from a family of viruses that circulates in bats. (Each branch is labeled with a strain code, followed by an abbreviation for its host: H = human, C = civet, B = bat. Adapted from Hon et al, 2008)

Human Petri Dishes

To learn about the workings of evolution, scientists have reared bacteria and viruses in the laboratory, observing how they change through mutation and selection (pages 117 and 134). But, when bacteria and viruses infect us, we become living petri dishes, and our pathogens evolve inside us.

Figure 13.4 shows the record of this evolution within nine people infected with HIV. Doctors took blood samples from them every few months, and James Mullins and his colleagues at the University of Washington isolated the HIV viruses from them. By sequencing a gene from the viruses, the scientists were able to draw the evolutionary tree. The viruses in all the patients descended from a common ancestor, but once they entered their hosts, they continued to evolve. Each time the viruses replicate in a host cell, many of them mutate. Most of the mutations cripple the viruses, but a few raise their fitness.

During the evolution of HIV, natural selection favors mutations that allow the viruses to copy themselves faster. The fitness of HIV can also be raised by mutations that allow the virus to escape destruction. An infected person's immune system can come to recognize parts of the surface proteins on HIV, enabling it to attack the viruses. As viruses with one kind of surface protein begin to suffer major assaults from our immune cells, other viruses with different proteins can

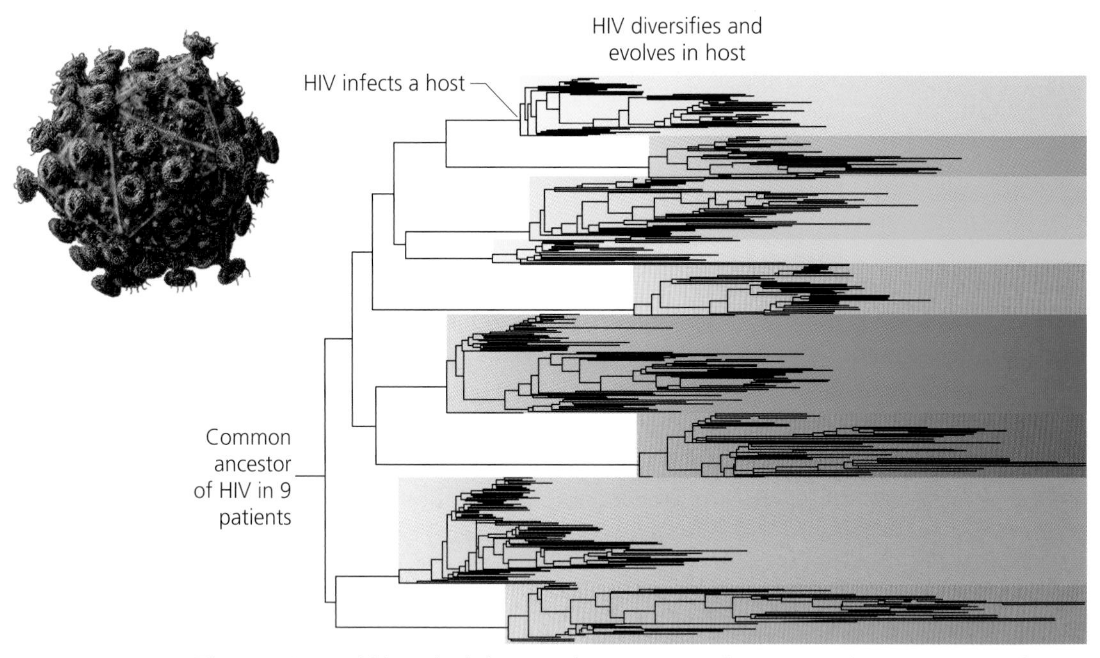

Figure 13.4 Within a single host, pathogens can replicate many times, mutate, and undergo natural selection. This diagram shows how related HIV viruses infected nine people (colored bands), and diversified inside them. These newly evolved viruses were better able to avoid the host's immune system and resist HIV-fighting drugs. (Adapted from Roualt, 2004)

reproduce more rapidly. As a result, the diversity of HIV within a single person blooms.

Pathogens don't make us sick and die because they enjoy it; this suffering—or, technically, virulence—is just a byproduct of the way in which pathogens grow and reproduce. Malaria, for example, is caused by single-celled parasites called *Plasmodium*. The parasite is carried by mosquitoes, which inject it into the bloodstream when they bite. *Plasmodium* can then invade red blood cells. If the infected blood cells enter the spleen, they are destroyed. *Plasmodium* avoids this fate by making sticky proteins that glue the infected red blood cells to the walls of blood vessels, so that they never pass through the spleen. Over time, these cells stick together in clumps, blocking the flow of blood and causing blood vessels to tear open.

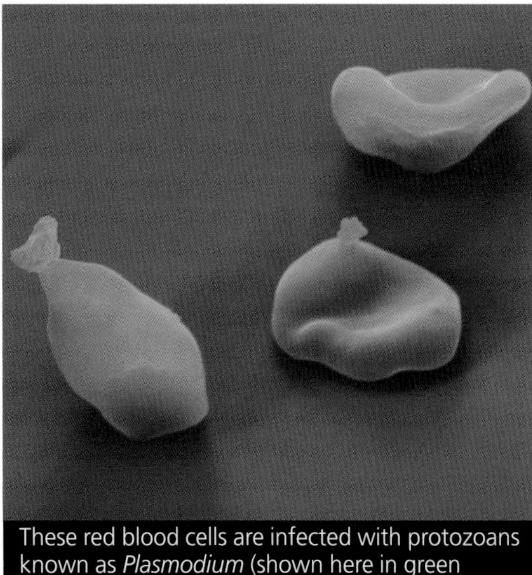

These red blood cells are infected with protozoans known as *Plasmodium* (shown here in green emerging from the cells). *Plasmodium* is the cause of malaria, one of the most devastating diseases in the world. They are carried from host to host by mosquitoes.

Some diseases are more virulent than others. There are pathogens that barely slow us down, and others that can kill us in a few days. Virulence varies not only between species but within them as well. *Toxoplasma gondii*, a relative of *Plasmodium*, can live peacefully, for the most part, inside people's brains. Billions of people are infected with *Toxoplasma* without even knowing it. In Brazil, however, many strains are more likely to cause dangerous fevers and inflammation.

A pathogen's virulence can evolve over time. Rabbits provide one of the best examples of this sort of evolution. Before 1859 there were no rabbits in Australia. They were introduced by Thomas Austin, a farmer who had enjoyed shooting rabbits in Scotland before immigrating to Australia. Without predators to control them, the rabbits exploded across the continent, eating so much vegetation that they began to cause serious soil erosion. In the 1950s, scientists deployed a biological counteroffensive, known as rabbit myxoma virus.

The rabbit myxoma virus, which was discovered in South America, causes deadly infections in hares. Scientists expected that the virus would also decimate the Australian rabbits. At first, the virus worked as expected, and rabbits died in droves. But then the virus gradually became less virulent. Rabbits would get sick, but then recover. In later years, the rabbit myxoma virus became more virulent again, but it never became as deadly as the original strain. Rabbits today continue to be an ecological blight on Australia.

Scientists can also observe the virulence of pathogens as they evolve in their labs. Kanta Subarro, a virologist at the National Institute of Allergy and Infectious Diseases, witnessed such an evolution while searching for an animal model she could use to study SARS. Mice did not get sick from the SARS virus, not even mice that had been genetically engineered so that they couldn't develop an immune system. So Subarro and her colleagues inoculated mice with the SARS virus, gave it a chance to replicate inside them, and then isolated the new viruses to infect new mice. As the virus replicated inside mice and then moved to new hosts, it evolved. Over the course of just 15 passages, it changed from a harmless virus into a fatal one. One sniff of SARS was now enough to kill a mouse.

Why did the myxoma virus and the SARS virus evolve to different levels of virulence? According to one hypothesis, the answer can be found in a trade-off that pathogens face. Their reproductive success depends on both their replication within hosts and their infection of new hosts. If a pathogen multiplies too quickly, it may kill its hosts before it can spread to a new one. If it's gentle enough to keep its hosts alive, it will be rapidly outcompeted by faster-growing strains. The details of how a particular pathogen replicates and spreads may determine its most adaptive level of virulence.

Dieter Ebert, an evolutionary biologist at the University of Fribourg, has tested the trade-off hypothesis using water fleas and bacteria. The bacteria (called *Pasteuria ramosa*) infect the water fleas and produce a vast number of spores inside the animals. The bacteria eventually kill the water fleas, whereupon the spores stream out of their dead bodies.

The particular way in which *Pasteuria* reproduces allowed Ebert's team to measure its reproductive success exactly (**Figure 13.5**). They infected a set of genetically identical water fleas with the bacteria and then counted the number of spores produced upon their deaths. The scientists found that the bacteria cut the life span of the water fleas roughly in half, from three months on average to a month and a half. Some bacteria killed their hosts in just 20 days, while others left them alive for more than 70 days.

At both extremes, the bacteria produced fewer spores than an intermediate host lifespan of 55 to 60 days. The fast-killing bacteria killed off their hosts before they could produce many spores. The slow killers, Ebert speculates, don't make good use of their hosts, which may get too old to provide a good supply of nutrients to their pathogens.

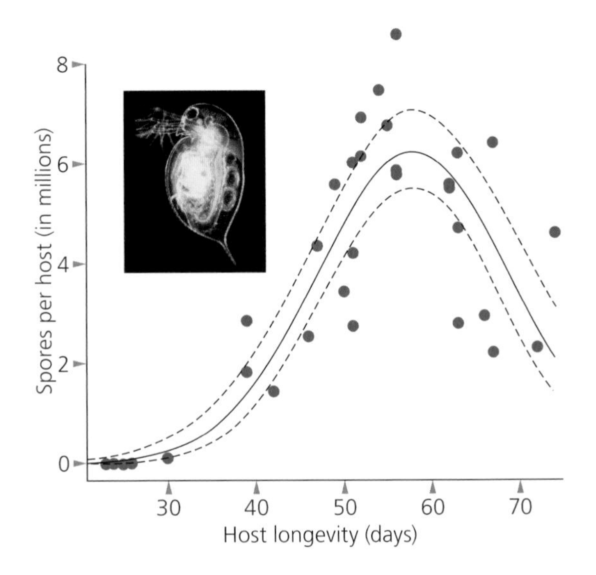

Figure 13.5 Like other organisms, parasites face an evolutionary trade-off. Parasites can boost their fitness by reproducing quickly inside their hosts. But if their hosts die too quickly, the parasites have less time to exploit them. Scientists documented this trade-off in bacteria that infect water fleas (inset), castrate their hosts, and then produce spores inside them that are released when the water fleas die. The bacteria that kill their host quickly or very slowly produce fewer spores than the ones that kill at an intermediate time. (Adapted from Jensen et al, 2006)

In a case like that of *Pasteuria*, natural selection will favor the pathogens at the top of the curve in **Figure 13.5**—in other words, the pathogens that reproduce the most. Exactly where the top of the curve ends up for a particular pathogen depends on its own set of conditions. Kanta Subarro's experiment may have led SARS to become more virulent because the virus no longer paid a cost for being deadly. Subarro ensured that the viruses would be able to get to another mouse, no matter how fast they replicated in their current host. Outside the laboratory, pathogens may become more virulent as it becomes easier for them to move between hosts. Pathogens can also become more virulent if several different strains infect the same host at once. If one strain of the pathogen reproduces slowly and causes little harm to its host, it will be outcompeted by the more virulent pathogens living alongside it.

Molded by Parasites

Pathogens evolve to adapt to their hosts, and their hosts, in turn, adapt to the pathogens. All living things have defenses against infections. Even bacteria can recognize invading viruses and chop up their DNA. Animals—and in particular, vertebrates—have evolved particularly elaborate immune systems, made up of many different types of cells that trade signals with one another, swallow up some pathogens, and manufacture chemical weapons to destroy others.

Despite the power and sophistication of the human immune system, our pathogens continue to thrive. One reason for their continued success is that they can evolve faster than we can. While humans may take two or three decades to reproduce, pathogens can divide in a matter of hours or less. Some pathogens, such as HIV, have very high mutation rates, which produces a great deal of genetic variation that speeds up their response to selection. Viruses and bacteria can also acquire genes through horizontal gene transfer, speeding up their evolution even more.

The high-speed evolution of pathogens means that natural selection is always favoring new defenses in their hosts. When evolutionary biologists measure the strength of selection on the human genome (page 148), immune-system genes consistently turn up near the top of the list.

Humans have evolved new strategies for fighting diseases in just the past few thousand years. One of the most devastating of these recent diseases is malaria. Scientists suspect that the strains of *Plasmodium* that cause the most harm today first evolved in Africa within the past 6,000 years as farmers began to clear forests. Malaria-spreading mosquitoes could breed in the standing water in their fields, and could then infect the farmers as they slept in their villages. The disease then spread wherever mosquitoes could carry it—even to England and the northern United States. In the mid-1900s, public-health workers eliminated malaria from the United States by eliminating breeding sites and spraying pesticides in people's homes. But in Africa and elsewhere it remains a major scourge, infecting 250 million people a year and killing 880,000.

In response, humans have evolved a number of defenses to malaria in just the past few thousand years. Sickle-cell anemia (page 119) is the byproduct of one of those defenses. People suffer from sickle-cell anemia when they inherit two copies of an allele called *HbS*. If they get one copy, however, they are only one-tenth as likely to get severe malaria. It's possible that *Plasmodium* parasites can't grow as fast inside red blood cells with the *HbS* allele, or that infected cells are eliminated from the body faster.

In any case, carrying a single *HbS* allele has provided a huge boost in fitness. Neil Hanchard, a geneticist at the Mayo Clinic in Minnesota, and his colleagues recently compared the DNA of people with the *HbS* allele in different parts of Africa, as well as people of African descent in Jamaica. The scientists found that the DNA surrounding the *HbS* allele was very similar from person to person. Such a similarity (known as linkage disequilibrium) is a sign of strong recent natural selection (page 127).

As people have moved around the world, they've taken their pathogens along for the ride. Their movements have brought diseases to people who had never been exposed to them before. In some cases, the encounter has been devastating. When Spanish conquistadors arrived in the New World, they brought with them an army of pathogens that had evolved in the unhealthy cities of Europe. The conquistadors themselves had some resistance to those diseases, thanks to a long coevolutionary relationship. But the people they met in the New World had left the Old World some 20,000 years ago. They had very little resistance to the diseases of the Spaniards and died in vast numbers. If not for the coevolution of humans and pathogens, world history would have taken a very different path.

Five centuries later, HIV moved into humans in Africa and then spread out to the rest of the world. Along the way, however, some people have proven remarkably resistant to the virus. They carry HIV in their blood for years, and yet their immune system never collapses. The reason appears to be that they are missing a receptor called CCR5, which HIV generally needs to make its way into immune cells.

The CCR5 mutation is remarkably common, especially in Europe. As many as 10% of people in some regions carry it. In Africa and elsewhere, however, it's practically nonexistent. The variations in DNA around the CCR5 mutation today suggest that the mutation arose about 700 years ago. That's long before HIV ever existed. Something else must have driven up the allele to its current levels in Europe.

Experiments suggest that the mutation may have been favored originally because it protected people against another major scourge. Some scientists have proposed the original defense was against bubonic plague, which killed off roughly a quarter of all Europeans in the fourteenth century. Others have argued that repeated epidemics of smallpox were more likely to have been the cause.

Parasite-driven variations in our species cannot be ignored in the search for new treatments for diseases. Scientists may find inspiration for new treatments by looking at the ways in which defenses have evolved naturally. However, they must also bear in mind the complexities of that evolution. While natural selection may create new adaptations, they often have many different effects, some of which are dangerous. If we could get rid of our CCR5 receptors, it's conceivable that the rate of HIV infection might go down. But we might discover an unexpected side effect. CCR5 also shows signs of protecting us from other diseases. People who carry the CCR5 mutation that protects them against HIV are more likely to suffer more from diseases such as West Nile fever.

Evolution's Drug War

Just as we are shaped by our pathogens, our pathogens are reshaped by us. Host and pathogen can evolve new adaptations to each other. One of the most remarkable sets of adaptations first came to light in the late 1980s. Michael Zasloff, then a research scientist at the National Institutes of Health, made the discovery almost by accident. At the time, he was using frog eggs to study how cells use genes to make proteins. He would cut open African clawed frogs, remove their eggs, and then stitch them back up. After he had put enough of the frogs in a tank, he'd take them to a nearby stream and let them go. Sometimes the tank water became murky and putrid, and yet his frogs—even with their fresh wounds—did not become infected.

Zasloff suspected that the frogs were defending themselves by making some kind of antibiotic. He ground up frog skin until he isolated short chains of amino acids known as peptides. These particular peptides could kill bacteria, thanks to their negative charge, which attracted them to the positively charged membranes of bacteria, but not to the frog's own positively charged cells. Once the peptides made contact with the bacteria, they punched holes in their membranes, causing the bacteria to burst open.

If frogs had these antimicrobial peptides, Zasloff reasoned, then it was possible that other animals had them too. He turned out to be right: these powerful molecules are produced in animals ranging from insects to sharks to humans. We produce many different antimicrobial peptides on our skin, in the lining of our guts, and in our lungs. Mutations that interfere with the ability to produce them can make people dangerously vulnerable to infections.

It turns out that bacteria have evolved counterdefenses against antimicrobial peptides. Some bacteria make a protein that can cut a host's antimicrobial peptide into pieces before it can do any damage. Animals, in turn, have responded by stiffening the peptides, making them harder to cut. But microbes have responded with even more counterdefenses: some species secrete proteins that grab the antimicrobial peptides and prevent them from entering the bacteria.

One of the most potent ways for animals to overcome all of these strategies is to make lots of different kinds of antimicrobial peptides. New ones can be produced by gene duplication, or by recruiting proteins that originally had other functions. The more antimicrobial peptides an animal makes, the harder it is for bacteria to evolve counterdefenses against them all. As a result, the genes for antimicrobial peptides have undergone more evolutionary change than any other group of genes found in mammals.

Silver Bullets

Scientists discovered the first antibiotics, made by bacteria and fungi, in the mid-1900s, and they soon ushered in a new chapter in the history of medicine. Infections that once almost certainly would have been lethal simply disappeared in a matter of days. Some optimists declared that infectious diseases would soon be a thing of the past.

But not long after antibiotics first became available, doctors began reporting that they sometimes failed. In the 1950s, Japanese doctors used antibiotics to battle outbreaks of dysentery caused by *E. coli*, only to watch the bacteria develop resistance to one drug after another.

The Japanese doctors had come face to face with one of evolutionary medicine's most sobering lessons: medicine itself can drive the evolution of disease. The microbes that breed in an infection also mutate, and some of those mutations may help the microbes to resist an antibiotic. In the absence of an antibiotic, those mutations may not increase their fitness. In fact, they may even make the microbes grow more slowly. As a result, those mutations will remain rare. But antibiotics can alter the evolutionary landscape in a flash. If a mutation provides a microbe with even a little resistance, it will have more reproductive success than vulnerable microbes that die off altogether.

Resistance can evolve in many ways. Some mutations can make it harder for antibiotics to attack their targets inside the microbe. Some alter membrane pumps so that the microbes can flush the antibiotics out quickly before they can cause serious harm. As a lineage of resistant microbes takes over a population, new mutations emerge, some of which can make them even more resistant. The evolutionary costs of these mutations can be eliminated by new mutations, which allow the microbes to reproduce just as quickly as vulnerable strains. These compensatory mutations allow resistant microbes to survive when there are no antibiotics to give them an edge.

Much of what scientists understand now about the evolution of resistance has come from watching it take place in their laboratories. When Michael Zasloff discovered antimicrobial peptides, he immediately wondered if they might prove to be resistance-proof antibiotics. He teamed up with Graham Bell, an evolutionary biologist at McGill University in Montreal, and Bell's student Gabriel Perron to test his new drugs.

The researchers began by exposing *E. coli* to very low levels of an antimicrobial peptide. A few microbes survived, which the scientists used to start a new colony. They then exposed the descendants of the survivors to a slightly higher concentration of the antimicrobial peptide. Again, most of the bacteria died, and they repeated the cycle, raising the concentration of the drug even more. After only 600 generations, 30 out of 32 colonies had done the impossible: they had become resistant to a full dose of antimicrobial peptides.

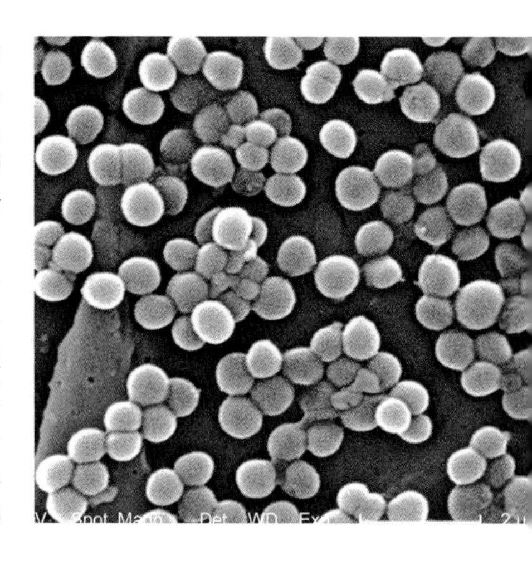

A human body can also serve as a living flask in which bacteria can become more resistant. Alexander Tomasz, a microbiologist at Rockefeller University, and his colleagues have tracked the evolution of resistance in a single patient, known only as JH (**Figure 13.6**). In 2000, JH developed an infection of *Staphylococcus aureus* bacteria in a heart valve. He was treated with an antibiotic called rifampin, which failed to work; his doctors

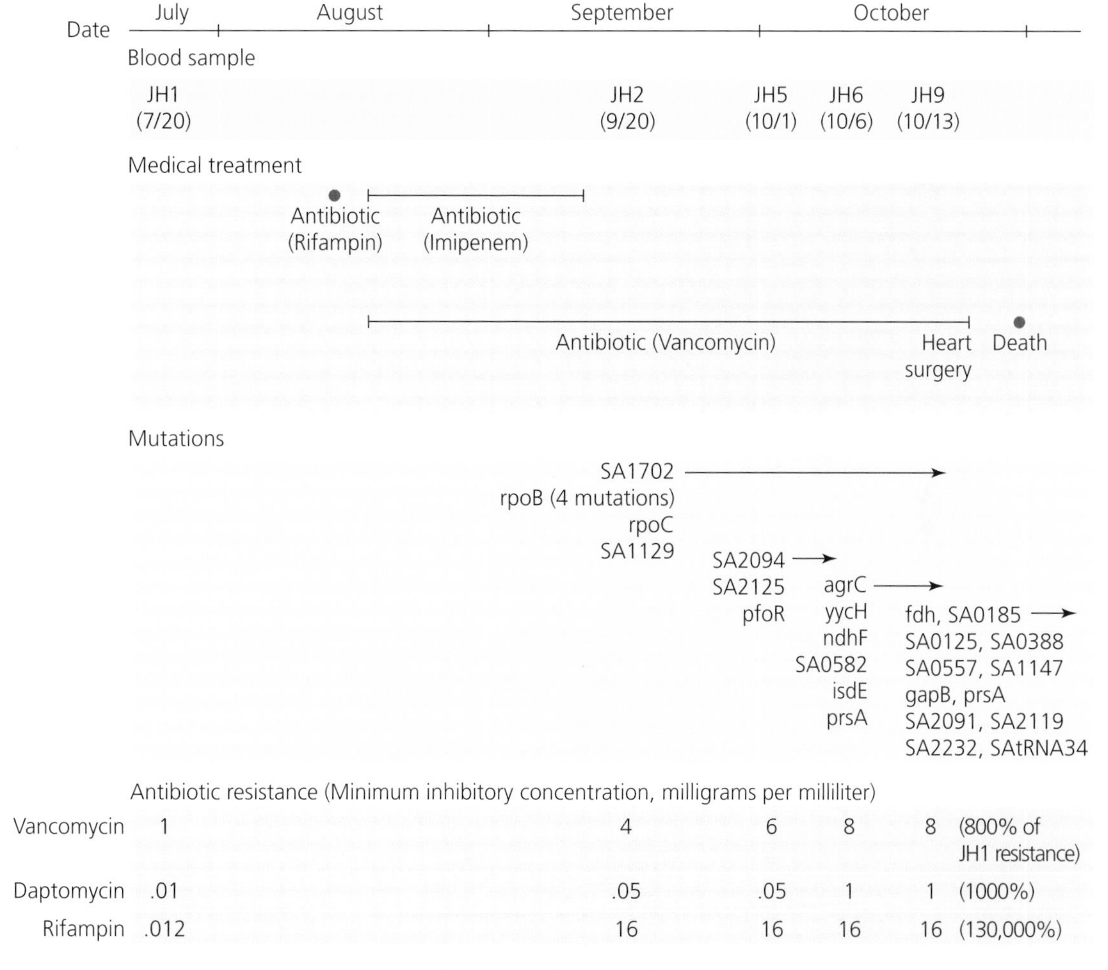

Figure 13.6 In 2000 a patient known as JH developed an infection of *Staphylococcus aureus* (top right). Doctors took a series of blood samples and identified the new mutations that arose in the bacteria over the course of the infection. Mutations that conferred more resistance to antibiotics were favored by natural selection. (Adapted from Mwangi et al., 2007)

then gave him heavy doses of more powerful antibiotics, such as vancomycin, which failed as well. After three months of treatment, surgeons replaced his heart valve, but he died two weeks later.

Tomasz and his colleagues were able to isolate the bacteria from a series of five blood samples that doctors had taken from JH over the course of the infection. They sequenced a *Staphylococcus aureus* genome from the first sample and then analyzed the DNA from later samples. The bacteria from the later samples shared a set of genetic markers with the original one, demonstrating that they were their descendants, rather than new arrivals from a separate infection.

Over the course of the infection, Tomasz and his colleagues found, the bacteria evolved increased resistance to three out of four antibiotics. The bacteria became eight times more resistant to vancomycin, for example, and 1,000 times more resistant to rifampin. These changes were the result of mutations that arose in the bacteria and were favored by natural selection. In all, Tomasz and his colleagues pinpointed 35 mutations that distinguished the bacteria in the last sample from those in the first. They could even see the mutations accumulate from one sample to the next. Some of the mutations are familiar to scientists from other resistant strains of bacteria; others are new, altering the bacteria in ways the scientists have yet to understand.

The bacteria in JH built up new mutations through vertical gene transfer. But bacteria can also acquire genes through horizontal gene transfer, which can speed up the evolution of antibiotic resistance dramatically. Many species of bacteria that live in the soil have genes that can, by coincidence, provide resistance to antibiotics; from time to time, they can pass those genes on to bacteria that cause human infections. Once genes evolve high levels of resistance, they can move from one species to another, either in the soil or in our bodies.

Resistance genes may start out being transferred individually, but over time they can be combined. Some of them are carried on ringlets of DNA called plasmids that bacteria exchange. Those plasmids sometimes mutate, splicing together their DNA so that resistance genes from separate plasmids can end up together on one. A plasmid resistant to two antibiotics may raise the fitness of bacteria much more than a plasmid with just one. And if it should pick up a third resistance gene, its fitness rises even more. This is probably what happened in Japan during the dysentery outbreaks: drugs fostered the evolution of *E. coli* resistant to several antibiotics.

Today, new strains of pathogens are emerging that are resistant to just about every antibiotic on the market. The only way to cope with the crisis is to treat it as an evolutionary phenomenon. It's not enough to recognize that evolution is taking place; doctors need to understand the complexities of that evolution. Some hospitals have tried to fight resistant bacteria by rotating their antibiotics over the course of a few months, so that the bacteria don't have much time to evolve increased resistance. That strategy hasn't worked, and mathematical models of evolution show why: a more promising strategy, according to the models, is to give different antibiotics to different patients, to slow down the transmission of bacteria between them.

Nature may be able to give scientists some clues to new strategies. It's striking, for example, that Zasloff, Bell, and Perron were able to observe bacteria evolve high levels of resistance to antimicrobial peptides in a matter of weeks, while bacteria have not evolved such resistance over millions of years in the natural world. That's because the scientists exposed *E. coli* to an intense dose of a single kind of molecule. Our bodies, on the other hand, have evolved many different antimicrobial peptides, all of which were modified into more effective forms.

Unfortunately, modern medicine tends to work more like Zasloff's experiment than like our own evolution. It may be time to find a new, more complex strategy. At the same time, however, some old-fashioned strategies still work well. Washing your hands with soap is a very effective way to avoid getting infected, and it does not foster the evolution of resistant strains.

Genetic Drift and Disease

In every generation, many harmful mutations spontaneously arise. They may produce life-threatening genetic disorders, or they may make people more prone to diseases like cancer later in life. Natural selection can eliminate many of them because they have such strong effects on reproductive success. But natural selection's broom doesn't sweep perfectly clean. Because natural selection favors sickle-cell heterozygotes, for example, it cannot eliminate sickle-cell anemia. Natural selection is also weak in small populations, where genetic drift can allow even harmful alleles to spread to high frequencies.

On the remote Pacific island of Pingelap, for example, 5% of the population is completely color-blind. By comparison, only 0.003% of the United States population suffers from this condition, called achromatopsia. Historical research indicates that, around 1775, a typhoon reduced an already small population down to just 20 survivors. One of those survivors carried one copy of the achromatopsia allele. Thanks to that fluke of history, 5% of the island's current residents now suffer from the disease because they carry two copies of the original allele, while 30% carry a single copy.

Islands are not the only place where genetic drift drives up genetic disorders. In the late 1600s, a small group of closely related farmers traveled from Switzerland to Germany and finally to the United States, where they became known as the Amish. Shunning intermarriage with other groups, they kept to themselves, and they still do today. In effect, they've created a genetic island in the middle of a continent. Not surprisingly, the Amish suffer high rates of certain genetic disorders, such as Ellis–van Creveld syndrome, which leads to extra fingers and dwarfism (**Figure 13.7**). Scientists have been able to trace that disorder to a single Amish couple who immigrated to Pennsylvania in 1744.

Genetic drift's biggest impact on human health probably occurred about 100,000 years ago. *Homo sapiens* evolved in Africa about 200,000 years ago (page 139). For at least 100,000 years, our species remained on that continent,

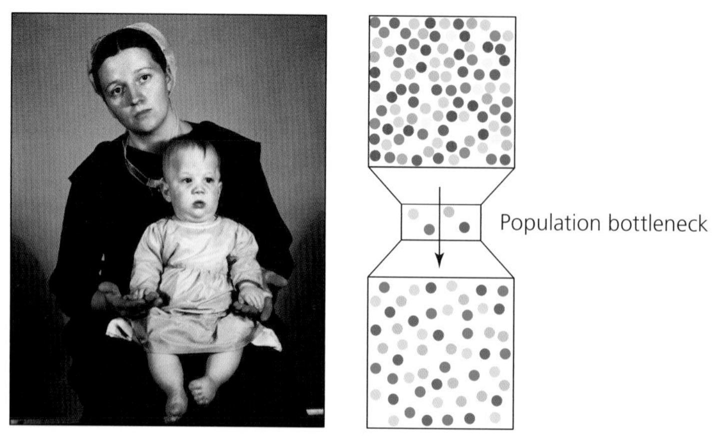

Figure 13.7 Some human populations suffer from high rates of certain genetic disorders. The Amish, for example, are unusually likely to suffer from Ellis–van Creveld syndrome, which causes deformities to the skeleton. The Amish descend from a small group of immigrants from Europe who settled in the United States. Because their population was so small, their genotypes were not a representative sample of the genotypes in their homeland. Even one person with genes for a rare genetic disorder could make it much more common in the population.

evolving an increasing level of genetic diversity. Then a very small number of Africans—probably originating in or around Ethiopia—migrated out of Africa. Their descendants would people Europe, Asia, and Australia. This migration out of Africa is recorded in the collective DNA of our species. The genetic variation among populations in Africa is much larger than the diversity found on the rest of the planet, even though Africans only make up 13% of the world's population.

The migrants who left Africa were not a representative sample of the continent's population. Like the settlers of islands, they had unusually high numbers of certain alleles, including some that cause diseases. One of these diseases is myotonic dystrophy, a genetic disorder that causes muscles to waste away slowly. It's triggered by a sequence of repetitive DNA near a gene for a muscle protein, known as DM. On rare occasions, the repetitive DNA is accidentally mutated into a longer segment of repetitive DNA. In later generations, that longer segment can become longer still. Eventually, it becomes so long that it disrupts the gene.

Myotonic dystrophy is rare in most of the world, but in sub-Saharan Africa it's practically nonexistent. To investigate why, Sarah Tishkoff (page 140) and her colleagues analyzed the stretch of DNA that includes the DM gene, comparing segments from 25 different populations around the world. They found that most Africans had only a few copies of the repetitive DNA. In Ethiopia, however, some populations had longer chunks of repetitive DNA, and these longer chunks are even more common out of Africa.

Tishkoff concluded that a mutation arose in an early Ethiopian, expanding the repetitive DNA near the DM gene. That Ethiopian's descendants were part of the migration out of Africa that led to the peopling of Europe and Asia. As the migrants rapidly expanded across the other continents, they passed down this repetitive DNA, which then mutated again in some people into a form that caused myotonic dystrophy.

Maladapted

Natural selection can adapt organisms to their changing environments, but it does not produce perfect forms of life. In many cases, natural selection actually leads not to adaptation but to maladaptation—the source of many of our medical problems.

Some of our maladaptations are the result of tinkerings with old body plans for new environments. Evolution can modify only what already exists, rather than ripping up old plans and creating new ones from scratch. As a result, many structures have glaring weaknesses. When male human embryos develop, for example, their testes start out high up in their bodies. Over time, the testicles descend. As they migrate, they push down on the body wall, creating a weak spot. It is here that a loop of intestine can slip through, creating an inguinal hernia.

Why should the testicles make such a strange journey? A comparison with a shark shows why. Shark testicles develop in the animal's abdomen and stay there. As a result, sharks don't get hernias, because the wall of muscle surrounding their bodies is intact. But when mammals evolved a warm-blooded metabolism, a problem emerged. Sperm are delicate, and overheating them can kill them off. Mammal testicles descended until they had formed a sac outside the body wall. There they could stay cooler than the core of the body.

Some diseases emerge from natural selection when it favors genes that come into conflict, as in the case of babies and their mothers. Preeclampsia in pregnant women is not the inadvertent side effect of some adaptation that benefits a woman's health; if David Haig is right, it is the direct effect of natural selection on the genes that babies use to get nutrients from their mothers.

The Natural Selection of Cancer

Darwin first recognized that natural selection could operate in a population of individuals. But selection can also operate at other levels. Cells, for example, can compete with each other within our bodies. We know this particular kind of natural selection as cancer.

Every time a cell divides, there's a tiny chance that it will mutate. In some cases, those mutations strike genes that control the rate at which cells divide. These gatekeeper genes ensure that cells divide only when they need to, and not when they shouldn't. For example, when you cut yourself, cells in the skin and other tissues rapidly divide to heal the wound. If they were to keep dividing, they'd create an expanding mass of flesh. Mutations to these gatekeeper genes allow cells to grow faster than their neighbors. In some cases, these runaway mutant cells can form a tumor.

Within a developing tumor, cells continue to mutate, and cells with mutations that speed up their growth come to dominate the population of cancer cells (**Figure 13.8**). Some genes, for example, normally only become active in sperm cells, helping them to grow rapidly throughout a man's adult life. Normally these sperm-growth genes are kept silent in other parts of the body. But mutations can switch them on in cancer cells, making them divide faster.

Natural selection not only allows cancers to become more aggressive, but also to become harder to treat. As tumors grow, some mutations may make them resistant to drugs. Much like bacteria, those resistant mutants will be able to grow faster than susceptible cancer cells. The same mathematical models that help evolutionary biologists to understand the rise of antibiotic-resistant bacteria are now being adapted to shed light on the evolution of resistance to anticancer drugs in tumors.

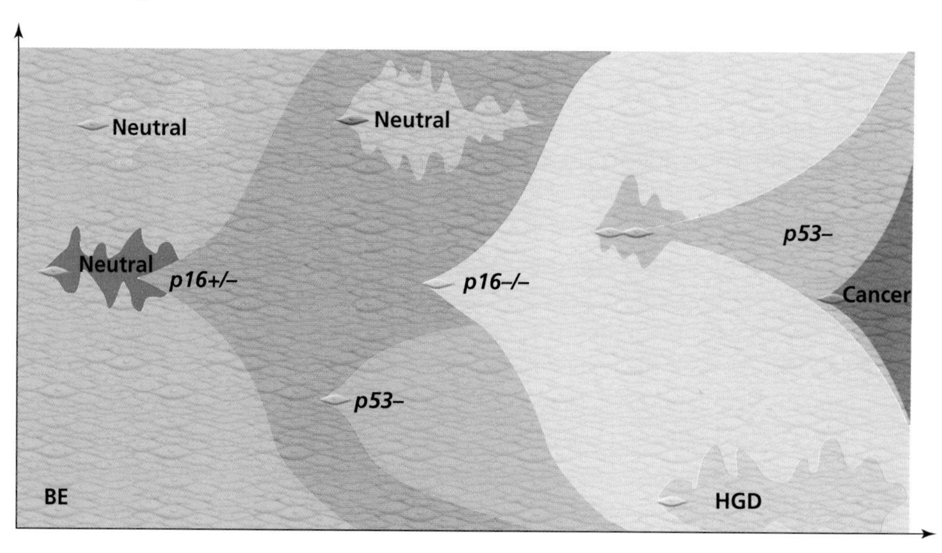

Figure 13.8 Cancer biologists have come to view tumors as the result of a special form of natural selection. This diagram shows how cancer can arise in the esophagus. In a condition known as Barrett's esophagus, a segment becomes precancerous (indicated here by the pink "BE" cells). Over time, as the esophagus cells divide, mutations arise. Many mutations are neutral, and some slow down a cell's rate of division. But every now and then mutations arise that speed up a cell's growth rate. The cell's descendants come to dominate the surrounding tissue. Later, a new mutation arises in one of its descendants, raising its fitness even further. Eventually a cell acquires enough mutations to develop into full-blown cancer of the esophagus. (Adapted from Crespi and Summers, 2005)

The fact that cancer cells follow the rules of natural selection suggests that it's important to start treating tumors as early as possible, because the longer they evolve, the more likely a resistant mutation may spontaneously arise in a cell. Evolutionary theory has led some researchers to explore using continuous, low-level treatments of cancers instead of a large pulse of drugs.

Defending Against Our Own Imperfection

Like hernias, cancer is a maladaptation of our bodies. It is the inescapable risk of being a multicellular animal. In order to make our bodies function properly, the cells inside them must cooperate. But cooperation always opens the way for cheating to evolve (page 250). Cancer, in other words, is a tragedy of the commons.

Because cancer can be lethal, natural selection has produced defenses against it. If a cell becomes precancerous, for example, proteins within the cell detect the change and trigger it to commit suicide. Some precancerous cells evade this checkpoint, however, but the vast majority of them are killed off by the immune system.

The very architecture of our bodies shows signs of having adapted to fight against cancer. One of the best defended organs is the colon, which is not surprising because it's so vulnerable to the disease. Every day, 100 billion cells lining the inner wall of the colon are shed, replaced by new cells generated from underlying stem cells. They continue to proliferate throughout our lifetime, and all that cell division provides a tremendous opportunity for a cell to turn cancerous.

Instead of a smooth sheet of cells, the colon's inner wall is made up of hundreds of millions of tiny pockets, called crypts. Each crypt contains about 2,000 cells. Stem cells at the bottom of each crypt continually divide, each time producing a differentiated cell and another stem cell. The new cells get pushed up the sides of the crypt until they reach the top and are shed.

Natalia Komarova, a mathematician at the University of California, Irvine, and her colleagues sought to understand the anatomy of the crypts by building a mathematical model of the colon. Different arrangements of cells in the crypts lead to different risks for cancer. If the cells go through many divisions over a person's life, they have a higher chance of mutating and turning cancerous. One alternative would be for stem cells to produce only a single generation of differentiated cells, which would then be shed. But that arrangement would create a risk of its own: it would require a vast number of stem cells, each of which would have to divide rapidly to produce a differentiated cell and a new stem cell.

Komarova found that the actual shape of crypts is a good solution that avoids these two extremes. Cells in crypts get to be only a few generations old at most before they are shed, and it takes relatively few stem cells to keep each crypt regenerating itself. Colon cancer is the third most common form of cancer in the United States, but, if Komarova is right, the architecture of the human colon keeps it from being far more common.

Old Age: Evolution's Side Effect

A child born in 1900 could expect to live, on average, for 47 years. A child born today can expect to live 78 years. Much of that change can be credited to clean drinking water, decent sewer systems, a steady supply of food, and improved medical care. But there's actually less to our longer lifespan than meets the eye. Most of the progress has come from saving the lives of children, as well as young mothers who used to die in childbirth. At the other end of life, people today still age in much the same way people did a hundred years ago. Their bones get brittle, they lose their stamina, and they lose their ability to fight infections. The decline begins in middle age and continues gradually for decades (**Figure 13.9**).

Most animals age in much the same way, although some get old faster than others. Blue whales may live 200 years, while fruit flies live for only a matter of weeks. Our closest relatives, the chimpanzees, rarely survive past 50 years, even under the best conditions. In the past, scientists speculated that the particular lifespan of each species was itself an adaptation, set by natural selection. But the evidence now shows that that's not the case. Aging is one of the unpleasant byproducts of natural selection, much like sickle-cell anemia.

The English biologist J. B. S. Haldane was the first to recognize the evolutionary nature of aging. In the 1930s he studied Huntington's disease, a genetic disorder that slowly destroys the nervous system. It is unusually common for a devastating genetic disorder: about one in 18,000 people suffers from the disease. If Huntington's disease resulted only from spontaneous mutations in sperm or eggs, Haldane realized, it should be much rarer that that.

Haldane proposed that the reason for Huntington's disease being so common was its late onset. It does not begin to cripple most people until they are in their forties. That would have provided them with plenty of time to raise children, passing on the disease-triggering alleles to the next generation before they died.

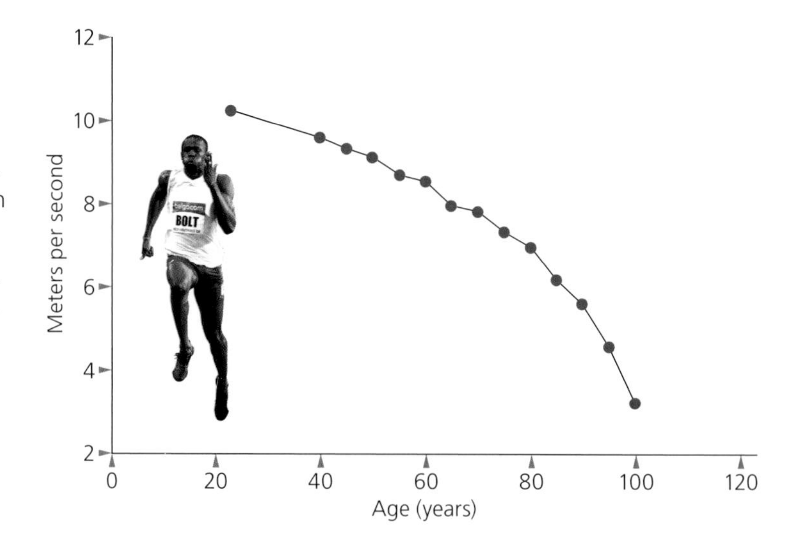

Figure 13.9 This graph illustrates how the human body declines with age. Scientists produced it by gathering the world records for the men's 100 meter run by age groups. They then calculated the average speed of each champion runner. The maximum speed starts dropping in the third decade of life. Evolutionary biologists seek to understand how this decline evolved. (Adapted from Koella and Stearns, 2008)

If Huntington's disease claimed its victims when they were teenagers, it would have been much rarer.

Haldane recognized an important feature of natural selection: it has different effects at different times in life. It acts weakly on mutations with effects in old age, and it acts strongly on mutations with effects in childhood and the prime reproductive years.

Since then, evolutionary biologists have recast Haldane's ideas in more precise terms, building mathematical models to calculate how natural selection fades with age. George Williams (page 279) also proposed that a mutation that benefits an organism while it's young, will be favored by natural selection, even if it becomes harmful in old age. Its benefit outweighs its cost. Thomas Kirkwood, an evolutionary biologist at Newcastle University in England, argues that organisms face a trade-off between reproduction and repairing their cells. If they put all their energy into repairing their cells, they will slow down the aging process, but they won't be able to produce as many offspring as other organisms. Fast-breeding individuals pass down more of their genes, and as a result the population cannot repair its cells very well. They get old.

To test these ideas, scientists have run experiments on animals. Stephen Stearns, an evolutionary biologist at Yale University, and his colleagues wondered if they could alter the longevity of fruit flies. They reared hundreds of fruit flies in two sets of vials. Twice a week, the scientists killed most of the flies in one set of vials and replaced them with some of their offspring. The other flies in the other vials could live longer, because, twice a week, the scientists only killed a small fraction of the flies.

The flies getting killed off faster didn't benefit as much from mutations that acted on them late in life. There were just too few of them left alive by then. In fact, mutations that caused harm late in life could be favored by natural selection if they boosted the reproductive success of the flies while they were still young.

That was the prediction, at any rate—and that's what Stearns and his colleagues discovered. After killing off the flies for generations, they relented and let the flies live out their natural lifespans. The flies that suffered high mortality evolved to grow up faster and to become ready to mate sooner. They were also more likely to die sooner than the flies that didn't face such high mortality.

This trade-off can explain how our bodies have ended up defending us from diseases in a less-than-perfect way. One of the most important controls against cancer is the p53 tumor-suppressor protein. It responds to stress inside cells, particularly to damaged DNA, which may signal the first steps towards cancer. It can cause a cell to die or to stop dividing. In either case, p53 prevents the cell from possibly growing into a tumor, but it takes a toll in the process. As the years pass, p53 can kill or stunt so many cells that tissues can no longer renew themselves. By forcing cells into early retirement, p53 may prevent them from becoming tumors, but the cells may damage surrounding tissue and release abnormal proteins that stimulate the growth of cancer cells.

In other words, p53 is a very effective stopgap defense against cancer. It helps keep young people relatively cancer free. But it also damages the body in the process. The damage accumulates slowly, over the course of many years. By the time it has an impact on us, we're so old that natural selection cannot alter it.

None of this means that we can't hope to extend the human lifespan. Some of the most promising results in the study of longevity have come from experiments in which scientists reduce the amount of food animals can eat. If they cut the normal number of calories in an animal's diet, it often lives much longer. Some experiments suggest that restricting an animal's diet triggers a special response in its cells. They begin to produce proteins that can repair the damage caused by the stress of not getting enough to eat.

This response appears to be an ancient strategy, given that the same stress-fighting genes can be found in animals ranging from nematode worms to mice. These genes may have evolved as a way to cope with short-term stress, such as famines and droughts. Scientists who study nematode worms have found that they can double the lifespan of the animals with mutations that keep these genes switched on. The genes may be able to continually repair damage to cells, fighting the effects of aging. Restricting calories may have the same effect, by keeping the genes switched on permanently.

David Sinclair, a Harvard biochemist, and his colleagues have discovered that certain chemicals can turn on these genes, and they have been trying to turn one of those compounds, resveratrol, into an anti-aging drug. Resveratrol may well do what they hope it will, but it may also have unexpected side effects. Nicole Jenkins, a biologist at the Buck Institute for Age Research in California, and her colleagues have found that mutations that bring long life to nematode worms actually lower their fitness. The scientists put 50 of the long-lived worms in a dish with 50 normal worms and then let them breed. Jenkins then randomly picked out 100 of the eggs and used them to rear the next generation. The long-lived worms were just as fertile as the normal worms, Jenkins found, and yet within just a few generations they had vanished from the dish. They had been outcompeted by the shorter-lived worms.

Natural selection, once again, did not favor long life simply for long life's sake. Jenkins's experiment raises the possibility that taking resveratrol or some other medication to prolong aging may bring with it some kinds of side effects. It would be nice to have an elixir of life, but an understanding of evolution's trade-offs shows why that's so unlikely.

Sick from Sexual Conflict

The best strategy for mating and rearing offspring is not the same for males and females. As a result, sexual conflicts can evolve, producing traits and behaviors that can seem downright destructive—such as the habit some birds have of abandoning their young (page 285). David Haig and other researchers are now investigating the impacts of sexual conflict on human health. Sexual conflict can help to explain preeclampsia as a tug-of-war between mothers and fetuses. In cases such as these, different genes come into conflict.

But sexual conflict can also leave a mark on a single gene. Scientists have identified dozens of genes that are imprinted (page 286). Animals inherit two alleles of a gene from their parents, but the copy from one parent is almost always silenced. Gene imprinting may be a strategy that has evolved to let one parent reduce the effect of a gene that benefits the other parent. In mice, for example, a gene called *Igf2* stimulates the growth of fetuses. The female's copy is silenced. Likewise, a gene called *Igf2r* limits the growth of fetuses. The father's copy, not the mother's, is silenced.

Scientists learned about the roles of these genes by disrupting them. If a mouse's paternal copy of *Igf2* is disabled, for example, the mouse develops to be 40% smaller at birth. If its maternal copy of *Igf2r* is disabled, it is 25% bigger than average. But nature, it turns out, has also produced this experiments in humans.

Sometimes a child is born expressing both copies of *Igf2*, instead of only the father's. The child will be born unusually big—weighing 50% more than an average baby. The child will also suffer a range of other symptoms of unchecked growth. Its heart and other organs are often enlarged, causing them to malfunction. Its tongue may grow so big that the child has trouble breathing, eating, and speaking. Sometimes one half of the body grows faster than the other, and the child may suffer tumors in the liver and kidneys. This condition, which strikes 1 in every 35,000 children, was recognized long before it was linked to *Igf2*. Doctors dubbed it Beckwith-Wiedemann syndrome.

One in 75,000 children suffers the opposite problem. In these children, the father's copy of *Igf2* is silenced, so that they produce no *Igf2* at all. This silencing leads to Silver-Russell syndrome. Children suffering this condition are born small, without much fat underneath their skin. After birth, they grow slowly, ending up far shorter than average at maturity.

Modern medicine may also be disrupting the delicate balance that has evolved between maternal and paternal genes. A growing number of couples who have trouble having children turn to in vitro fertilization (IVF): doctors inject sperm directly into eggs in a dish. Once the eggs start dividing, they implant them in the woman's uterus. Jane Halliday, a geneticist at the University of Melbourne, and her colleagues surveyed 1.3 million birth records from the Australian state of Victoria, comparing children conceived through IVF with other children. They found that children conceived by IVF were nine times more likely to develop Beckwith-Wiedemann syndrome than normal children.

IVF may raise the risk of Beckwith-Wiedemann syndrome because it bypasses the normal process by which genes are imprinted in fertilized eggs. When a sperm delivers its DNA into an egg, the methyl groups that cap both male and female genes are stripped away. Only then do special proteins in the fertilized egg put methyl groups back on its DNA.

Eamonn Maher, a geneticist at the University of Birmingham in England, has suggested that this imprinting requires a special set of compounds made by the uterus, compounds that are missing from the medium in which IVF eggs are kept.

Another possible explanation is that the risk of Beckwith-Wiedemann syndrome may have something to do with the infertility of the women who resort to IVF. Whatever makes it hard for them to conceive may also impair their eggs' ability to imprint genes.

In either case, it's clear that gene imprinting is a delicately balanced process, and disrupting it can have lifelong effects on people. Evolutionary medicine can explain why this part of development is so fragile and so susceptible to malfunctioning: because evolution did not find an optimal solution. In fact, thanks to sexual conflict, it *could* not.

Mismatched with Modern Life

For people in the United States and other developed countries, it's nearly impossible to imagine the suffering infectious diseases brought a century ago. Along with tremendous outbreaks of such scourges as influenza, many other diseases also steadily killed off people year in and year out. Since then, the death rate from infectious diseases has dropped dramatically (**Figure 13.10**).

Yet, in the countries that have experienced these drops in infectious diseases, people have seen an increase in another kind of disease. Rather than being caused by pathogens, these rising diseases are caused by our own immune sys-

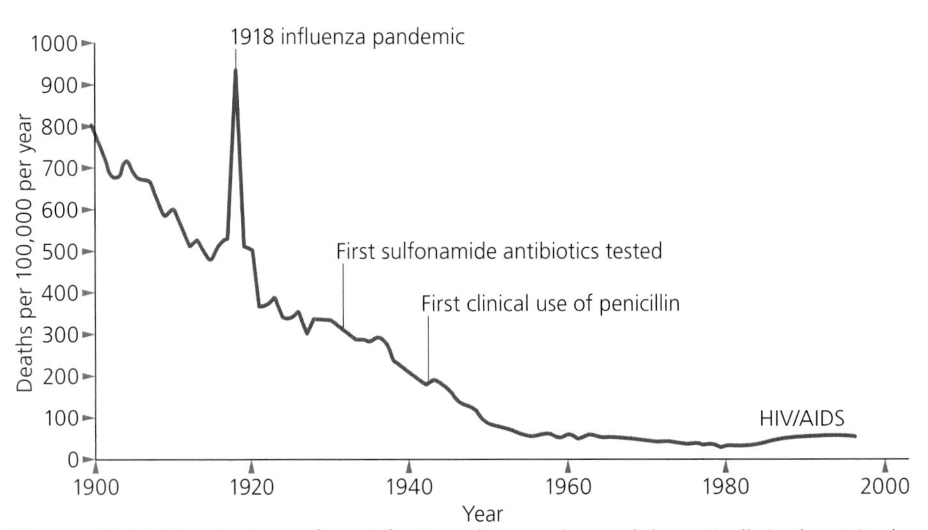

Figure 13.10 The death rate from infectious diseases dropped dramatically in the United States over the twentieth century. It began to decline thanks to better hygiene, clean drinking water, and better food. The invention of antibiotics in the mid-1900s helped push the death rate even lower. The HIV epidemic that began in the late 1980s raised the death rate, although it remained far lower than at the beginning of the century. (Adapted from Koella and Stearns, 2008)

tems. Crohn's disease, for example, occurs when the immune system attacks the lining of the intestines. Asthma is caused by inflammation in the lungs. Type 1 diabetes occurs when immune cells attack insulin-producing cells in the pancreas. All these diseases were once rare in the United States and other developed countries, but now are increasingly common.

It's intriguing to note how these diseases seemed to have followed in the wake of affluence. They first emerged in countries such as England and the United States, which were among the first nations to go through an industrial revolution and then improve their public health. Later, when other countries went through the same transition, they also saw a rise in autoimmune diseases. Within countries, a similar pattern can be found. In Venezuela, for example, the population is split mainly between cities and farms, with some Indians still living in isolated villages in the rain forest. Venezuelan city dwellers have higher rates of allergies than Venezuelan farmers, and Venezuelan Indians have no allergies to speak of.

In 1989, David Strachan, an epidemiologist at the London School of Hygiene and Tropical Medicine, suggested that these autoimmune diseases were breaking out because children had become too clean. They were not being exposed to dirt and dust, and they were not being infected by bacteria and parasitic worms. As a result, their immune systems were attacking themselves, rather than pathogens. Today, scientists refer to this as the hygiene hypothesis.

The hygiene hypothesis builds on the fact that life in the twenty-first century is very different from what it had been for most of our 200-million-year history as mammals. Our ancestors were under constant assault from pathogens. They could not respond with antibiotics or bandages. Instead, natural selection responded by fine-tuning the immune system, which became better able to cope with the infections. But if the immune system became too aggressive in fighting infections, it could actually do more harm than good. If it attacked any peculiar-looking cell, it might wipe out the intestinal bacteria we depend on for our nutrition. It might even damage our own tissues by releasing too many toxic chemicals.

It appears that the immune system has evolved to an uneasy truce with pathogens. When the immune system encounters certain kinds of bacteria and intestinal worms, it becomes subdued. In effect, it learns to tolerate our inner companions. This truce is an ancient one, shared by other vertebrates. If you dissect a healthy opossum or gorilla, you'll find not just bacteria but several species of worms living inside it without triggering a major reaction from the immune system.

When our own species first emerged 200,000 years ago, our ancestors inherited this tolerant immune system. But suddenly, in just the past few decades, a large fraction of our species has entered into an entirely new relationship with the life within us. Many children do not get exposed to many microbes, playing inside houses rather than out in the dirt. When they get sick from bacteria, they are treated with antibiotics that kill many harmless bacteria along with the pathogenic ones. Intestinal worms are now a thing of the past.

According to the hygiene hypothesis, our bodies are maladapted to this kind of upbringing. The immune system is primed to receive calming signals from bacteria and worms, but it gets none. In some individuals, it begins to overreact, either to their own bodies or to normally harmless substances such as peanuts or cat dander.

Scientists have been testing the hygiene hypothesis from many different directions, and their results are encouraging. Martin Blaser and Yu Chen, microbiologists at New York University, have been looking at the effects of one particularly important species of bacteria called *Helicobacter pylori*. *H. pylori* lives in the stomach, and it's had a particularly intimate relationship with our species. Its phylogeny mirrors our own. *H. pylori* strains that live in Native Americans are more closely related to the ones in Asian people; African strains of *H. pylori* are more diverse than strains on other continents.

H. pylori was ubiquitous in our species before the advent of antibiotics, but it's now on the decline. Only one in five American children now carries it. Blaser and Chen analyzed the medical histories of more than 7,400 people who took part in a nutrition survey. As part of the survey, researchers collected stool samples. Blaser and Chen checked the samples for signs of *H. pylori*. In 2008, they reported that children between three and 13 years old who carried *H. pylori* were 59% less likely to have asthma than children who were free of the bacteria. They were also less likely to have hay fever or eczema. Other experiments suggest that *H. pylori* triggers the body to make certain kinds of immune cells, known as Th17 cells, which regulate the way the body will respond to invading bacteria.

On the whole, the public is healthier, thanks to antibiotics and better hygiene, but that's small comfort to people who suffer from the autoimmune diseases that may have been produced as a side effect. Some researchers are using the hygiene hypothesis to figure out ways to treat these diseases of modernity. Some doctors have dispensed parasitic worms to people suffering from Crohn's disease, and they've found that the parasites tend to reduce the symptoms. These experiments are just a proof of principle, however. Ideally, doctors would be able to prescribe drugs that would trigger the same response as the worms and bacteria, without the harm that they can cause. (*H. pylori*, for example, may protect children from asthma, but it also increases the chance of developing stomach cancer.) It's possible that someday people will swallow pills containing surface proteins from parasitic worms or bacteria to teach their immune systems how to behave themselves.

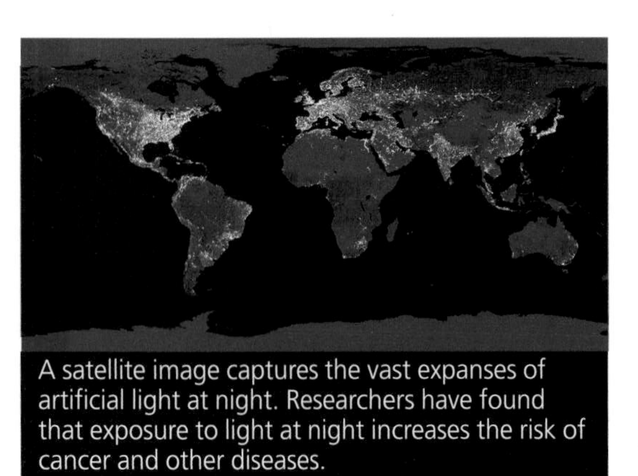

A satellite image captures the vast expanses of artificial light at night. Researchers have found that exposure to light at night increases the risk of cancer and other diseases.

Modern medicine is not the only sudden change our bodies face. Animals, plants, and many other organisms have evolved an internal clock that controls how their bodies function at different times of day. The human body clock controls the expression of many genes and the release of hormones. It creates 24-hour fluctuations in our appetite, our body temperature, and our wakefulness. When sunlight enters our eyes each day, it resets the clock. Today, however, we are confusing our bodies. We keep lights on well past dark, for example, and we sometimes fly from one time zone to another. Lights at night and long-distance plane travel have both been linked to an increased risk of cancer, possibly because they disrupt our hormone cycles.

The sudden mismatches between our bodies and modern life have changed the fitness of our alleles. As we've seen in previous chapters, the fitness of a mutation is not some absolute value. It depends on the other genes with which a mutant gene shares its genome, and it also depends on the environment in which its owner lives. This perspective changes the way we think about genetic disorders. What is healthy in one century may not be healthy in another century.

Evolutionary Medicine: Gloom and Hope

Evolutionary medicine is a sobering science. It reveals the many obstacles we face in fighting diseases, from the rapid evolution of antibiotic resistance to the deep vulnerabilities that make us susceptible to illnesses. But evolutionary medicine can also offer researchers inspiration and guidance in the struggle to improve our health. Models of evolution can show which antibiotic treatments will lead to more resistance, and which treatments will lower it. Scientists are also building similar models to help avoid the evolution of tumors that are resistant to chemotherapy and the evolution of HIV strains that resist antiviral drugs.

Evolutionary biology also points the way to potential new treatments. We humans are particularly vulnerable to a form of blood poisoning called hyperuricemia—an excess of a chemical called uric acid. All organisms, from bacteria to humans, break down DNA in their food with a set of proteins so that they can use its fragments to build new molecules. Most species break it down until all that's left is a relatively harmless waste product called allantoin (**Figure 13.11**). But humans can't reach the end of this pathway. Instead, we stop with an intermediate molecule called uric acid. We excrete the uric acid out of our kidneys in our urine.

Stopping with uric acid brings special risks for us. If we can't flush it out of our bodies fast enough, it can cause all kinds of damage. Uric acid can crystallize into painful kidney stones, or it can cause gout, a disease caused by sand-like grains of uric acid that build up in the extremeties, such as the joints of the toes. Uric acid can also contribute to hypertension and cardiovascular disease. When cancer treatments kill off tumor cells, they dump their DNA into the bloodstream. If patients can't handle the surge of uric acid that follows, they may suffer kidney failure.

Exactly how other animals turn uric acid into allantoin has long been a mystery. The first step requires a protein called urate oxidase, which turns uric acid into an intermediate called HIU. But until recently scientists didn't know how HIU turns into allantoin.

A

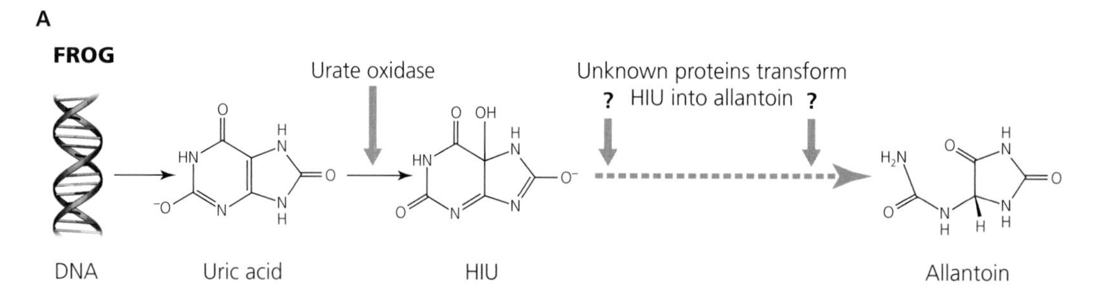

B

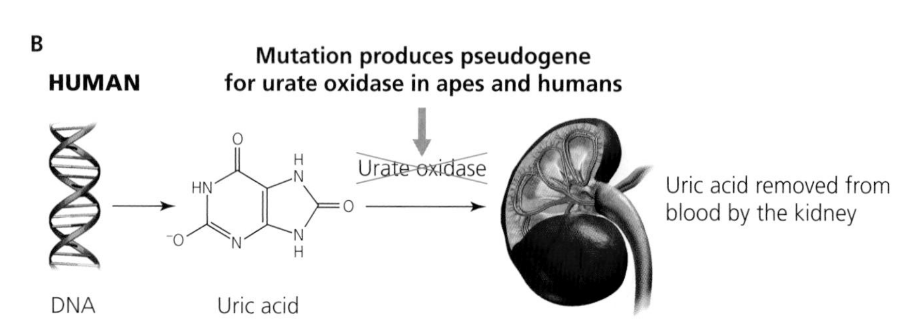

C

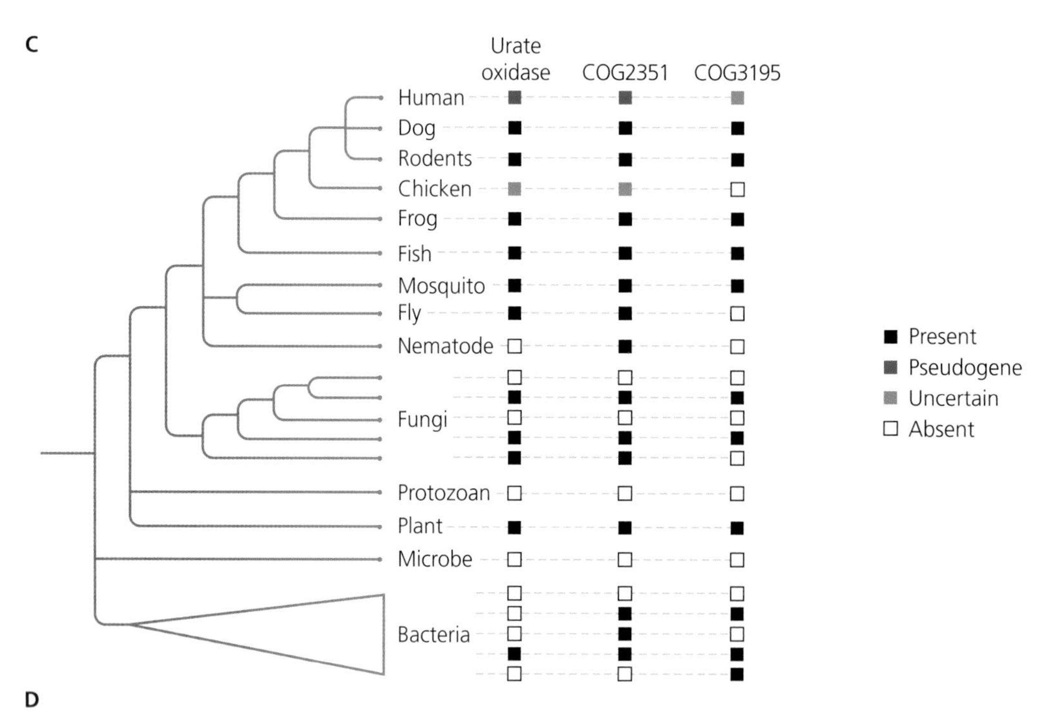

D

FROG

Urate oxidase

COG2351

COG3195

DNA

Uric acid

HIU

Allantoin

Our ancestors lost the ability to make allantoin a long time ago—about 20 million years ago, in fact. In 2002, Naoyuki Takahata and colleagues at Kyoto University showed that chimpanzees, gorillas, orangutans and humans all share the same pseudogene for urate oxidase. That protein must have been disabled in the common ancestor of humans and other great apes. Losing urate oxidase raised levels of uric acid in the blood of primates. It's possible that change had some benefit—researchers have suggested that it may have helped to protect primates from dangerously low levels of salt. Whatever the advantage, however, primates also became vulnerable to dangerously high levels of uric acid, a danger we still face today.

Doctors have tried treating gout and other uric-acid–triggered diseases by giving patients urate oxidase, in the hopes of breaking uric acid down into a less dangerous form, but this treatment has not been very successful. Some researchers suspect its failure is due to the fact that urate oxidase is just the first of a series of proteins that turn uric acid into harmless allantoin. To treat high levels of uric acid, some scientists suggest, we must re-create the lost chemical reactions of our ancestors.

Recently, Riccardo Percudani, a molecular biologist at the University of Parma in Italy, and his colleagues decided to find the missing proteins. They reasoned that they must have coevolved with urate oxidase. If the urate oxidase gene was disabled and became a pseudogene in a lineage, for example, the missing proteins were likely to have become pseudogenes as well. In other cases, all of the proteins might have been completely deleted. The scientists also took advantage of the fact that genes that function together are often located close to one another in a genome, sometimes even fused together. It was possible, they reasoned, that the genes for the missing proteins were close to the gene for urate oxidase.

The researchers searched online databases of genomes noting which ones contained working versions of uric oxidase genes or pseudogenes and which lacked any uric oxidase gene at all. They then looked for other genes that showed associations with the uric oxidase gene from one species to another. Their search yielded two genes. No one had figured out what those two genes were for. Percudani and his colleagues synthesized proteins according to the sequence of the two genes and mixed them with uric acid. The scientists discovered that each protein made a precise cut in the bonds of the uric acid, and together they could turn it into allantoin. The discovery of these two missing proteins opens the possibility of an effective treatment for hyperuracemia. In effect, doctors

Figure 13.11 Opposite: Humans are vulnerable to diseases such as gout because they cannot completely break down uric acid, a byproduct of DNA. A: Most animals, such as frogs, can transform uric acid into a harmless molecule called allantoin. They use a series of proteins to carry out the transformation, starting with urate oxidase. B: In humans, the urate oxidase gene is disabled, and so uric acid must be removed by the kidney. C: A team of Italian scientists set out to find the genes that allow frogs and other animals to break down uric acid. They searched for genes that tended to be present or absent along with urate oxidase. They discovered two such genes. D: By combining these proteins with uric acid, the researchers were able to transform it into allantoin. This discovery promises to help in the search for new treatments for diseases involving uric acid. (Adapted from Ramazzina, 2006)

may be able to cure a disease by giving patients back the proteins our ancestors lost 20 million years ago.

Most of the medical drugs in use today were not synthesized by researchers from scratch. For the most part, researchers discovered molecules that had been made by living things. The breast cancer drug taxol, for example, was originally discovered in a yew tree. Most antibiotics on the market today were discovered in fungi or bacteria. Scientists have barely begun to survey the world's diversity of biological molecules, and so there's good reason to expect that there are many valuable new compounds out there that are waiting to be discovered (page 238).

Insights from evolution are helping scientists to zero in on potential new drugs. Snake venoms, for example, have been transformed into drugs to treat blood clots, high blood pressure, and other ailments. Like any kind of drug, snake-venom molecules have some drawbacks. One is their size: snake-venom molecules are relatively large for drug compounds. The larger a drug molecule, the more easily it can be recognized by the immune system, which may then destroy it before it can have an effect.

Bryan Fry of the University of Melbourne has been tracing the evolution of snake venom through the duplication and co-option of old genes (page 160). His research led him back through time to discover when snake-venom genes first evolved. When he looked in the mouths of the lizards most closely related to snakes, he discovered proteins that he believes also function as venom. Molecule for molecule, the lizard proteins appear to be just as powerful as snake venoms. There is one important difference, however—the lizard proteins Fry has identified so far are smaller than snake-venom molecules. Their small size may make them even more promising as drugs than snake venom.

Medical researchers are like treasure hunters, wandering across a continent in search of jewels. A real treasure hunter wouldn't even start the search without a map. Evolutionary biology offers medical researchers a map of life.

TO SUM UP...

- **Evolutionary medicine is the study of the evolutionary roots of health and sickness.**
- **Some diseases are caused by pathogens that have been infecting our ancestors for millions of years. Some have begun infecting humans in just the past few decades.**
- **Reconstructing the evolutionary trees of emerging pathogens allows scientists to determine their origins.**
- **Human pathogens are always rapidly evolving, adapting to their human hosts.**
- **The virulence of pathogens emerges from an evolutionary trade-off between replicating quickly and moving easily from host to host.**
- **Pathogens have driven the evolution of defenses in human populations.**

- Pathogens evolve resistance to antibiotics.
- Genetic drift can lead to harmful mutations becoming common in a population.
- Our evolutionary history has given the human body many flaws.
- Cancer develops in an evolutionary process.
- Aging is an evolutionary byproduct of natural selection for traits that help animals survive during their reproductive years.
- Sexual conflict can lead to gene imprinting disorders.
- Our bodies are adapted to a preindustrial life. Some diseases are the result of the mismatch between our bodies and modern life.
- Evolutionary biology reveals the great obstacles to treating some diseases, but it also opens new avenues of research for potential new cures.

Minds and Microbes

14

The Evolution of Behavior

The standard uniform for paleontologists is casual. T-shirts, cut-offs, and floppy old hats are common sights around most fossil digs. But in a Spanish cave called El Sidron, the dress code is decidedly more formal. The fossil hunters there are dressed in white coveralls, surgical masks, and sterile gloves. And rather than ordinary rock hammers and chisels, the paleontologists at

Left: Neanderthals, the most closely related species to humans, became extinct 28,000 years ago. Above: Excavating in a cave in Spain, researchers have discovered DNA in Neanderthal fossils. They are comparing the genes to ours to gain clues about how human behavior evolved.

El Sidron use sterilized blades to dig at the bones, which they quickly put into a freezer. They look less like fossil hunters than characters out of a science-fiction movie about a killer plague.

The scientists at El Sidron have gone to these extremes because they are conducting no ordinary fossil hunt. They have been searching for Neanderthal fossils in order to extract their DNA, and by examining their DNA, to learn how much Neanderthals behaved like we do. To succeed at this audacious task, they have to take every possible precaution to ensure that not even a flake of their skin or a drop of sweat contaminated the fossils with human DNA.

In 2006, the researchers excavated a 48,000-year-old Neanderthal fossil from the cave and shipped samples from 22 bones to Svante Pääbo at the Max Planck Institute for Evolutionary Anthropology. Pääbo and his colleagues took even more care with the bones, preparing them in an ultra-clean laboratory. They ground the bones into a powder and gradually removed all the minerals and organic matter until only DNA was left. They had prepared special molecules known as primers that could rapidly make many copies of a particular gene they wanted to find. They were searching for *FOXP2*, a gene that is known to be crucial for communication in mammals, and for language in humans (page 148).

In 2002, Pääbo and his colleagues had demonstrated that *FOXP2* experienced exceptionally strong natural selection in our ancestors after they branched off from other apes seven million years ago. The fact that *FOXP2* helps in the development of speech suggests that its evolution in the human lineage was a crucial step in the evolution of full-blown language. A question naturally arose: When did *FOXP2* evolve into the form found in humans today? Searching for the gene in Neanderthal DNA might give Pääbo a clue to the answer. Using a molecular clock, Pääbo estimated that the last common ancestor of humans and Neanderthals lived 800,000 years ago. If *FOXP2* changed after that split, the Neanderthals would not be expected to have our version of the gene. If it changed before the split, they would probably share it.

In 2008, Pääbo and his colleagues announced the result of their hunt. They had found the Neanderthal *FOXP2* gene. When they read its sequence, they found that it was the same as human *FOXP2*, and not like that of chimpanzees or other apes. So the transformation of this gene—possibly a key step on the way to full-blown language—took place after our ancestors diverged from other apes seven million years ago, but before our ancestors diverged from those of Neanderthals, an estimated 800,000 years ago.

To understand the evolution of human beings—or of any species—scientists cannot just look at anatomy or physiology. Humans are distinctive not just because we stand on two legs or have unique sweat glands embedded in our naked skin. Our behavior—what we do, and how we do it—is also unique. Unlike other species, we can use language, think in symbols and concepts, and get inside other people's heads. In this chapter we'll explore how behavior evolves, and see how human nature has emerged from hundreds of millions of years of evolutionary history. And we'll consider the many kinds of evidence scientists can study to understand the forces that have shaped behavior—from psychological experiments on humans and apes to DNA that's been lying in a cave in Spain for 48,000 years.

Behavior Evolves

The evolution of human behavior has a deep history. Today we use our brains to take in information from our surroundings and to come up with decisions about how to act. But long before any brain existed, living things were making decisions, and their behavior was being shaped by evolution. Living bacteria offer clues to how our single-celled ancestors behaved.

Scientists have carried out some particularly revealing experiments on the evolution of behavior in *Myxococcus xanthus*, a species of soil bacteria. *M. xanthus* travels by spraying out a slippery sheet of goo over which it can glide, pulling itself forward with long hairs that sprout from its front end. Hundreds of thousands of *M. xanthus* will join together into huge swarms to chase after other microbes. In a pack, each *M. xanthus* can travel faster than it could on its own. When the *M. xanthus* pack runs down their prey, they release lethal proteins that tear their victims apart. The pack of *M. xanthus* can then feast together.

This chain of events requires each microbe in the swarm to respond to signals from its environment with actions. *M. xanthus* does not make its decisions based on a perfect knowledge of its surroundings. Some bacteria can sense light, for example, in order to carry out photosynthesis. But *M. xanthus* does not get its food that way, and so it lacks light-sensitive receptors. Instead, it depends on a microbial version of taste. When certain molecules flow by *M. xanthus*, they are snagged on its receptors. That snagging triggers a cascade of chemical reactions that can cause the bacteria to respond in certain ways, such as switching on their lubricating nozzles, reversing direction, or releasing their deadly digestive proteins.

The behavior of *M. xanthus* emerges from the interactions of their 7,457 genes with the environment. Mutations that alter some of those genes can lead to changes in that behavior, and, as a result, their behavior can evolve. Richard Lenski (page 118) has observed the evolution of *M. xanthus's* behavior in an experiment he conducted with Kristina Hillesland and Gregory Velicer.

The scientists reared populations of *M. xanthus* in petri dishes in which they placed *Escherichia coli*, one of the species of bacteria *M. xanthus* hunts. In one dish, the *E. coli* were arranged in a closely-packed set of blocks. It took relatively little time for *M. xanthus* to slither from one block to the other. In the other, the blocks were more widely spaced. For 14 days, the scientists allowed *M. xanthus* to hunt for *E. coli* in each dish and to reproduce. They then removed some of the *M. xanthus* and put them in a new dish with the same arrangement of food. They repeated this cycle for a year.

During that year, *M. xanthus* evolved. The bacteria in both dishes became faster hunters. The *M. xanthus* in the dishes with the sparse supply of *E. coli* became fastest of all, with some populations evolving to swim ten times faster than their ancestors. The *M. xanthus* in the sparse dishes also underwent another change: once they encountered their prey, it took them less time to eat it.

The evolution of *M. xanthus's* behavior in this experiment was governed by a trade-off. In the dishes where *E. coli* was packed close together, *M. xanthus* didn't have to swim far to find food. Swimming fast did not bring as much reward to them than it did to the bacteria in the other dishes. Instead, the bacteria that channeled some of their energy into growing faster had more reproductive success.

When a swarm of *M. xanthus* wipes out all the nearby prey, they sometimes make a collective decision about how to survive starvation. They come together and form a mound. A small fraction of the bacteria in the mound develop hard cases and become spores. They can be blown away in the wind or carried off by water. When they eventually reach a more hospitable place, they can emerge and begin to grow, divide, and hunt again. Biologists don't yet fully understand why certain *M. xanthus* in a mound become destined to form spores. It appears to be a random luck of the draw. Biologists also don't know much about how a free-living *M. xanthus* turns itself into a spore, although it's clear that they can only do so inside a mound, rather than on their own. Because only a few percent of the bacteria in a mound become spores, the vast majority must stay behind, to face almost certain death from lack of food. They die, in other words, so that others might live.

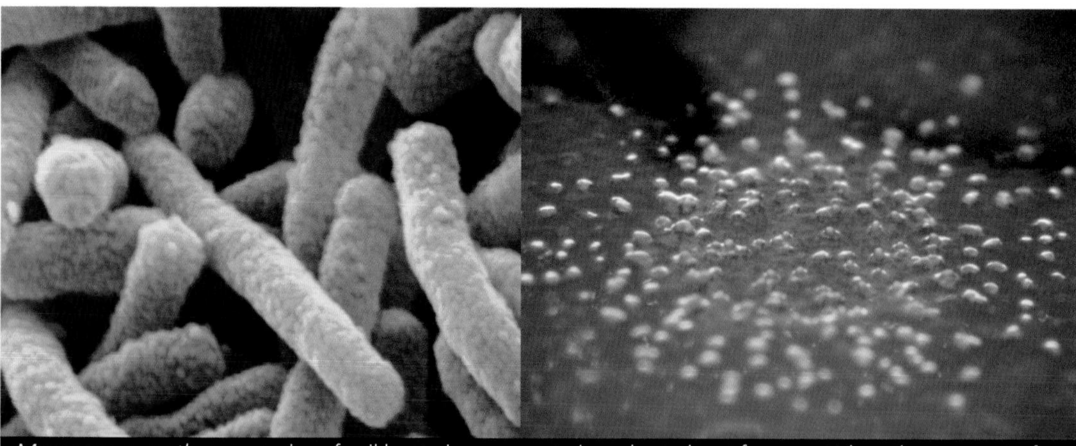

Myxococcus xanthus, a species of soil bacteria, cooperate in order to hunt for prey. When they run out of food, they come together into mounds (right). A few percent of the bacteria in each mound become spores which can be transported away and survive; the other bacteria stay behind and die.

In other words, the behavior of *M. xanthus* includes not only strategies for finding food, but also social interactions with other bacteria. And just as its food-searching behavior can evolve, its social behaviors can evolve as well. Gregory Velicer was able to observe this social evolution by letting *M. xanthus* evolve for 1,000 generations in a rich broth. Under these conditions, the bacteria never faced starvation and could always get plenty of food on their own. Velicer discovered that at the end of the experiment, most of the lines of bacteria had lost the ability to swarm, or to form spores, or both. Mutations to the genes for swarming and spore formation did not harm the bacteria; they may have even been better off without these behaviors.

Surprisingly, some of the newly evolved bacteria were not just asocial—they were positively antisocial. Velicer found that if he starved a population made up only of these cheaters, they could not form mounds. But if he mixed some cheaters in with ordinary *M. xanthus*, the cheaters could join mounds. When Velicer looked at the spores produced by these mounds, he was surprised to find that the cheaters were far more common than you'd expect if the spores were randomly selected from the bacteria in the mound. Somehow, the cheaters had found a way to exploit the spore-selection process, so that they were ten times more likely to form a spore as a normal *M. xanthus*. It was as if the crew of a sinking ship were drawing straws for spaces on a lifeboat, and a few of them figured out how to make sure they didn't draw a short straw.

Velicer wondered what would happen to a mixed population of cooperators and cheaters if they passed through several rounds of mound-forming. Since the cheaters would be overrepresented among the spores, they might gradually become more common, while the cooperators became rarer. Velicer set up a new experiment in which *M. xanthus* alternated between a rich broth and a dish with no food. As he had predicted, the cheaters became more common. In fact, if they became too common an entire population could get wiped out, because there were no longer enough cooperating *M. xanthus* left to make the mounds during famines.

Like many researchers who carry out evolution experiments, Velicer has also had his share of surprises. As he and his colleagues were studying the evolution of cheating in *M. xanthus*, they discovered that a strain of cheaters had given rise to a cooperator that could form mounds on its own again. Velicer and his colleagues sequenced the genome of the new cooperator and discovered a single mutation. The new mutation did not simply reverse the mutation that had originally turned the microbe's ancestors into cheaters. Instead, it struck a new part of the genome. Yet Velicer has yet to figure out how it brought about this remarkable return to altruism.

These experiments help to demonstrate how long behavior has been evolving. Billions of years before our ancestors had a brain, their behavior was being shaped by natural selection, as genes and binding sites in microbes evolved to produce new responses to signals. Even after animals evolved brains, however, the evolution of their behavior still followed many of the same rules that governed their single-celled ancestors.

The Origin of Nerves

Some of the cells that make up the human body still behave like free-living microbes. An immune cell that detects an invading bacterium, for example, may respond by crawling after it and devouring it. But the behavior of the *entire* human body is governed by a special set of cells adapted specifically for processing information. Those cells are neurons.

A neuron generates signals with pulses of electric charge that move from one end of the cell to the other. The signals can move from one neuron to the next at a special junction of the two cells called a synapse. The neuron sending the signal dumps chemicals known as neurotransmitters into the synapse, and they're taking up by the receiver. If you press an elevator button, the sensation travels from your finger along a series of neurons to the brain, a fantastically dense, complex organ, made of roughly 100 billion neurons interlinked by 100 trillion connections. The brain uses this sensory information to make decisions and send out commands to the body.

Most other animal species have nervous systems as well. Some have much smaller brains, and some have no recognizable brain at all. The tiny roundworm *Caenorhabditis elegans*, which measures only a millimeter long, has a nervous system that contains just 302 neurons. Yet even the nervous system of *C. elegans* shares many homologies with our own, from the channels it uses to generate electric charges to the neurotransmitters that let signals jump across synapses.

These homologies indicate that neurons evolved long ago, in the common ancestor of humans, *C. elegans*, and all other animals with nervous systems. In fact, in recent years, scientists have pushed back the origins of the animal nervous system to before there were even animals. The closest living single-celled relatives of animals are known as choanoflagellates. These protozoans, which live in streams and ponds, feed on bacteria they gather by beating their tails. Some choanoflagellate species form colonies, while some collect bacteria on their own. In 2008, Nicole King of the University of California, Berkeley, led a team of scientists that sequenced the choanoflagellate genome. Choanoflagellates turned out to have a number of genes that had previously been found only in animals. Among those genes are some that make proteins in neurons.

When young neurons move through the brain to find their final location, for example, they switch on a gene called reelin that makes a protein that helps guide them. Choanoflagellates turn out to have a reelin gene. Neurons make receptors to grab signaling molecules from other parts of the body; choanoflagellates make the same kind of receptors. In order to produce electric pulses, neurons open up special channels to let charged calcium atoms flow across their membranes. Choanoflagellates are the first nonanimals ever found that make the same kinds of calcium channels. In other words, some of the building blocks for our brains—the genes that would be essential for building and running neurons—already existed in single-celled creatures some 700 million years ago. Back then the genes had other functions, just as they do today in choanoflagellates (**Figure 14.1**).

Sponges left the earliest evidence of any animal in the fossil record. And in many studies on animal phylogeny, sponges branched off very early from all other animals. Sponges are exceptional in the animal kingdom for having no nervous system. One hypothesis to explain all these findings is that sponges diverged before the nervous system evolved in the common ancestor of the other animals.

But even without a nervous system, sponges have some cells that bear a striking resemblance to true neurons. Sponges start out in life as tiny, free-floating larvae that drift through the water. Some of the cells on the surface of the larvae develop a deep pit, out of which grows a long hair. This arrangement is a common one among cells that are specialized for sensing their surroundings. No one knows whether these cells actually sense anything, but some scientists have suggested that they detect changes in the water that may tell the sponge that it is time to settle down and find a place to anchor in order to grow into an adult. These so-called globular cells may be able to spread their messages through the larva's body by secreting hormones.

In 2008 Bernard Degnan, a biologist at the University of Queensland in Australia, and his colleagues analyzed the network of genes that guide the development of globular cells. They discovered that each of the genes is closely related to a gene in the network that guides the development of neurons. Meanwhile, Ken Kosik of the University of California, Santa Barbara, has been examining some of the proteins that act as a skeleton inside globular cells. It turns out that they create an intricate scaffolding that is strikingly similar, protein for protein,

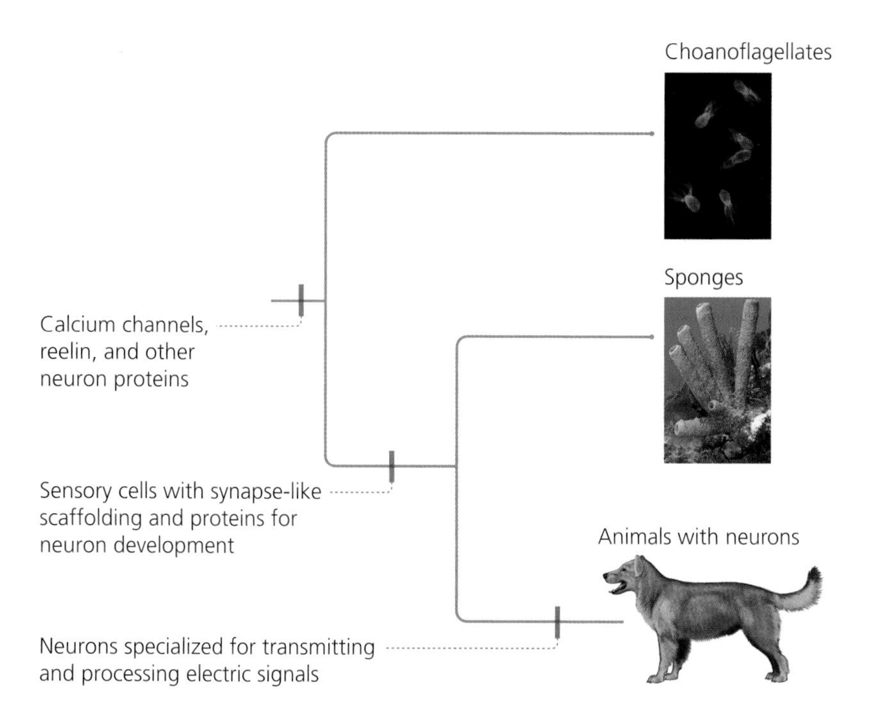

Figure 14.1 The evolution of neuron-associated genes began long before the emergence of neurons, as indicated by studies on sponges and choanoflagellates.

to the scaffolding inside a synapse. Both Kosik's and Degnan's research suggests that globular cells in sponges share a common ancestry with full-blown neurons.

These studies on sponges and choanoflagellates point to the same working hypothesis. The neurons found in millions of animals today, from humans to flies to jellyfish, are the result of a long, stepwise evolution, in which genes for sensing the environment and communication were modified to produce a new kind of cell. Once the neuron emerged, animals could do something their ancestors never could do: they could learn.

Learning New Tricks

The behavior of an animal is partly a result of the genes that control the development of its neurons. A baby gull that has never seen another gull in its life will react to its parents by pecking on their bills to beg for food. This behavior is not a result of the baby gull taking in everything it can see and hear, weighing all that information, and then choosing to peck on a bill. Instead, gulls automatically peck in response to something bill-shaped with some red on it. Even a red-and-white stick is enough to trigger it to peck (**Figure 14.2**).

There's no single gene "for" bill-pecking in the gull genome. The response emerges as the bird's nervous system develops, as its eyes begin to recognize shapes and colors, and as it gains control of the muscles in its neck and head. Nevertheless, gulls reared in a normal environment will almost always begin to peck at red and white bill shapes from the first time they see them. Many genes work together to produce this behavior, and they are passed down from gulls to their chicks.

The genes that encode these automatic behaviors evolve just like other genes. Stevan Arnold, a biologist at the University of Oregon, has documented the evolution of new behaviors in terrestrial garter snakes. These snakes, which live in the western United States, generally feed on animals like fish and frogs. In some

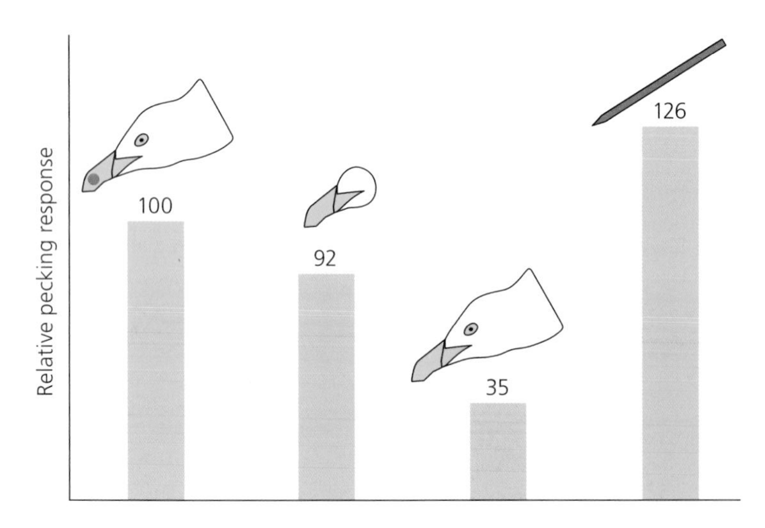

Figure 14.2 Herring gull chicks respond to the sight of their parents by begging for food. To determine what the chicks respond to, researchers build cardboard cutouts of adult gull heads. A cutout of the beak alone is almost as effective at triggering a response from a chick as a full head. On the other hand, a head without a red spot is less likely, while a red stick with white stripes is much more effective. These experiments show that the red beak spot is mainly responsible for a chick's response. (Adapted from Alcock, 2004)

areas near the California coast, however, they eat the slugs that are abundant in the damp coastal forests. To see how the snakes had acquired these tastes, Arnold collected female garter snakes carrying embryos and brought them to his lab. When the baby garter snakes were born, Arnold began to offer them food. The coastal snakes refused chunks of fish, but they eagerly feasted on pieces of slug. Even a cotton swab that smelled of slugs caused them to attack. Most inland snakes refused the slugs, and many refused to eat them until they were on the verge of starvation. When Arnold interbred the inland and coastal snakes, their hybrid offspring showed more interest in slugs than their inland parents, but less than their coastal parents.

Arnold's results demonstrated that populations of the same species can evolve different behaviors, and it's likely that this evolution happens rapidly. Studies on the DNA of terrestrial garter snakes show that they were restricted to a few refuges during the last Ice Age and rapidly moved into new territories as the glaciers retreated. While the inland and coastal populations have evolved different behaviors, they share a recent common ancestor.

While some behaviors are fixed, many behaviors can change during an animal's lifetime. Animals, in other words, can learn. Learning is not just what we do in school, and it often happens without our awareness that it's happening. Learning is the development of new behaviors and the modification of old ones. Fears, for example, can be learned. Psychologists can train people to have new fears by having them listen to a series of sounds. Most of the sounds are just ordinary tones, but, from time to time, they play a loud, painful sound that triggers an anxious response from the subjects, which can be measured with electrodes on the skin. If the psychologists consistently play a particular note—say, middle C—before the painful sound, their subjects will unconsciously learn that middle C means trouble ahead. Just hearing middle C will trigger a fear response, even if no painful sound comes afterwards.

Nervous systems make learning possible. No one has found any evidence that bacteria or protozoans can learn, but even the most humble nervous system seems capable of learning. A fruit fly, for example, does not simply react to its environment in a fixed way. It can learn things about potential dangers, potential rewards, and potential mates. Male fruit flies can mate successfully only with receptive females, for example, but young males often try to mate with unreceptive ones. It takes them a few encounters to learn the signs of a receptive female.

Scientists can give fruit flies lessons as well. Tadeusz Kawecki and his colleagues at the University of Fribourg in Switzerland have run experiments in which the flies learn clues about where to find good food. They offer the insects a choice of orange or pineapple jelly to eat. Both smell delicious to the flies. But the ones that land on one of the two flavors of jelly will discover that it is spiked with bitter-tasting quinine. Within a few hours, the flies will show a strong preference for the quinine-free jelly. They learn to associate the sweet smell of the other jelly with the nasty surprise in store if they have a taste.

Humans and flies can learn thanks to the underlying biology their nervous systems share. Learning changes the connections between neurons. The odds that a particular neuron will be able to pass on a particular signal to another neuron depend on how strongly they are connected by synapses. As animals learn, the strength of synaptic connections changes. Neurons may grow new contacts, or they may pare old ones back. As a result, new pathways emerge for signals traveling through the nervous system. The new pathways can link the perception of a particular stimulus (the smell of oranges, for instance) to neurons that produce a particular action (the impulse to fly away).

All of the molecules that are involved in learning are encoded in genes, and variations in those genes can lead to variations in how animals learn. Kawecki and his colleagues have found, for example, that some flies can learn to associate quinine with jelly flavors faster than other flies. They then took advantage of that variation to design an experiment to see if learning in flies can evolve.

Kawecki and his colleagues gave the flies three hours to learn to avoid the dish of jelly laced with quinine. Then they let the flies lay eggs on either dish, but they collected only the eggs from the quinine-free dish. Many of the flies that laid those eggs had learned which dish tasted good. The scientists reared the eggs and allowed the new flies to feed on the jelly dishes. This time, however, the scientists switched the quinine to the other flavor, and then in the next generation they switched back. By switching the quinine, Kawecki and his colleagues hoped that their experiment would foster the evolution of general learning, rather than an instinctive attraction to one particular flavor.

After 15 generations, the scientists tested the flies for their ability to learn. They gave the flies two dishes of jelly to choose from and let them take as long as they needed to learn which one to avoid. Ordinary flies needed several hours to figure this out. But the evolved flies needed less than an hour.

But Kawecki and his colleagues discovered that the flies pay a price for fast learning. The scientists pitted smart fly larvae against a different strain of flies, mixing the insects and giving them a meager supply of yeast to see who would survive. About half the smart flies survived; 80% of the ordinary flies did.

Reversing the experiment showed that being smart does not ensure survival. The scientists reared flies for 30 generations on a meager diet. The flies that adapted to surviving on the poor food left more offspring. When the scientists tested the learning abilities of the flies in the thirtieth generation, they found that the insects did a worse job than their ancestors. In effect, survival had made them stupid.

The ability to learn does not just harm the flies in their youth. Kawecki and his colleagues discovered that the average lifespan of their fast-learning flies was 15% shorter than those of flies that had not experienced selection on the quinine-spiked jelly. When the scientists selected flies for long life, the insects evolved to be up to 40% worse at learning than ordinary flies.

Kawecki's experiments demonstrate that learning, like many other adaptations, can impose a cost on an organism. It's not clear what is creating that cost. Forming a lot of connections between neurons may cause harmful side effects, possibly by producing cell-damaging molecules. It's also possible that the proteins used by neurons during learning interfere with other chemical reactions, such as the ones that repair cells.

Natural selection can favor increases in learning and memory only if their costs are outweighed by their benefits. That balance is different for each species. Learning may be favored when a species cannot rely on automatic responses—that is, when its environment becomes less predictable. Some bee species, for example, feed on a single flower species. They can find plenty of nectar using automatic cues. Other bees are adapted to many different flowers, each with a different shape and a different flowering time. Learning may be a better strategy in such cases. Scientists have carried out only a few studies to test this idea. One study, published in 2008 by scientists at the University of London, showed that fast-learning colonies of bumblebees collected up to 40% more nectar than slower-learning colonies.

Each species may evolve until it reaches the equilibrium between the costs and the benefits of learning. Kawecki's experiments demonstrate that flies have the genetic potential to become significantly smarter in the wild, but only under his lab conditions did evolution actually move in that direction. Outside a laboratory, this kind of improvement may impose too much of a cost.

The Vertebrate Brain: Not an All-Purpose Computer

Our behavior is similar in some ways to the behavior of insects, roundworms, and other animals because we all inherited a nervous system from a common ancestor. After our own lineage branched off from theirs, though, our nervous system evolved many new features. It is organized around a large brain protected by a bony skull, supplied with information from powerful eyes and other sensory organs, and controlling a big, fast-moving body. So it shouldn't come as a surprise that our behavior is more like the behavior of other animals that also have this same kind of nervous system—in other words, our vertebrate relatives: fishes, amphibians, reptiles, birds, and other mammals.

The oldest signs of this kind of brain can be found in 530-million-year-old rocks in China. Those rocks contain hundreds of fossilized impressions of a tiny creature called *Haikouichthys*. Measuring about three centimeters long, it has many (but not all) of the hallmarks of living vertebrates. Its spinal canal is surrounded by vertebrae, which are supported by a notochord. It has a series of pouches and arches to support gills. Two dark spots at the front of its body appear to be simple eyes. It has holes on the side of its head where sound-sensitive nerves probably grew, and it has another cavity up front that paleontologists suspect was a nostril for smelling. Its head contains a mass of cartilage that appears to have surrounded a primitive brain.

Over the next 100 million years, fishes evolved from such humble creatures into the biggest animals in the world. They became predators that could search for prey, in many cases chasing other animals down. Many of the major features

A key step in the evolution of vertebrate behavior was the evolution of a brain. The earliest fossil evidence of a brain is from the 530-million-year-old *Haikouichthys*.

of our nervous systems evolved during this transition. As the bodies of fishes became longer, their fossils show that their neurons did as well. This was made possible by the evolution of myelin, an oily sleeve for neurons that acts like the insulation around a wire, preventing the loss of electrical signals over long distances. The growing neurons of fishes began to supply the brain with information from larger sense organs, and new motor neurons allowed fishes to steer their bodies in complex ways.

Controlling those neurons was an evolving brain. The brain evolved into distinct regions, which can still be seen in the brains of all living vertebrates today, known as the cerebellum, the optic tectum, and the cerebrum (**Figure 14.3**). Each of these regions took on different functions. The cerebellum, located at the base

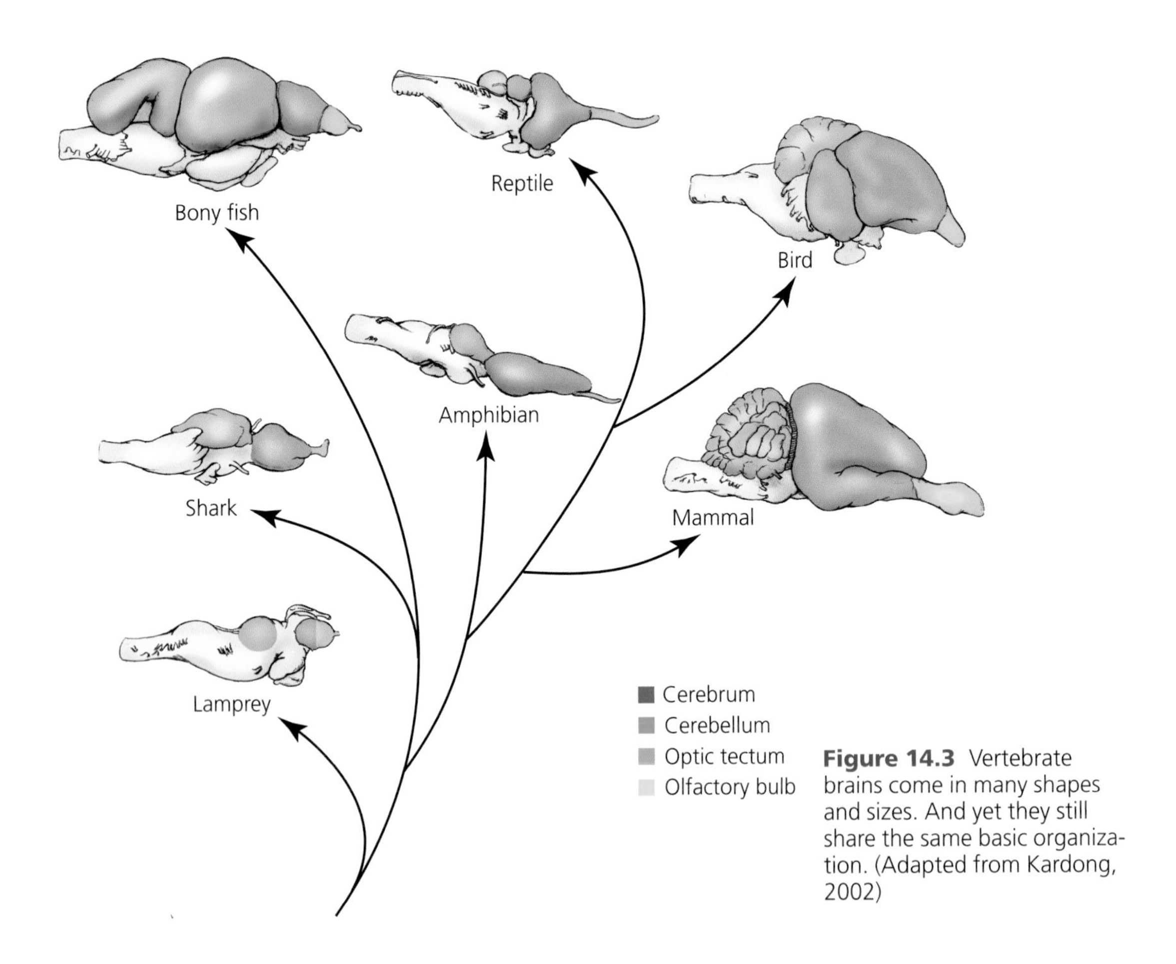

Figure 14.3 Vertebrate brains come in many shapes and sizes. And yet they still share the same basic organization. (Adapted from Kardong, 2002)

of the brain, is especially important for balance, for example. People who suffer damage to their cerebellum have trouble walking. Fish use their cerebellum to stay balanced in water.

As vertebrates diverged from a common ancestor, their brains diverged as well. In some sharks, the cerebellum is larger than the other sections, while in salmon the optic tectum is larger. The vertebrates that moved onto land—the tetrapods—tended to evolve larger cerebrums. In mammals, the outer layer of the cerebrum, known as the cerebral cortex, expanded drastically. In humans, the cerebral cortex now takes up 90% of the brain (**Figure 14.4**).

The cerebral cortex is especially important for our most sophisticated kinds of thinking—for recalling memories, for language, for making tough choices. As it develops, it becomes parceled into some 200 distinct areas. The neurons in a particular area are densely interconnected, and together they help with certain men-

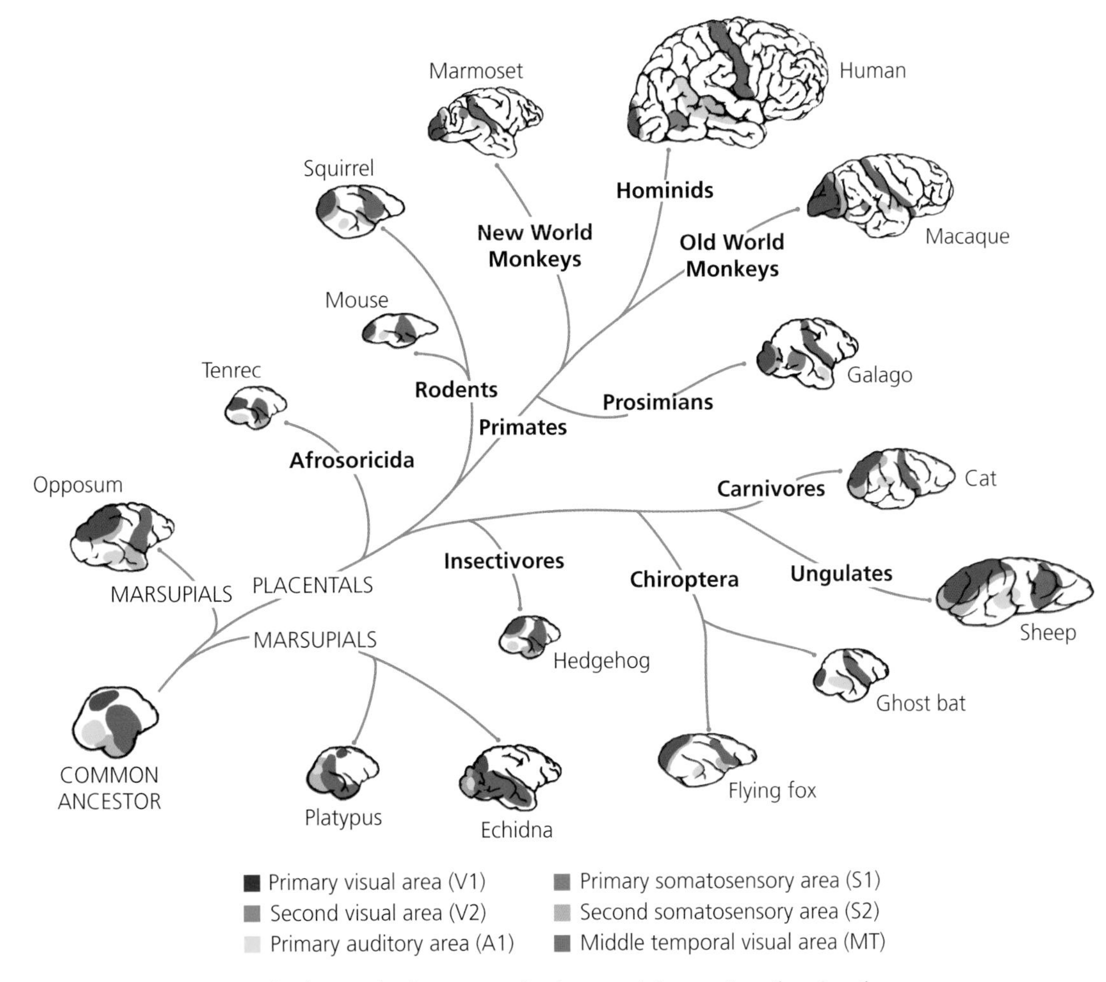

Figure 14.4 In mammals, the cerebral cortex evolved to much larger sizes than in other vertebrates. Different regions of the cerebral cortex are dedicated to special tasks, such as processing sight or sound. Some of these regions are highlighted in this evolutionary tree. (Adapted from Krubitzer, 2007)

tal tasks. One area, called the somatosensory cortex, receives signals from touch-sensitive neurons in the skin. Another area, the visual cortex, maps the signals from the eyes and then sends them to other areas of the brain (**Figure 14.5**).

Because our brains are organized in this way, they cannot work like an all-purpose computer that processes all kinds of information in the same way. They are biased in the information they extract from the senses. The somatosensory cortex offers a stark illustration of how biased vertebrate brains can be. Neuroscientists map the somatosensory cortex of animals by touching parts of their skin and recording the responses of neurons in the brain. Some parts of the body are better represented in this body map than others. Neuroscientists chart these patterns by making drawings of an animal's body, making some parts bigger and some smaller. The body map for one species is distorted in different ways than that of another.

A few burrowing animals are mapped out in **Figure 14.6** An eastern mole has extremely sensitive front feet, nose, and whiskers. Those are the parts of the body that it uses to dig through the dirt and to sense the presence of insects that it can eat. A star-nosed mole, by contrast, has long, fingerlike extensions around its nose that it uses to probe the mud around streams. That tiny patch of skin is more sensitive than the rest of its body. Naked mole-rats also burrow through the ground, but they live in arid regions of Africa, where they use their teeth to dig tunnels. For them, the teeth rather than the nose dominate their somatosensory cortex.

We humans are also biased about the information we receive, but in a different way. A map of our somatosensory cortex (page 339) shows our hands, lips, and tongues to be exquisitely sensitive. That pattern is a result of our own ecology—particularly our adaptations for using tools with our hands.

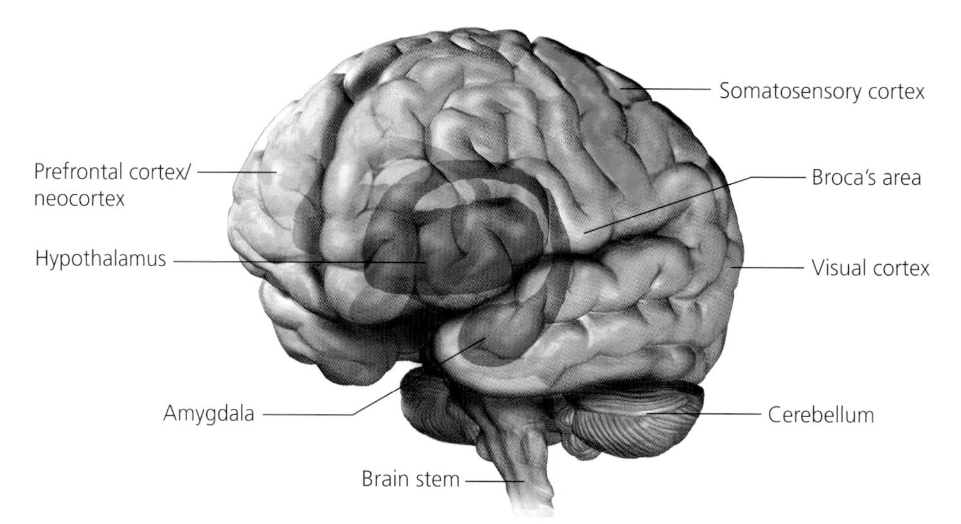

Figure 14.5 The human brain, like other vertebrate brains, is divided into many specialized regions, each of which helps carry out certain functions. The cerebellum, for example, is important for balance. The somatosensory cortex organizes sensory information from the skin. Broca's area is involved, among other things, in language. Neuroscientists can find homologs to many of these areas in other animals.

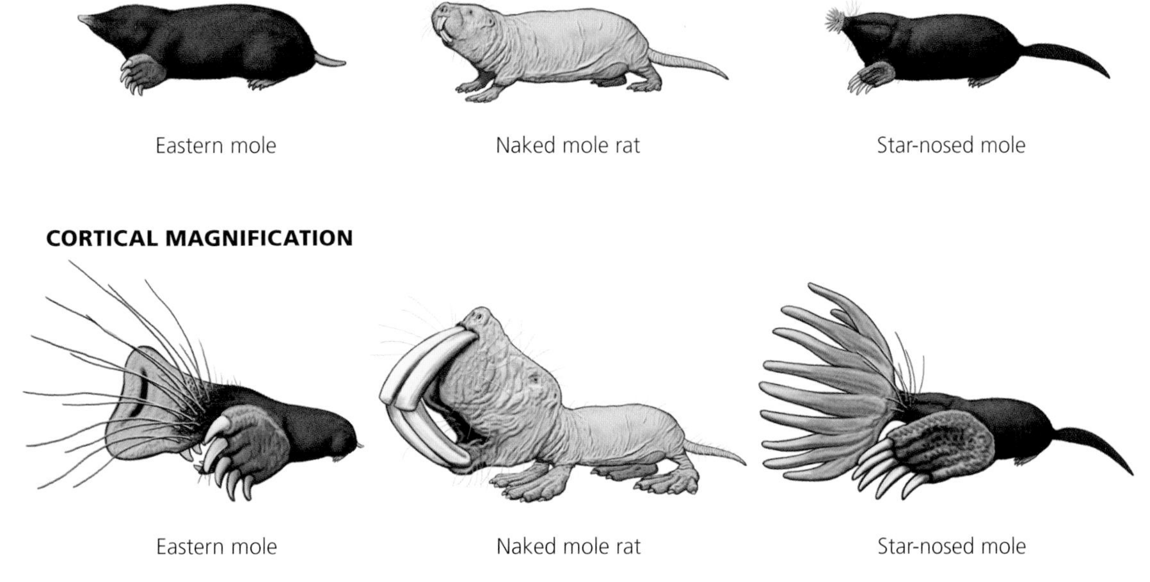

Figure 14.6 The brains of mammals are adapted to the ecological niches in which they live. For example, each mammal has some regions of the body that are very sensitive, and others that are not. These images show the relative concentration of neurons dedicated to each part of the body in the somatosensory cortex of the brain. Left: Moles use their forelegs to burrow, and their noses and whiskers to sense prey. These regions are the biggest in their body map. Middle: The naked mole rat, which spends its life digging tunnels through dry earth, has a body map dominated by its teeth, mouth, and feet. Right: Star-nosed moles, on the other hand, have fleshy appendages on their noses, and these "stars" dominate their body maps. Below: This sculpture represents the human body map in the somatosensory cortex. Unlike rodents, our hands and mouths are most sensitive—reflecting our adaptations for using tools and language. (Rodents adapted from Alcock, 2004)

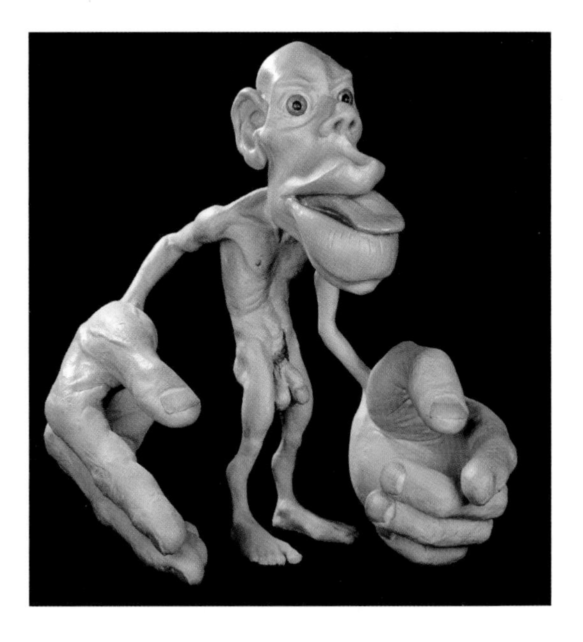

Once the brain receives this biased information, it looks for specific patterns. Our species communicates with complex sounds from our throats, and the sensitivity of our hearing is tuned to the frequencies used most in speech. Parts of our brain specialize in reading lips, and that information is merged with the information from the ears to predict what people are saying. If you were to listen to a recording of someone saying "My bab pop me poo brive" while you watch a video of someone mouthing "My gag kok me koo grive," your brain would combine those streams of information so that you'd think you were hearing, "My dad taught me to drive."

Experiments like this one demonstrate that our brains are not just cameras that passively film the outside world and then make a perfect replica. Instead, our brains filter out all but a limited number of signals and then use that information to make predictions about what the world is like.

The Origin of Emotions

There's more to our minds than sensing and acting. We also have feelings. To trace the evolutionary roots of our moods, desires, and emotions, scientists are pinpointing the activities in the brain that make them possible. Many of those activities bear striking similarities with what's going on in animal brains, particularly in the brains of our fellow mammals. These comparisons show that, 100 million years ago, our ancestors had already evolved the basic systems that are essential for many of our emotions.

Fear, for example, triggers responses from an almond-shaped region on the underside of the brain called the amygdala. The amygdala becomes active at the sight of fearful things, such as a picture of a gun or an angry face. In fact, neuroscientists can observe activity in the amygdala when people see these pictures for just a tenth of a second, which is not enough time for them to become aware they've seen anything. Our brains have shortcuts that can relay information from our eyes and ears to the amygdala without passing through the cerebral cortex. The amygdala, in turn, sends signals to other parts of the brain that produce changes in the body, such as a rapid heartbeat and heightened attention.

Some of the earliest insights about the human amygdala came from studies on the brains of mice and rats. In the mid-1900s, a number of scientists began to experiment on rodents to understand the biology of fear. Like all other mammals, rats and mice have amygdalas, which are linked to other parts of the brain in much the same arrangement as in the human brain. By implanting electrodes in the brains of the animals, scientists could observe how rodents learned fear as synapses in the amygdala became stronger.

The fear we feel is not identical with the fear experience of a rat. The human amygdala can become active when people read threatening words, such as *poison* or *danger*, an experience a rat obviously will never have. But our reactions to those words have evolved from the same underlying circuitry we share with our rodent cousins. This capacity to feel fear was one of the most important that early mammals could have. It allowed them to respond quickly to threats. If they suddenly saw a predator about to attack, for example, those early mammals could freeze, flee, or retaliate. Some of the dangers that mammals faced were reliable enough that natural selection could produce instinctive fears. Some studies suggest, for example, that we are born with an innate fear of snakes, which would have threatened our primate ancestors for millions of years. But mammals did not have to rely only on hard-wired fears. They could also learn a healthy fear for any new dangers they might encounter.

There's more to life than being scared, though. Motivations help mammals reach important goals such as finding food or mates. The most important region for generating these motivations is a small cluster of neurons in the brain stem. If a rat, for example, should be searching for food and unexpectedly get a whiff of something delicious, those neurons will release a tiny surge of a neurotransmitter called dopamine. The dopamine-producing neurons have a vast number of connections to many networks in the brain, and so they can quickly alter how the entire brain functions. Dopamine arouses an animal's attention, and also makes it easier for neurons to form stronger connections with other neurons. A rat's brain can begin to associate cues like odors with its long-term goals, such as finding food.

The power of dopamine over the mammal brain is astonishing. One way to demonstrate its importance is by genetically engineering mice so that they can't produce it. These dopamine-free mice are in many ways perfectly normal. They still prefer the taste of sucrose over other foods, and they can learn where food is located. But they lose the motivation to pursue any goals. They will simply starve from that lack of motivation less than a month after they're born. If scientists give them injections of dopamine, however, they will feed for about 10 hours, until the motivation disappears again.

Too much dopamine can be just as dangerous as too little. Scientists often reward rats by giving them food if they press a lever in response to the right signal—in response to a green light but not a red one, for example. The rat's brain produces surges of dopamine as it learns the rule. If the scientists give the rats an injection of dopamine each time they press the lever, however, something else happens. The rats will keep pressing the lever again and again. They will do nothing else, not even eat. Ultimately, they may die of starvation.

Humans have inherited the same dopamine-delivery system. It doesn't make us feel happy so much as eager with anticipation. The rewards that can trigger a release of dopamine are, like our fears, more sophisticated than those that occur in a rat's brain. It can be triggred by winning at a slot machine, the sight of an attractive face, or even hearing a joke. Unfortunately, the reward system can also be hijacked by substances that cause the brain to release unnaturally large amounts of dopamine. Cocaine and other drugs do just this, and it's the reason that they can become so addictive.

Our relationships with other people—particularly with our family—trigger many of our most intense emotions. These bonds also have an ancient history. Our reptilian ancestors probably gave their young little care after they hatched from their eggs. But the bond became more intimate in early mammals. Mothers evolved the ability to make milk for their young. To nurse their offspring, they needed to interact with their offspring for weeks, months, or even years.

All living mammals use the same hormone, known as oxytocin, to foster mothering. A gland in the brain called the hypothalamus releases oxytocin late in pregnancy. Some of the hormones latch onto receptors in the mammary glands, causing them to begin producing milk. Some oxytocin latches onto neurons in the brain, altering a mother's behavior. In sheep, for example, oxytocin causes a ewe to bond with her lamb just after birth. She will be able to recognize the smell and bleat of her own lamb for the weeks that she nurses it. If scientists block the uptake of oxytocin in a ewe, she will reject the lamb. An injection of oxytocin into a ewe that's not pregnant will make her act like a mother to an unrelated lamb.

Oxytocin is released by the brains of women both during pregnancy and after birth. The touch of a baby during nursing is enough to trigger an increase of the hormone. Oxytocin tends to cause women to bond more with their babies, as measured by the sounds they make, the number of times they check in on the children, and how much they gaze at them. But experiments in recent years suggest that oxytocin shapes our dealings outside the family as well. Scientists have fashioned oxytocin sprays that can deliver the hormone into the nose. It enters the blood and then eventually reaches the brain, where it reduces the activity of the amygdala. People given oxytocin become more trusting of others, and more willing to forgive. They even do a better job of empathizing with other people simply by looking at their facial expressions.

What's striking about these results is not just that oxytocin seems to have taken on many new social roles in our species, but also that it is stimulated by different mechanisms. For sheep, rats, and most other mammals, oxytocin is triggered mainly by smell. That's not the case in humans; the release of oxytocin into the bloodstream and other emotion-related responses depend much more on our sense of sight. The reason we are different is that we are not just any mammal: we are primates.

Primates: Eyes, Tools, and Societies

About 100 million years ago, our ancestors probably looked vaguely like shrews, with four short legs, a long tail, eyes facing out to either side, and a tiny brain with a very tiny cerebral cortex. But then our ancestors began to evolve into a distinctive new form. They became primates.

All living primates share a number of traits. Their thumbs are free enough from their other fingers to swing around and create a powerful grip. Most primates also have opposable big toes, although humans lost that particular trait as they evolved flat feet for walking. Primates have flat nails instead of claws, and their eyes point forward rather than to the sides. And along with these anatomical traits, primates share a number of molecular ones, such as a distinctive collection of mobile elements called *Alu* elements (page 138).

The oldest fossils of mammals more closely related to living primates than to other living mammals date back about 65 million years. Simon Tavare, a molecular evolutionist at the University of Southern California, and his colleagues have tried to estimate the origin of primates based on the molecular clock recorded in their DNA. They estimate that the last common ancestor of living primates lived 81 million years ago. If they're right, there's a 16 million year gap between the oldest fossil primates yet found and the origin of the primate lineage. Future generations of paleontologists can test Tavare's hypothesis by searching for those early fossils.

Our most distant living primate relatives include the lemurs of Madagascar. We share a more recent common ancestor with monkeys and apes. Apes branched off from the ancestors of Old World monkeys about 30 million years ago. Among the apes, our closest living relatives are the chimpanzees and bonobos of Africa. Our common ancestor with them lived roughly seven million years ago.

Over the course of primate evolution, the brains of our ancestors underwent some extraordinary changes. It was during this time, for example, that vision, rather than smell, became the dominant sense in our ancestors. Fossil braincases of early mammals bulge at the front, where a smell-processing brain region called the olfactory bulb was located. Many mammals today still rely mainly on their sense of smell. Shrews dedicate 60% of their cerebral cortex to processing information from their noses. Primates dedicate just a few percent of their brains to smell, and that figure drops to about 1% in apes.

This shift is also recorded in our genes. All tetrapods use the same family of genes to produce odor receptors on the ends of neurons that grow inside their noses. Like other genes, these olfactory receptor genes have sometimes been accidentally duplicated. Afterwards, some of the duplicated genes have evolved to become sensitive to different kinds of odors, while others have been disabled by a mutation. At first, disabled olfactory receptor genes linger on as pseudogenes, and eventually many of them are deleted from the genome altogether. Mammals that depend on their sense of smell have large families of olfactory receptor genes. Mice, for example, have 1,391 olfactory receptor genes, 508 (about 36%) of which are pseudogenes. Primates, by contrast, have far fewer olfactory genes and a higher proportion of pseudogenes. A macaque, for example, has 606 olfactory genes, of which 46% are pseudogenes. Humans have 802 olfactory genes, but 52% of them—more than half—are pseudogenes.

A number of studies indicate that our primate ancestors shifted from smell to sight thanks to a shift in their diet. Old World monkeys and apes eat mainly leaves and fruit, and they depend on their ability to see the colors to judge when a particular fruit or leaf is ready to be picked. It turns out that Old World monkeys and apes also share a duplicated opsin gene that other primates lack, and this gene duplication gives us better vision in the red and orange region of the light spectrum. It's exactly these colors that are most important for a primate to see in leaves or fruits to judge if they're edible. All this evidence indicates that the ancestor of apes and Old World monkeys came to rely more on vision to find food, and less on smell.

This shift towards vision also altered the social lives of primates. Many other mammals communicate to each other with a language of odors. The molecules that waft from a newborn lamb enter its mother's nose and trigger changes in her brain. She will recognize her lamb by smell until she has finished nursing it. Primates evolved to respond emotionally more to the sight of their fellow primates than to their smell. This transition brought with it the evolution of a new kind of face. Primates evolved new arrangements of facial muscles that allowed them to make a much wider range of expressions. They also evolved new regions of the brain that specialized in recognizing the faces of other primates and understanding what kind of face they were making.

These new adaptions helped to turn primates into intensely social animals. Most species of primates live their entire lives in a group. They sleep together, search for food together, escape predators together, and sometimes even do battle with other groups of primates together. Living in groups may be better for most primates than living alone, but these groups create a new arena within which competition arises. Primates struggle to reach the top of the hierarchy in a group, and with that rise often comes reproductive success. Joan Silk, a primatologist at the University of California, Los Angeles, has observed baboons in Kenya for two decades, and she has found that high-ranking female baboons grow faster, produce healthier infants, give birth at shorter intervals, and generally have a higher lifetime reproductive success than lower-ranking baboons. For baboons, and for many other primates, achieving a high rank is not a matter of having the biggest teeth or the loudest scream. It's more a matter of social connections, of alliances that can last for decades. The most socially integrated baboons, Silk has found, have the highest reproductive success in a baboon society.

These conditions have driven the evolution of primates, making them keen social observers. They can recognize individual members of their own group and remember who is related to whom, and who is allied with whom. They can tell what other members of their group are looking at, and they can use that knowledge to deceive them. A female gorilla will sometimes sneak off with a low-ranking male to mate, and the pair will make only quiet sounds instead of their normally loud mating calls. Monkeys and apes have even evolved a kind of social economy. In 2007, for example, a student of Silk's named Kimberly Duffy reported that a top-ranking male chimpanzee gave lower-ranking males the opportunity to mate with females in his group in exchange for their alliance with him against other males.

In 1992 Robin Dunbar, a psychologist at the University of Oxford, hypothesized that the social complexity of primates drove the evolution of their brains. Compared with other mammals, primates have very big brains (and we humans have the biggest brains of all, relative to our bodies). Dunbar proposed that natural selection favored big brains in species that had complicated social lives. To test this idea, Dunbar began to measure different parts of primate brains to see what fraction was taken up by the neocortex, the region of the cerebral cortex where the most sophisticated information processing takes place. He found that the bigger the average social group in a primate species, the bigger the fraction of the brain that was taken up by the neocortex (**Figure 14.7**). Dunbar and others have also found similar correlations between the neocortex and other measures of social complexity, such as how often primates deceive one another.

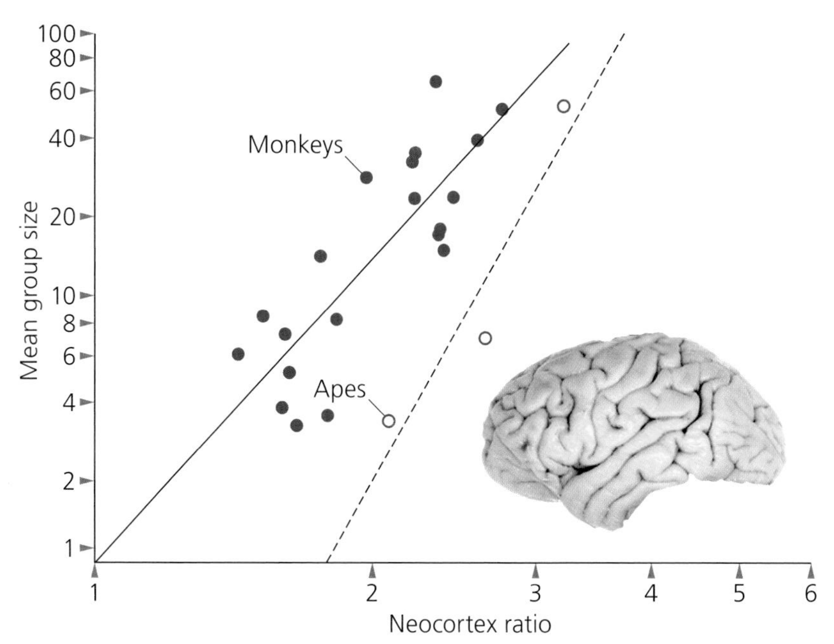

Figure 14.7 The social lives of primates may have had a profound effect on the evolution of their brains. This graph shows how primate species that live in large groups tend to have a proportionately larger neocortex. (Adapted from Dunbar, 2007)

Dunbar's so-called social-brain hypothesis could explain how primates—and humans, in particular—broke out of the trade-off between the costs and benefits of learning that Kawecki and his colleagues documented with *Drosophila* flies. If the only benefit an animal gets from learning is how to get better food, it may evolve until it reaches the point at which the benefits of learning balance out the cost. But, for primates, learning is important for succeeding in a society—a society that itself is evolving. If reproductive success depends on social intelligence, then the average intelligence of the population will go up. Now natural selection will favor even bigger brains for better social cognition. Primate societies may thus create something akin to a coevolutionary arms race (page 249), in which individuals drive each other towards bigger and bigger brains.

It's also possible that social complexity isn't the only force that's driven the evolution of big brains in primates. Big brains may also be favored because they allow mammals to innovate and to adapt to new conditions. Daniel Sol, a biologist at Universitat Autònoma de Barcelona, and his colleagues examined 400 cases in which people accidentally introduced mammals to new habitats. Some species died off, some established small populations, and others thrived. In 2008, Sol reported that the mammals with bigger brains tended to thrive, while the small-brained ones had a harder time. Similarly, Simon Reader and Kevin Laland found that, among primates, bigger brains tended to be found in species that showed more signs of being able to learn from one another and to come up with innovative new ways of finding food.

Apes in particular are very innovative, being able to eat foods that other primates have to pass up. Richard Byrne, a primatologist at the University of Saint Andrews in Scotland, has documented how gorillas manage to eat stinging nettles by first carefully removing the spines from the plants. Chimpanzees in Tanzania feed on the fruit of the vine *Saba florida* by first breaking open its hard shell to expose its interior, dividing it into smaller sections to eat, removing the pieces of shell still attached to the sections, and finally extracting the sweet flesh inside. At many stages in this long process, the chimpanzee's two hands are doing different jobs, and sometimes it may dedicate individual fingers to different tasks.

Apes don't just use their fingers in innovative ways; they also make tools. Orangutans, for example, use branches to ward off bees or wasps. They dip leaves into deep holes to obtain drinking water, and they strip twigs in order to stick them into ant nests or spiny fruits. Chimpanzees make insect probes as well. They also place nuts on flat rocks and smash them with pebbles. They sometimes lay out several leaves on wet ground to keep their backsides dry. In Cameroon, female chimpanzees have been observed impaling small primates called bush babies with spears they fashioned from sticks.

These innovations are not the result of simple hard-wired behavior. Not all chimpanzees who have nuts and rocks at hand actually use the rocks to smash the nuts. Many populations of apes have unique combinations of tools and techniques that neighboring populations lack. Some primatologists argue that this pattern is evidence that apes have culture—something once thought to be unique to humans. One ape may invent a new kind of tool, and other apes—particularly young ones—will learn how to make it by observing the inventor.

Uniquely Human (or, at Least, Uniquely Hominid)

We now reach the final branching point that divides us from all other living things. Seven million or so years ago, the ancestors of chimpanzees and bonobos diverged from the ancestors of human beings. There are many things about human behavior that set us apart from the rest of life. The fact that you have almost finished reading this book is one of them. No other species can communicate with full-blown language—a system of sounds, gestures, or written symbols that can convey information not just about what's immediately in front of us, but what lies in the distant past, in the far-off future, or in a world that never will be.

Language, many scientists argue, is at the core of human nature. In order to use it, we have to be able to understand abstract concepts, instead of just using labels for obvious things like snakes or birds. Language lets us do things other animals cannot do, such as make complex plans together and gain a deep understanding of the inner lives of other humans.

There are more than six thousand languages on Earth, each the product of a particular culture with a particular history. But underlying this staggering diversity, languages share a lot in common. All spoken languages are based on sets of sounds, which can be combined into thousands of words, which, in turn, can be combined into an infinite number of sentences. All languages share some basic rules of grammar. They also all have some kind of syntax. The order in which words are arranged in a sentence can change their meaning.

Despite language's complexity, children don't need to attend a linguistics class to learn how to speak. They quickly pick up the rules of grammar for themselves in the first few years of life. Even if a child can't hear, the drive to develop language doesn't stop. Deaf children who watch other people talk in sign language will start to "babble" with their hands. In the 1980s, a group of deaf children in Nicaragua even invented a sign language of their own, complete with grammar and syntax.

Along with having a capacity for learning the rules of language, our brains are also well adapted to hearing speech. When we listen to someone speaking, the network of brain regions that becomes active is not the same as the one we use to listen to ordinary sounds. Certain kinds of brain damage can cause "word deafness," leaving people unable to understand speech

It was not until the discovery of *FOXP2* in 1990 that scientists began to uncover some of the genes underlying our capacity for language. British researchers pinpointed the *FOXP2* gene while studying a family in which many members had trouble speaking and understanding grammar. Some of the family members took part in a study run by neuroscientist Frédérique Liégeois and her colleagues at University College London. As Liégeois scanned their brains with a functional magnetic resonance imager, the family members listened to nouns and thought of verbs to go with them. The family members with the defective version of *FOXP2* had less activity in a region called Broca's area (**Figure 14.8**). Broca's area, it turns out, plays a pivotal role in processing language.

It would be a mistake to think of *FOXP2* as "the" language gene, just as it would be a mistake to look for a single beak-pecking gene in gulls. Doubtless there are

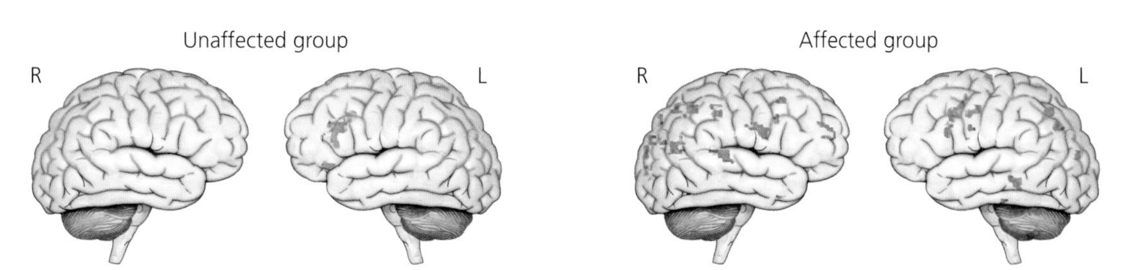

Figure 14.8 To investigate the role of *FOXP2* in language, neuroscientists scanned the brains of people with defective versions of the gene and their relatives who have working versions. The scans picked up areas of the brain that became active when the subjects listened to words and then thought of other words. In the brains of the subjects with a working version of *FOXP2*, Broca's area became especially active. (It's the large orange spot on the left side of the unaffected brain shown here.) In the brains of affected people, on the other hand, Broca's area was much less average than normal, while other parts of the brain became more active. (Adapted from Liégeois, 2005)

many other genes that work together to produce the human capacity for language, but studies on *FOXP2* add to the evidence that language has evolved as an adaptation in our species.

Many adaptations evolved from old parts that were modified for new uses (Chapter 6). While full-blown language may indeed be unique to our species, scientists can see its foundations by looking at our primate relatives. Broca's area and many other regions of the cortex that help process language are all joined together by a bundle of nerve fibers. If this bundle, called the arcuate fasciculus, is damaged, the ability to speak is damaged, too. People may have a hard time reading words aloud, and may jumble the syllables in words as they speak. James Rilling of Emory University and his colleagues recently made the first detailed comparison between the human arcuate fasciculus and the connections inside the brains of other primates (**Figure 14.9**).

Before Rilling's research, scientists had dissected brains to map these nerve fibers. Rilling was able to get a much more detailed picture with a technique called diffusion tensor imaging, which detects water molecules in nerve fibers.

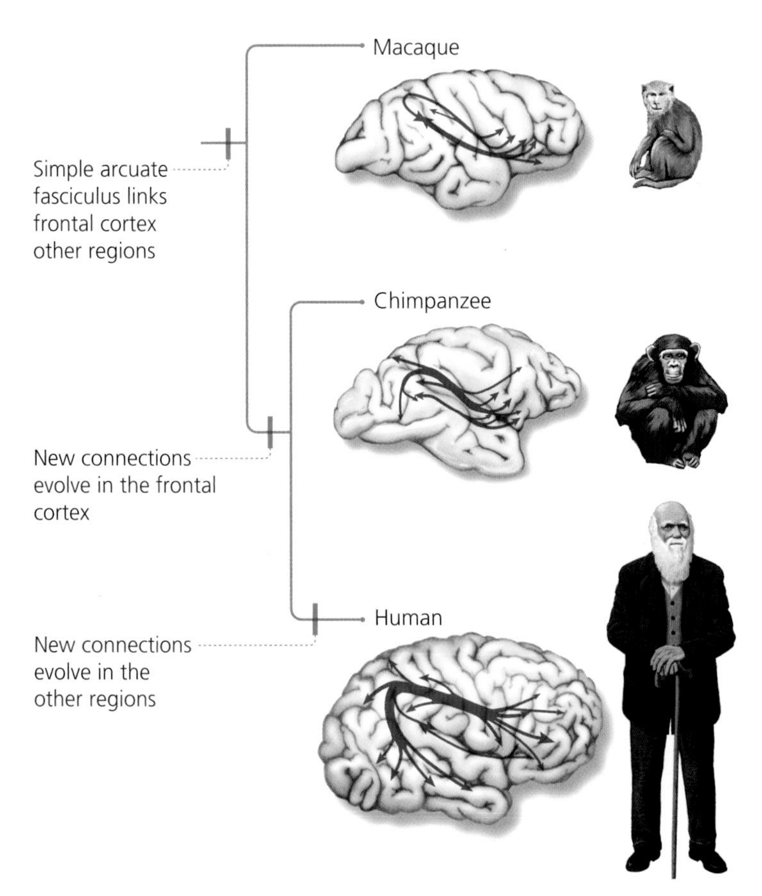

Figure 14.9 The arcuate fasciculus is a bundle of nerve fibers essential to human language. Scientists have discovered how it has expanded over the course of primate evolution. (Adapted from Rilling, 2008)

Rilling and his colleagues found that, over the course of primate evolution, the connections between the frontal and temporal lobes became denser. Chimpanzees have many more connections in the frontal lobe than macaque monkeys, and humans, in turn, have more connections to the temporal lobes than chimpanzees.

The regions that the arcuate fasciculus join together also have a long evolutionary history. Monkeys use Broca's area to interpret the calls of other monkeys. While monkeys don't have a huge vocabulary of words like we do, they can make several different calls, each of which has a distinct meaning. They may make one alarm call when snakes attack, for example, and another when a leopard attacks. Allen Braun and his colleagues at the National Institutes of Health recently investigated whether the brains of monkeys respond to these calls differently than they do to other noises. They had macaque monkeys listen to random noises, notes from musical instruments, and the coos and screams of other macaques. The scientists discovered that the coos and screams from macaques triggered regions of the brain that the other sounds did not. Those regions turn out to be homologs to language-processing regions in the human brain, such as Broca's area. Braun and his colleagues proposed that these regions of the brain processed simple sounds from other primates before they evolved into language centers.

By the time the apes evolved, this communication system had become more complex. Chimpanzees can adjust their gestures and calls in subtle ways, depending on who's listening. When wild chimpanzees are attacked by other chimps, for example, they give a distinctive scream. Sometimes, bystander chimps will respond by intervening and breaking up the fight. Katie Slocombe and Klaus Zuberbuhler, two primatologists at University of St. Andrews in Scotland, have observed that, when attacks are more severe, the victims make higher-pitched screams that last longer and come in longer bouts. But when high-ranking chimpanzees were nearby, the victims would scream as if the attacks were much worse than they really were. Slocombe and Zuberbuhler suggest that the victims are trying to take advantage of the power of high-ranking chimpanzees in their area.

Chimpanzees and other apes also communicate with gestures, something not observed in other primates. They may wave their arms, reach out their hands, or slap the ground. Chimp gestures generally convey some kind of request—to play, for example, or to share some food—but there's not any deep biological impulse linking one gesture to one meaning. In fact, gestures vary from population to population among chimpanzees, and they come to look like rituals. Some scientists

Chimpanzees use many different gestures to communicate with each other. Two populations of chimpanzees may give two different meanings to the same gesture. Some researchers have suggested that gestures might have played a critical part in the evolution of human language.

Between 1.5 million years ago and 200,000 years ago, hominids in Africa, Asia, and Europe made large teardrop-shaped stone tools. They probably used these so-called hand-axes to butcher animals, cut wood, and gather roots and other plant foods.

have suggested that the first steps that early hominids made towards human language might have been through gestures, rather than sounds.

Seven million years ago, the earliest hominids likely communicated as living apes do today, with a combination of gestures and simple calls. They had some of the building blocks of full-blown language, but other building blocks would evolve later. Spoken words do not fossilize, and so scientists have to find other lines of evidence to understand the evolution of language. As we saw in Chapter 4, early hominids lived in open woodlands, where they evolved into bipeds. By two million years ago, however, the African habitats of hominids had become much more arid and were dominated by grasslands.

The skulls of hominids allow scientists to estimate the sizes of their brains (**Figure 14.10**). For the first few million years of hominid evolution, their brains were about the size of chimpanzees' brains. About two million years ago, however, their brains began to expand, and they continued to do so until about 100,000 years ago. Paleoanthropologists have also documented the rise of hominid tools, starting with the earliest known stone tools, which were made about 2.5 million years ago. For perhaps two million years, hominids used these tools to carve meat off of dead wildebeests and other big game. Only in the last few hundred thousand years did our ancestors begin to hunt game with weapons.

In general, primates with big brains tend to have complex social lives. Some scientists argue that the human brain is no exception to this rule. As hominids evolved into open-country apes, they evolved into a highly cooperative species, in which the individuals in a group worked together far more closely than their ancestors had. Mutations that provided hominids with better social skills—the ability to learn from others, to make friends, to manipulate rivals—boosted reproductive success.

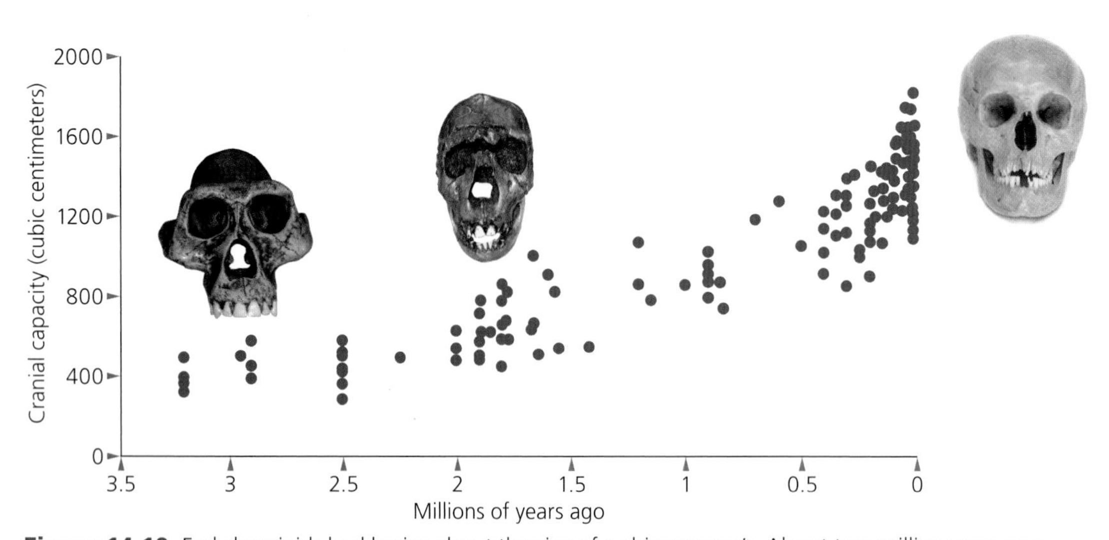

Figure 14.10 Early hominids had brains about the size of a chimpanzee's. About two million years ago hominid brains began to increase, and after one million years their growth accelerated. Today humans have brains about three times the size of the earliest hominids. (Courtesy of Nicholas Matzke, www.ncse.org)

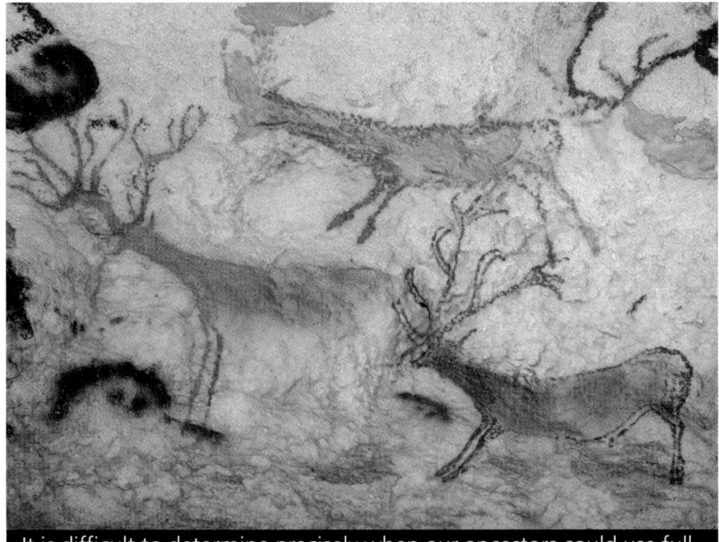

It is difficult to determine precisely when our ancestors could use full-blown language. But most researchers agree that humans could understand symbols by the time they were making figurative art. Archaeologists have found 35,000-year-old sculptures of women in Germany. They have also found cave paintings, such as these images painted 20,000 years ago in Lascaux, France.

In 2007, Esther Herrmann and her colleagues at the Max Planck Institute for Evolutionary Anthropology tested this "ultrasocial" hypothesis by testing the mental skills of 105 two-year-old children against 106 chimpanzees and 32 orangutans. Herrmann gave the children and the apes an identical series of tests. Some of the tests measured their understanding of space, quantities, or physical causes and effects. To test their spatial memory, for example, she put a toy or a piece of food under one of three cups and then let her subjects try to pick the right cup. Herrmann also tested the children and apes for social cognition. For instance, she showed her subjects how to get a toy or a piece of food out of a plastic tube, and then she gave them another tube. If they could learn by observing, they would be able to open it by themselves.

If social forces have driven human brain evolution, then children should rapidly develop social skills, such as learning from others and understanding what other people know or don't know. They shouldn't be particularly adapted for understanding math—at least no more than other apes. Of course, children can go on to understand math far better than apes, but only after they have developed the social skills that they will then use to learn. When Herrmann and her colleagues had finished running the test and had tallied the results, they discovered that, as predicted, children did no better than apes on the physical tests. On the social tests, however, they did significantly better.

As the hominid brain grew, perhaps thanks to social pressure, it changed in other ways. Scientists have found that, as a rule, the bigger a mammal's brain is, the more areas it has in its cerebral cortex. The extra areas that evolved in the hominid cortex could have been put to use processing information in new ways; and as the hominid brain expanded, it had to be rewired. All the areas of the cerebral cortex had to be linked together by white matter stretching from one end of the brain to the other. As Rilling's research on the arcuate fasciculus shows, hominids evolved many new connections. One possibility scientists are exploring is that the new connections that formed in the expanding brain allowed hominids to mix together information from different regions of the mind in new ways. This merging could have had a revolutionary effect on language. Now words could summon up vivid images and distant memories. Now they could represent more than just leopards and trees—they could represent concepts and beliefs, stories and dreams.

We are only at the beginning of this particular chapter in the history of science. Having identified a clear-cut gene involved in language, *FOXP2*, scientists are now looking closely at other candidates. They'll eventually look for signs that these genes have experienced natural selection, and they'll learn how they work in our closest primate relatives. They may even find versions of those genes in Neanderthal fossils. In the scheme of life, the evolution of language is just one of many fascinating episodes in the biological history of Earth, from the rise of feathered dinosaurs, to the origins of snake venom, to the greening of the land.

But we can be excused for being especially interested in it. Evolution helps show us who we are, and how we got this way.

TO SUM UP...

- **Behavior is shaped by evolution. Scientists can observe the evolution of behavior in laboratory experiments.**
- **The evolution of nervous systems allowed animals to develop more sophisticated behaviors.**
- **The building blocks of neurons evolved millions of years before neurons did.**
- **Learning emerged when animals evolved nervous systems. Learning evolves according to a trade-off between its costs and benefits.**
- **Vertebrates evolved large brains that evolved further to solve different ecological problems.**
- **The basic systems essential for human emotions are found in other mammals.**
- **The behavior of primates is shaped by their highly social lives.**
- **Humans are ultrasocial primates.**

Acknowledgments

Throughout the creation of this book, I relied on the generosity and expertise of many people. I would like to thank Joel Kingsolver, Kevin Padian, Greg Wray, and Marlene Zuk for serving as my advisors, helping me from my first outlines to the final draft. I would also like to thank all the scientists who offered their thoughts on the topics to include in this book, who read drafts to check for accuracy, who evaluated artwork, and who helped in many other ways: Fred W. Allendorf, Abigail Allwood, Walter Alvarez, Lyall Anderson, Mark Batzer, William Bemis, Jeffrey Bennett, Martin Brasier, Patricia Brennan, Rick Brenneman, Graham Budd, Jennifer Clack, Michael Coates, Shelley Copley, Ted Daeschler, David Duffy, Douglas Emlen, Douglas Erwin, David Fitch, Carles Lalueza Fox, Matt Friedman, Bryan Grieg Fry, Anne Gaskett, Scott Gilbert, George Gilchrist, Robert Guralnick, Joel Hirschhorn, Aaron Hirsh, Thomas Holtz, Gene Hunt, Douglas Irschink, Jacob Koella, Thomas Lackner, Trevor Lamb, Clarence Lehman, Richard Lenski, Sara Lewis, Jonathan Losos, Zhe-Xi Luo, Louise Mead, Axel Meyer, David Mindell, Christopher Norris, Todd Oakley, Karen Ober, Robert Pennock, Kenneth Petren, David Pfennig, Richard Prum, David Reznick, Beth Shapiro, Ruth Shaw, William Shear, Neil Shubin, David Strait, Douglas Taylor, Hans Thewissen, John Thompson, Joseph Thornton, Sarah Tishkoff, Paul Turner, Frances Westall, and Michael Whitlock.

Many thanks also to Carl Buell, Betty Gee, Lynn Golbetz, Gunder Hefta, Mark Ong, Emiko Paul, Ben Roberts, Laura Roberts, Tom Webster, and Michael Zierler for their invaluable help during the production of this book. I'm deeply grateful to Ben Roberts for recognizing the opportunity for a book like this and for inviting me to write it. I thank my daughters Charlotte and Veronica for their patience with their erstwhile dance partner. And most of all I thank my wife Grace, who, as always, kept my life in its proper balance for the past year (as she does every year).

Glossary

adaptive radiation: A rapid diversification of a lineage, accompanied by the evolution of species adapted to new ecological habits and environments.

allele: A version of a gene found in a population with a different sequence of bases than is found in other alleles.

allopatric speciation: The evolution of new species through geographical isolation.

allopolyploidy: The evolution of new species by the duplication of genomes.

amino acid: A building block of proteins.

antibiotics: Molecules produced by bacteria, fungi, and other organisms that can kill or inhibit the growth of bacteria.

Archaea: One of the three major lineages of life. Archaea are single-celled microbes similar to bacteria.

asexual reproduction: Reproduction without sex.

background extinctions: Extinction rates that are not significantly higher or lower than the average rate of extinction.

Bacteria: One of the three major lineages of life, bacteria are single-celled organisms that are common in soil, in water, and inside larger organisms.

base: Part of a nucleotide of DNA. There are four bases, which serve as an "alphabet" for genes.

bilaterian: An animal belonging to a lineage that includes vertebrates, insects, and worms. Bilaterian animals share many traits in common, including left–right symmetry.

biogeography: The study of the geographical distribution of species.

biological species concept: A concept of species defined by Ernst Mayr as "actually or potentially interbreeding populations which are reproductively isolated from other such groups."

BMP4: Bone morphogenetic protein 4, a protein that regulates the development of many structures in animals, including bird beaks and the dorsal–ventral axis of bilaterians.

body plan: A set of anatomical features found in all species in a lineage. The vertebrate body plan, for example, includes left–right symmetry, a skeleton, a skull, and a brain.

chromosome: A threadlike package of DNA and associated proteins. Humans have 46 pairs of chromosomes.

clade: A group of organisms that share a common ancestor. In some cases, a clade corresponds to a traditional classification, such as a species, an order, or a class.

cladogram: A diagram of the distribution of shared derived characteristics among a group of related organisms.

codon: Three consecutive bases that specify an amino acid in a protein.

coelacanth: A lobe-fin closely related to lungfishes and tetrapods.

coevolution: The evolution of two or more ecological partners as they adapt to each other.

coevolutionary arms race: An antagonistic form of coevolution in which species evolve stronger toxins, weapons, or defenses against each other.

commensal: Describes a species that depends on another species for its survival, without lowering or raising the fitness of its partner.

continuous variation: Variation in a trait, such as height, that varies smoothly through a population.

cospeciation: The parallel speciation of coevolutionary partners.

derived trait: Describes a trait that is shared only by species within a clade. Hair is a derived trait of mammals. All mammals have eyes, but eyes are not a derived trait of mammals because other animals have eyes as well.

discontinuous variation: A trait that is found in two or more discrete forms, such as the wrinkled or smooth surfaces of peas.

disparity: The morphological variation among a group of organisms.

dispersal: The spread of species to new ranges, such as when birds colonize distant islands.

DNA: Deoxyribonucleic acid, which stores genetic information in almost all organisms (with the exception of RNA viruses).

Ediacarans: Organisms that left some of the earliest known animal fossils. Some Ediacarans probably belong to the same lineages as living animals. Others may belong to lineages that have since became extinct.

eukaryotes: One of the three major branches of life. Eukaryotes include animals, plants, fungi, and single-celled protozoans. All eukaryotes share certain traits, such as a nucleus.

evolution: Descent with modification.

extinction: The termination of a species or a lineage.

fitness: The success of an organism in its environment, which allows it to spread its genes in the next generation.

fixation: The process by which one particular allele spreads through a population over the course of generations, until no other allele of the same gene remains.

gametes: Sex cells; egg cells and sperm cells.

genes: Segments of DNA that together encode a protein or an RNA molecule.

gene duplication: A mutation that produces an extra copy of a gene.

gene flow: The spread of genes across the geographical range of a population, or from one population to another.

gene recruitment: An evolutionary transition in which a gene begins to be expressed in another organ, in another pathway, or in some other context.

genetic drift: A change in the frequency of an allele owing to random sampling errors in a population.

genome: The full set of genetic information in a cell.

genotype: The genetic makeup of an organism; usually contrasted with the phenotype.

germ-line mutation: A mutation that occurs during the development of gametes, which can be passed down to offspring.

Hardy–Weinberg principle: The frequencies of genotypes will be constant in a population from generation to generation unless they are disturbed by natural selection or some other influence, such as drift, immigration, emigration, and nonrandom mating. A population in this constant state is said to be in Hardy–Weinberg equilibrium.

heritability: The fraction of variation in a trait found in a population that is due to genetic factors.

heterozygote: An individual that carries two different alleles of the same gene.

hominid: A human or another species of primate that was more closely related to humans than to other primates.

***Homo sapiens*:** The human species.

homology: Similarities in phenotype or genotype that are the result of common descent.

homozygote: An individual carrying two copies of the same allele of a gene.

horizontal gene transfer: The transfer of genetic material from one individual to another through a process other than heredity.

***Hox* genes:** A set of genes that regulates the development of animals. They are important for defining the head-to-tail axis of bilaterians.

hybrid swarm: A group of highly interbreeding species.

inclusive fitness: A measure of the fitness of an individual that takes into account the extra reproductive success that relatives may have as a result of the phenotype of the individual. For example, individuals that help their siblings raise offspring may have a higher inclusive fitness than individuals that try and fail to raise offspring of their own.

lobe-fins: A lineage of vertebrates that includes tetrapods and their closest living aquatic relatives, including *Tiktaalik*, lungfishes, and coealacanths.

lungfish: A freshwater lobe-fin found in Africa, Australia, and South America. Lungfishes are the closest living relatives of tetrapods.

mammal: An animal such as a human, a kangaroo, or a platypus, with a set of shared traits, including hair or fur, and the ability of females to produce milk to nourish their young

mass extinctions: Widespread extinctions of species that occur in a large pulse lasting for a few million years or less.

meiosis: Cell division that gives rise to sperm cells or egg cells. During meiosis, the number of chromosomes per cell is cut in half.

mitochondria: Sausage-shaped structures that generate most of the energy in eukaryotic cells. Mitochondria probably evolved from free-living bacteria.

mobile elements: Segments of DNA that be copied and pasted into other parts of the genome. Many mobile elements evolved from viruses.

monogamy: A mating system in which a male animal mates with only one female.

morphology: The form, shape, or structure of an organism.

mutualism: An ecological relationship in which two species depend on each other, such as sap-feeding insects and the bacteria inside them that synthesize amino acids that the insects can use.

natural selection: The process by which individuals better adapted for their way of life in their environment preferentially survive to leave more offspring with their traits to future generations.

Neanderthals: A population of *Homo sapiens* (or possibly a closely related species of hominids) that lived in Europe and Asia from about 200,000 to 28,000 years ago. Neanderthals were the closest relatives of living humans.

negative selection: The elimination of harmful alleles from a population through natural selection.

nonsilent substitution: A point mutation that changes the amino acid that is encoded by a codon.

nucleotide: A unit of DNA or RNA.

p53: A protein that suppresses tumors by causing cells with abnormal cell cycles to die or to stop dividing.

parasitism: An ecological relationship in which a parasite lowers the fitness of a host in order to raise its own fitness.

phenotype: The manifestation of a genotype.

phylogenetic species concept: The concept of a species as the smallest clade of organisms sharing a set of distinctive traits.

phylogeny: The evolutionary history of a group of organisms.

plastid: A structure found in plants, algae, and other organisms that carries out photosynthesis. Plastids evolved from free-living bacteria.

pleiotropy: The influence of a single gene on several phenotypic traits.

polyandry: A mating system in which females mate with multiple males.

polygamy: A mating system in which males mate with multiple females.

population genetics: The study of alleles in populations and how they change over time.

positive selection: The process by which new advantageous genetic variants sweep a population.

postzygotic isolation: The isolation of two populations through mechanisms that act after eggs are fertilized. For example, hybrids between some species are sterile.

prezygotic isolation: The isolation of two populations through mechanisms that act before eggs are fertilized. For example, females of one species may not be attracted to the songs of males from another species.

promiscuous proteins: Proteins that can carry out more than one kind of chemical reaction.

protein: A molecule assembled from amino acids.

pseudogene: A gene that has mutated so much that it can no longer encode a protein.

punctuated equilibria: A model of evolution in which stasis and random walks are most common, with rare bursts of significant change.

random walk: A series of random steps. The changes that occur in many lineages are random walks, as opposed to directional change or stasis.

recombination: The shuffling of chromosome segments during meiosis. Recombination is a major source of genetic variation in populations of sexually reproducing organisms.

relative fitness: The ratio between the average number of offspring produced from one genotype and the average number produced from another genotype.

retrovirus: A virus that encodes its genes in RNA and can insert its genetic material into a host's DNA. In some cases, retroviruses become integrated into the genomes of their hosts for millions of years.

ribosome: A set of RNA molecules and proteins that assemble amino acids into proteins.

ring species: A connected set of populations with a range that brings the two end populations into contact without interbreeding.

RNA: A single strand of nucleotides. RNA molecules serve as templates of genes for making proteins. They also play other roles, such as shutting down the expression of other genes.

***Sahelanthropus*:** The oldest known fossil of a hominid, estimated to have lived between six and seven million years ago.

sexual conflict: A situation in which the optimal reproductive strategy for males comes into conflict with the optimal strategy for females. Sexual conflict can lead to males and females evolving strategies that lower the fitness of their mates.

sex ratio: The ratio of males to females in a population.

sexual reproduction: A mode of reproduction in which gametes are combined. While bacteria are asexual, some researchers consider horizontal gene transfer a form of sexual reproduction.

sexual selection: Changes in the gene frequencies of a population caused by different levels of mating success of different genotypes.

silent substitution: A point mutation that does not change the amino acid encoded by a codon.

somatic mutation: A mutation that occurs in a cell in the body of an organism, as opposed to a mutation in a gamete.

speciation: The evolution of new species through splitting from an ancestral lineage.

stasis: A period of thousands or millions of years in which a lineage undergoes little or no directional change in a particular trait.

sympatric speciation: The divergence of individuals in a single area into two species.

synapsids: A clade of tetrapods that emerged 300 million years ago and that includes the mammals.

taxonomy: The science of classifying organisms by the use of reliable characteristics.

teleosts: A large clade of bony fishes that includes the most familiar species, such as trout and goldfish.

tetrapods: A clade of vertebrates that includes mammals, birds, reptiles, and amphibians. Tetrapods evolved from lobe-fins about 360 million years ago, acquiring limbs and other traits that allowed them to live on land.

vicariance: The splitting of a species or a group of species within a common ancestral geographic range. Vicariance can occur through the splitting of continents, the formation of new mountains, and other geological events.

virus: An infective agent consisting of DNA or RNA enclosed in a protein coat. Viruses can reproduce only by invading a host cell.

References

CHAPTER 1: EVOLUTION: AN INTRODUCTION

General references

These books are introductions to evolution intended for the general public:

Coyne, Jerry A., 2009. *Why Evolution Is True*. New York: Viking.

Mayr, Ernst, 2001. *What Evolution Is*. New York: Basic Books.

Zimmer, Carl, 2006. *Evolution: The Triumph of an Idea*. New York: HarperPerennial.

In January 2009, *Scientific American* published a special issue entitled *The Evolution of Evolution*, an excellent collection of articles surveying the science, written by leading evolutionary biologists.

The best Web site for learning about evolution is "Understanding Evolution": http://evolution.berkeley.edu

Here are three textbooks on evolutionary biology that are intended for undergraduates majoring in biology:

Barton, Nicholas H., et al., 2007. *Evolution*. Cold Spring Harbor, N.Y: Cold Spring Harbor Laboratory Press.

Freeman, Scott, and Jon C. Herron, 2007. *Evolutionary Analysis*, 4th ed. Upper Saddle River, N.J.: Pearson Prentice Hall.

Futuyma, Douglas J., 2005. *Evolution*. Sunderland, Mass.: Sinauer Associates.

Detailed references

A Case of Evolution: Why Do Whales Have Blowholes?

Marino, Lori, 2007. Cetacean brains: How aquatic are they? *Anatomical Record* 290: 694–700.

Nummela, Sirpa, J. G. M. Thewissen, Sunil Bajpai, Taseer Hussain, and Kishor Kumar, 2007. Sound transmission in archaic and modern whales: Anatomical adaptations for underwater hearing. *Anatomical Record* 290: 716–733.

Thewissen, J. G. M, ed., 1998. *The Emergence of Whales: Evolutionary Patterns in the Origin of Cetacea*. Advances in Vertebrate Paleobiology. New York: Plenum Press.

Thewissen, J. G. M., E. M. Williams, L. J. Roe, and S. T. Hussain, 2001. Skeletons of terrestrial cetaceans and the relationship of whales to artiodactyls. *Nature* 413: 277–281.

Thewissen, J. G. M., M. J. Cohn, L. S. Stevens, et al., 2006. Developmental basis for hind-limb loss in dolphins and origin of the cetacean bodyplan. *Proceedings of the National Academy of Science* 103: 8414–8418.

Thewissen, J. G. M., Lisa Noelle Cooper, Mark T. Clementz, Sunil Bajpai, and B. N. Tiwari, 2007. Whales originated from aquatic artiodactyls in the Eocene epoch of India. *Nature* 450: 1190–1194.

Zimmer, Carl. 1998. *At the Water's Edge: Macroevolution and the Transformation of Life*. New York: Free Press.

CHAPTER 2: BIOLOGY: FROM NATURAL PHILOSOPHY TO DARWIN

General references

Bowler, Peter J., 2003. *Evolution: The History of an Idea*, 3rd ed. Berkeley: University of California Press.

Magner, Lois N., 2002. *A History of the Life Sciences*, 3rd ed. New York: Marcel Dekker.

Young, David, 2007. *The Discovery of Evolution*, 2nd ed. Cambridge: Cambridge University Press, in association with Natural History Museum, London.

Detailed references

Nature before Darwin

Cutler, Alan, 2003. *The Seashell on the Mountaintop: A Story of Science, Sainthood, and the Humble Genius Who Discovered a New History of the Earth*. New York: Dutton.

Koerner, Lisbet, 1999. *Linnaeus: Nature and Nation*. Cambridge, Mass.: Harvard University Press.

Evolution before Darwin

Roger, Jacques, 1997. *Buffon: A Life in Natural History*. Ithaca, N.Y: Cornell University Press.

Fossils and the Death of Species

Oldroyd, D. R., 2006. *Earth Cycles: A Historical Perspective*. Westport, Conn.: Greenwood Press.

Rudwick, Martin J. S., 1985. *The Meaning of Fossils: Episodes in the History of Paleontology*. Chicago: University of Chicago Press.

Rudwick, Martin J. S., 1997. *Georges Cuvier, Fossil Bones, and Geological Catastrophes: New Translations & Interpretations of the Primary Texts*. Chicago: University of Chicago Press.

Winchester, Simon, 2001. *The Map That Changed the World: William Smith and the Birth of Modern Geology*. New York: HarperCollins.

Evolution as Striving

Appel, Toby A., 1987. *The Cuvier–Geoffroy Debate: French Biology in the Decades Before Darwin*. New York: Oxford University Press.

Burkhardt, Richard W., 1977. *The Spirit of System: Lamarck and Evolutionary Biology*. Cambridge, Mass.: Harvard University Press.

The Unofficial Naturalist

Browne, E. J., 1995. *Charles Darwin: A Biography*. New York: Knopf.

Desmond, Adrian J. and J. Moore, 1991. *Darwin*. New York: Penguin.

Padian, Kevin, 2008. Darwin's enduring legacy. *Nature* 451: 632–634.

Common Descent

Browne, E. J., 2006. *Darwin's Origin of Species: A Biography*. New York: Atlantic Monthly Press.

Darwin, Charles, 1998. *The Origin of Species by Means of Natural Selection, or, The Preservation of Favored Races in the Struggle for Life*. New York: Modern Library.

Box: What Is Science?

Okasha, Samir. 2002. *Philosophy of Science: A Very Short Introduction* (Very short introductions 67). Oxford: Oxford University Press.

CHAPTER 3: WHAT THE ROCKS SAY

Radioactive Clocks

Dalrymple, G. Brent, 1991. *The Age of the Earth*. Stanford, Calif: Stanford University Press.

"The Crust of the Earth Is a Vast Museum"

Prothero, Donald R., 2003. *Bringing Fossils to Life: An Introduction to Paleobiology*, 2nd ed. Boston: McGraw-Hill.

Prothero, Donald R., 2007. *Evolution: What the Fossils Say and Why It Matters*. New York: Columbia University Press.

Raff, Elizabeth C., Kaila L. Schollaert, David E. Nelson, et al., 2008. Embryo fossilization is a biological process mediated by microbial biofilms. *Proceedings of the National Academy of Sciences of the United States of America* 105: 19360-9365.

Traces of Vanished Life

Gaines, Susan M., 2008. *Echoes of Life: What Fossil Molecules Reveal About Earth History*. New York: Oxford University Press.

Lee-Thorp, Julia, and Matt Sponheimer, 2006. Contributions of biogeochemistry to understanding hominid dietary ecology. *American Journal of Physical Anthropology* Suppl. 43: 131–148.

Life's Earliest Marks

Allwood, Abigail C., Malcolm R. Walter, Balz S. Kamber, Craig P. Marshall, and Ian W. Burch, 2006. Stromatolite reef from the Early Archaean era of Australia. *Nature* 441: 714–718.

Allwood, Abigail C. John P. Grotzinger, Anfrew H. Knoll, Ian W. Burch, Mark S. Anderson, Max L. Coleman, and Isik Kanik, 2009. Article: Controls on development and diversity of Early Archean stromatolites. *Proceedings of the National Academy of Sciences* (June 10). doi: 10.1073/pnas.0903323106.

Brasier, Martin, Nicola McLoughlin, Owen Green, and David Wacey, 2006. A fresh look at the fossil evidence for early Archaean cellular life. *Philosophical Transactions of the Royal Society of London. Series B, Biological Sciences* 361: 887–902.

Knoll, Andrew H., 2003. *Life on a Young Planet: The First Three Billion Years of Evolution on Earth*. Princeton, N.J: Princeton University Press.

Menneken, Martina, Alexander A. Nemchin, Thorsten Geisler, Robert T. Pidgeon, and Simon A. Wilde. 2007. Hadean diamonds in zircon from Jack Hills, Western Australia. *Nature* 448: 917–920.

Rosing, Minik T., and Robert Frei, 2004. U-rich Archaean sea-floor sediments from Greenland—indications of >3700 Ma oxygenic photosynthesis. *Earth and Planetary Science Letters* 217: 237–244.

Schopf, J. William. 1999. *Cradle of Life: The Discovery of Earth's Earliest Fossils*. Princeton, N.J: Princeton University Press.

Life Gets Big

Budd, Graham E., 2008. The earliest fossil record of the animals and its significance. *Philosophical Transactions of the Royal Society of London. Series B, Biological Sciences* 363: 1425–1434.

Knoll, A. H., E. J. Javaux, D. Hewitt, and P. Cohen, 2006. Eukaryotic organisms in Proterozoic oceans. *Philosophical Transactions of the Royal Society of London. Series B, Biological Sciences* 361: 1023–1038.

Love, Gordon D, Emmanuelle Grosjean, Charlotte Stalvies, David A Fike, John P Grotzinger, Alexander S Bradley, Amy E Kelly, et al. 2009. Fossil steroids record the appearance of Demospongiae during the Cryogenian period. *Nature* 457: 718-721.

Marshall, Charles R., 2006. Explaining the Cambrian "explosion" of animals. *Annual Review of Earth and Planetary Sciences* 34 : 355–384.

Shu, D-G., H-L. Luo, S. Conway Morris, et al., 1999. Lower Cambrian vertebrates from south China. *Nature* 402: 42–46.

Xiao, Shuhai, and Marc Laflamme, 2009. On the eve of animal radiation: phylogeny, ecology and evolution of the Ediacara biota. *Trends in Ecology & Evolution* 24: 31-40.

Yin, Leiming, Maoyan Zhu, Andrew H. Knoll, et al., 2007. Doushantuo embryos preserved inside diapause egg cysts. *Nature* 446: 661–663.

Climbing Ashore

Berbee, Mary L, and John W. Taylor, 2007. Rhynie chert: a window into a lost world of complex plant–fungus interactions. *The New Phytologist* 174: 475–479.

Clack, Jennifer A., 2002. *Gaining Ground: The Origin and Evolution of Tetrapods*. Bloomington, Ind: Indiana University Press.

Labandeira, Conrad C., 2005. Invasion of the continents: cyanobacterial crusts to tree-inhabiting arthropods. *Trends in Ecology & Evolution* 20: 253–262.

Wellman, Charles H, Peter L. Osterloff, and Uzma Mohiuddin, 2003. Fragments of the earliest land plants. *Nature* 425: 282–285.

Wilson, Heather M., and Lyall I. Anderson, 2004. Morphology and taxonomy of Paleozoic millipedes

(Diplopoda: Chilognatha: Archipolypoda) from Scotland. *Journal of Paleontology* 78: 169–184.

Recent Arrivals

Brunet, Michel, Franck Guy, David Pilbeam, et al., 2002. A new hominid from the Upper Miocene of Chad, Central Africa. *Nature* 418:145–151.
Chiappe, Luis M., 2007. *Glorified Dinosaurs: The Origin and Early Evolution of Birds.* Hoboken, N.J.: John Wiley.
Johanson, Donald C., Blake Edgar, and David Brill, 2006. *From Lucy to Language: Revised, Updated, and Expanded.* New York: Simon and Schuster.
Luo, Zhe-Xi, 2007. Transformation and diversification in early mammal evolution. *Nature* 450: 1011–1019.
McDougall, Ian, Francis H. Brown, and John G. Fleagle, 2005. Stratigraphic placement and age of modern humans from Kibish, Ethiopia. *Nature* 433: 733–736.
Piperno, Dolores R., and Hans-Dieter Sues, 2005. Dinosaurs dined on grass. *Science* 310: 1126–1128.
Soltis, Douglas E., Charles D. Bell, Sangtae Kim, and Pamela S. Soltis, 2008. Origin and early evolution of angiosperms. *Annals of the New York Academy of Sciences* 1133: 3–25.
Wallace, David Rains, 2004. *Beasts of Eden: Walking Whales, Dawn Horses, and Other Enigmas of Mammal Evolution.* Berkeley: University of California Press.

Box: The past and present in science

Powers, Catherine M., and David J. Bottjer, 2007. Bryozoan paleoecology indicates mid-Phanerozoic extinctions were the product of long-term environmental stress. *Geology* 35: 995–998.

CHAPTER 4: THE TREE OF LIFE

General references

Dawkins, Richard, 2004. *The Ancestor's Tale: A Pilgrimage to the Dawn of Evolution.* Boston: Houghton Mifflin.
Gee, Henry, 1999. *In Search of Deep Time: Beyond the Fossil Record to a New History of Life.* New York: Free Press.
Lecointre, Guillaume, 2006. *The Tree of Life: A Phylogenetic Classification.* Cambridge, Mass: Belknap Press of Harvard University Press.
Shubin, Neil, 2008. *Your Inner Fish: A Journey into the 3.5-Billion-Year History of the Human Body.* New York: Pantheon Books.

Detailed references

How to Build a Tree

Cracraft, Joel, and Michael J. Donoghue, eds., 2004. *Assembling the Tree of Life.* Oxford: Oxford University Press.
Felsenstein, Joseph, 2004. *Inferring Phylogenies.* Sunderland, Mass: Sinauer Associates.
Gregory, T., 2008. Understanding Evolutionary Trees. *Evolution: Education and Outreach* 1: 121–137

From Fins to Feet

Clack, Jennifer A., 2002. *Gaining Ground: The Origin and Evolution of Tetrapods.* Bloomington, Ind.: Indiana University Press.
Daeschler, Edward B., Neil H. Shubin, and Farish A. Jenkins, 2006. A Devonian tetrapod-like fish and the evolution of the tetrapod body plan. *Nature* 440: 757–763.
Downs, Jason P., Edward B. Daeschler, Farish A. Jenkins, and Neil H. Shubin, 2008. The cranial endoskeleton of *Tiktaalik roseae. Nature* 455: 925–929.
Ruta, Marcello, Peter J. Wagner, and Michael I. Coates, 2006. Evolutionary patterns in early tetrapods. I. Rapid initial diversification followed by decrease in rates of character change. *Proceedings of the Royal Society B: Biological Sciences* 273: 2107–2111.
Shubin, Neil H., Edward B. Daeschler, and Farish A. Jenkins, 2006. The pectoral fin of *Tiktaalik roseae* and the origin of the tetrapod limb. *Nature* 440: 764–771.
Zimmer, Carl, 1998. *At the Water's Edge: Macroevolution and the Transformation of Life.* New York: Free Press.

Evolution as Tinkering

Jacob, Francois, 1977. Evolution and tinkering. *Science* 196: 1161–1166.
Luo, Zhe-Xi. 2007. Transformation and diversification in early mammal evolution. *Nature* 450: 1011–1019.
Rich, Thomas H., James A. Hopson, Anne M. Musser, Timothy F. Flannery, and Patricia Vickers-Rich, 2005. Independent origins of middle ear bones in monotremes and therians. *Science* 307: 910–914.
Rowe, T., 1996. Coevolution of the mammalian middle ear and neocortex. *Science* 273: 651–654..

The Feathered Dinosaurs Take Flight

Chiappe, Luis M., 2007. *Glorified Dinosaurs: The Origin and Early Evolution of Birds.* Hoboken, N.J.: John Wiley.
Dial, Kenneth P,, Brandon E. Jackson, and Paolo Segre, 2008. A fundamental avian wing-stroke provides a new perspective on the evolution of flight. *Nature* 451, no. 7181 (February 21): 985–989.
Norell, Mark A., and Xing Xu, 2005. Feathered Dinosaurs. *Annual Review of Earth and Planetary Sciences* 33 (January 1): 277–299.
Shipman, Pat, 1998. *Taking Wing:* Archaeopteryx *and the Evolution of Bird Flight.* New York: Simon & Schuster.
Turner, Alan H., Diego Pol, Julia A. Clarke, Gregory M. Erickson, and Mark A. Norell, 2007. A basal dromaeosaurid and size evolution preceding avian flight. *Science* 317: 1378–1381.

A New Ape

González-José, Rolando, Ignacio Escapa, Walter A. Neves, Rubén Cúneo, and Héctor M. Pucciarelli, 2008. Cladistic analysis of continuous modularized traits provides phylogenetic signals in *Homo* evolution. *Nature* 453: 775–778..

Harcourt-Smith, W. E. H., and L. C. Aiello, 2004. Fossils, feet and the evolution of human bipedal locomotion. *Journal of Anatomy* 204: 403–416.

Johanson, Donald C., 2006. *From Lucy to Language: Revised, Updated, and Expanded*. New York: Simon and Schuster.

Lieberman, Daniel E,, and Dennis M. Bramble, 2007. The evolution of marathon running: capabilities in humans. *New Zealand Journal of Sports Medicine* 37, no. 4–5: 288–290.

Richmond, Brian G., and William L. Jungers, 2008. *Orrorin tugenensis* femoral morphology and the evolution of hominid bipedalism. *Science* 319: 1662–1665.

David Strait, Frederick E. Grine, and John G. Freagle, 2007. Analyzing Hominid phylogeny. In *Handbook of Paleoanthropology*. New York: Springer.

Stringer, Chris, and Peter Andrews, 2005. *The Complete World of Human Evolution*. London: Thames & Hudson.

Zimmer, Carl, 2005. *Smithsonian Intimate Guide to Human Origins*. Washington, D.C.: Smithsonian Books.

CHAPTER 5: EVOLUTION'S MOLECULES

Introduction

Weedon, Michael N., and Timothy M. Frayling, 2008. Reaching new heights: insights into the genetics of human stature. *Trends in Genetics*, 24: 595–603.

Proteins, DNA, and RNA

Birney, Ewan, John A. Stamatoyannopoulos, Anindya Dutta, et al., 2007. Identification and analysis of functional elements in 1% of the human genome by the ENCODE pilot project. *Nature* 447: 799–816.

Feschotte, Cédric, and Ellen J. Pritham, 2007. DNA transposons and the evolution of eukaryotic genomes. *Annual Review of Genetics* 41: 331–368.

Gerstein, Mark B., Can Bruce, Joel S. Rozowsky, et al., 2007. What is a gene, post-ENCODE? History and updated definition. *Genome Research* 17: 669–681.

Zimmer, Carl, 2008. Now: the rest of the genome. *The New York Times*, November 11. http://www.nytimes.com/2008/11/11/science/11gene.html.

Sexual Reproduction

Levy, Samuel, Granger Sutton, Pauline C. Ng, et al., 2007. The diploid genome sequence of an individual human. *PLoS Biology* 5:, e254.

Mendel

Bhattacharyya, M. K., A. M. Smith, T. H. Ellis, C. Hedley, and C. Martin, 1990. The wrinkled-seed character of pea described by Mendel is caused by a transposon-like insertion in a gene encoding starch-branching enzyme. *Cell* 60: 115–122.

Henig, Robin Marantz, 2000. *The Monk in the Garden: The Lost and Found Genius of Gregor Mendel, the Father of Genetics*. Boston: Houghton Mifflin.

The Complex Path from Genotype to Phenotype

Hartl, Daniel L., 2009. *Genetics: Analysis of Genes and Genomes*, 7th ed. Sudbury, Mass: Jones and Bartlett.

Bogin, B, P. Smith, A. B. Orden, M. I. Varela Silva, and J. Loucky, 2002. Rapid change in height and body proportions of Maya American children. *American Journal of Human Biology* 14: 753–761.

Ridley, Matt, 2003. *Nature Via Nurture: Genes, Experience, and What Makes Us Human*. New York, N.Y: HarperCollins.

CHAPTER 6: THE WAYS OF CHANGE: MUTATION, DRIFT, AND SELECTION

General references

Here are four books on the latest research on population genetics, genetic drift, and natural selection. *The Making of the Fittest* is intended for the general public; the rest are textbooks:

Bell, Graham, 2008. *Selection: The Mechanism of Evolution*, 2nd ed. Oxford: Oxford University Press.

Carroll, Sean B., 2006. *The Making of the Fittest: DNA and the Ultimate Forensic Record of Evolution*. New York: W.W. Norton & Co.

Hartl, Daniel L., 2007. *Principles of Population Genetics*, 4th ed. Sunderland, Mass: Sinauer Associates.

Lynch, Michael, 2007. *The Origins of Genome Architecture*. Sunderland, Mass: Sinauer Associates.

Detailed references

Introduction

Herrel, Anthony, Katleen Huyghe, Bieke Vanhooydonck, et al., 2008. Rapid large-scale evolutionary divergence in morphology and performance associated with exploitation of a different dietary resource. *Proceedings of the National Academy of Sciences of the United States of America* 105: 4792–4795.

Mutations: The Origin of Variation

Lynch, Michael, Way Sung, Krystalynne Morris, et al., 2008. A genome-wide view of the spectrum of spontaneous mutations in yeast. *Proceedings of the National Academy of Sciences of the United States of America* 105: 9272–9277.

Turning Biology into Equations

Cavalli-Sforza, L. L., 2000. *Genes, Peoples, and Languages*. New York: North Point Press.

Genetic Drift

Hartl, Daniel L., 2007. *Principles of Population Genetics*. 4th ed. Sunderland, Mass: Sinauer Associates.

Small Differences, Big Results

Herring, Christopher D., Anu Raghunathan, Christiane Honisch, et al., 2006. Comparative genome sequencing of *Escherichia coli* allows observation of bacterial evolution on a laboratory timescale. *Nature Genetics* 38: 1406–1412.

Ostrowski, Elizabeth A., Robert J. Woods, and Richard E. Lenski, 2008. The genetic basis of parallel and divergent phenotypic responses in evolving populations of *Escherichia coli*. *Proceedings of the Royal Society(London), Series B, Biological Sciences* 275: 277–284.

The Speed of Evolution

Visscher, Peter M., William G. Hill, and Naomi R. Wray, 2008. Heritability in the genomics era—concepts and misconceptions. *Nature Reviews Genetics* 9: 255–266.

Natural Selection All Around

Cheptou, P.-O., O. Carrue, S. Rouifed, and A. Cantarel, 2008. Rapid evolution of seed dispersal in an urban environment in the weed *Crepis sancta*. *Proceedings of the National Academy of Sciences of the United States of America* 105: 3796–3799.

Grant, Peter R., and B. Rosemary Grant, 2008. *How and Why Species Multiply: The Radiation of Darwin's Finches*. Princeton Series in Evolutionary Biology. Princeton, N.J.: Princeton University Press.

Drinking Milk: A Fingerprint of Natural Selection

Enattah, Nabil Sabri, Tine G. K. Jensen, Mette Nielsen, et al., 2008. Independent introduction of two lactase-persistence alleles into human populations reflects different history of adaptation to milk culture. *American Journal of Human Genetics* 82: 57–72.

Sabeti, P. C., S. F. Schaffner, B. Fry, et al., 2006. Positive natural selection in the human lineage. *Science* 312: 1614–1620.

The Geography of Fitness

Harper, George R., and David W. Pfennig, 2008. Selection overrides gene flow to break down maladaptive mimicry. *Nature* 451: 1103–1106.

The Limits of Natural Selection

Tabashnik, Bruce E., Aaron J. Gassmann, David W. Crowder, and Yves Carriére, 2008. Insect resistance to Bt crops: evidence versus theory. *Nature Biotechnology* 26: 199–202.

CHAPTER 7: THE HISTORY IN OUR GENES

Introduction

Campbell, Michael C., and Sarah A. Tishkoff, 2008. African genetic diversity: implications for human demographic history, modern human origins, and complex disease mapping. *Annual Review of Genomics and Human Genetics* 9: 403–433.

The Genetic Archive

Hillis, D. M., J. J. Bull, M. E .White, M. R. Badgett, and I. J. Molineux. 1992. Experimental phylogenetics: generation of a known phylogeny. *Science* 255: 589–592.

Species Trees and Gene Trees

Hobolth, Asger, Ole F. Christensen, Thomas Mailund, and Mikkel H. Schierup, 2007. Genomic relationships and speciation times of human, chimpanzee, and gorilla inferred from a coalescent hidden Markov model. *PLoS Genetics* 3: e7.

Salem, Abdel-Halim, David A. Ray, Jinchuan Xing, et al., 2003. *Alu* elements and hominid phylogenetics. *Proceedings of the National Academy of Sciences of the United States of America* 100: 12787–12791.

Morphology and Molecules

Hallstrom, Bjorn M., and Axel Janke, 2008. Gnathostome phylogenomics utilizing lungfish EST sequences. *Molecular Biology and Evolution* (November 24): msn271. doi:10.1093/molbev/msn271

Humans, Finches, and HIV: Clues From the Genome

Gonder, Mary Katherine, Holly M. Mortensen, Floyd A. Reed, Alexandra de Sousa, and Sarah A. Tishkoff, 2007. Whole-mtDNA genome sequence analysis of ancient African lineages. *Molecular Biology and Evolution* 24: 757–768.

Grant, Peter R., and B. Rosemary Grant, 2008. *How and Why Species Multiply: The Radiation of Darwin's Finches*. Princeton, N.J.: Princeton University Press.

Underhill, Peter A., and Toomas Kivisild, 2007. Use of Y chromosome and mitochondrial DNA population structure in tracing human migrations. *Annual Review of Genetics* 41: 539–564.

Van Heuverswyn, Fran, Yingying Li, Elizabeth Bailes, et al., 2007. Genetic diversity and phylogeographic clustering of SIVcpzPtt in wild chimpanzees in Cameroon. *Virology* 368: 155–171.

Wain, Louise V., Elizabeth Bailes, Frederic Bibollet-Ruche, et al., 2007. Adaptation of HIV-1 to its human host. *Molecular Biology and Evolution* 24: 1853–1860.

Natural Selection versus Neutral Evolution

Nei, Masatoshi, 2005. Selectionism and neutralism in molecular evolution. *Molecular Biology and Evolution* 22: 2318–2342.

The Molecular Clock

Grant, Peter R., and B. Rosemary Grant, 2008. *How and Why Species Multiply: The Radiation of Darwin's Finches*. Princeton, N.J.: Princeton University Press.

Kumar, Sudhir, 2005. Molecular clocks: four decades of evolution. *Nature Reviews Genetics* 6: 654–662.

Rambaut, Andrew, David Posada, Keith A. Crandall, and Edward C. Holmes, 2004. The causes and consequences of HIV evolution. *Nature Reviews Genetics* 5: 52–61.

Wible, J. R., G. W. Rougier, M. J. Novacek, and R. J. Asher, 2007. Cretaceous eutherians and Laurasian origin for placental mammals near the K/T boundary. *Nature* 447: 1003–1006.

Ancient Selection

Enard, Wolfgang, Molly Przeworski, Simon E. Fisher, et al., 2002. Molecular evolution of FOXP2, a gene involved in speech and language. *Nature* 418: 869–872.

Studer, Romain A., Simon Penel, Laurent Duret, and Marc Robinson-Rechavi, 2008. Pervasive positive selection on duplicated and nonduplicated vertebrate protein coding genes. *Genome Research* 18: 1393–1402.

Deciphering the Genome

Pennacchio, L. A., M. Olivier, J. A. Hubacek, et al. 2001. An apolipoprotein influencing triglycerides in humans and mice revealed by comparative sequencing. *Science* 294: 169–173.

Pollard, Katherine S., Sofie R. Salama, Nelle Lambert, et al., 2006. An RNA gene expressed during cortical development evolved rapidly in humans. *Nature* 443: 167–172.

Sawyer, Sara L., Lily I. Wu, Michael Emerman, and Harmit S. Malik, 2005. Positive selection of primate TRIM5alpha identifies a critical species-specific retroviral restriction domain. *Proceedings of the National Academy of Sciences of the United States of America* 102: 2832–2837.

CHAPTER 8: ADAPTATIONS: FROM GENES TO TRAITS

General references

Carroll, Sean B., 2005. *Endless Forms Most Beautiful: The New Science of Evo Devo and the Making of the Animal Kingdom.* New York: Norton.

Dawkins, Richard, 1996. *The Blind Watchmaker: Why the Evidence of Evolution Reveals a Universe Without Design.* New York: Norton.

Detailed references

New Genes from Old

Blount, Zachary D., Christina Z Borland, and Richard E Lenski, 2008. Historical contingency and the evolution of a key innovation in an experimental population of *Escherichia coli*. *Proceedings of the National Academy of Sciences of the United States of America* 105: 7899–7906.

Copley, S. D., 2000. Evolution of a metabolic pathway for degradation of a toxic xenobiotic: the patchwork approach. *Trends in Biochemical Sciences* 25: 261–265.

Can I Borrow A Gene?

Fry, Bryan G., 2005. From genome to "venome": molecular origin and evolution of the snake venom proteome inferred from phylogenetic analysis of toxin sequences and related body proteins. *Genome Research* 15: 403–420.

Fry, Bryan G., Nicolas Vidal, Janette A. Norman, et al., 2006. Early evolution of the venom system in lizards and snakes. *Nature* 439: 584–588.

Piatigorsky, Joram, 2007. *Gene Sharing and Evolution: The Diversity of Protein Functions.* Cambridge, Mass: Harvard University Press.

Zimmer, Carl, 2005. Open Wide: Decoding the secrets of venom. *The New York Times*, April 5, 2005 sec. Science. http://www.nytimes.com/2005/04/05/science/05veno.html

Sculpting a Beak

Abzhanov, Arhat, Winston P. Kuo, Christine Hartmann, et al., 2006. The calmodulin pathway and evolution of elongated beak morphology in Darwin's finches. *Nature* 442: 563–567.

Recycled Feathers

Harris, Matthew P., Scott Williamson, John F. Fallon, Hans Meinhardt, and Richard O. Prum, 2005. Molecular evidence for an activator–inhibitor mechanism in development of embryonic feather branching. *Proceedings of the National Academy of Sciences of the United States of America* 102: 11734–11739.

Prum, Richard O., and Jan Dyck, 2003. A hierarchical model of plumage: morphology, development, and evolution. *Journal of Experimental Zoology Part B: Molecular and Developmental Evolution* 298: 73–90.

From Flies to Mice

Carroll, Sean B., 2005. *Endless Forms Most Beautiful: The New Science of Evo Devo and the Making of the Animal Kingdom.* New York: Norton.

De Robertis, E. M., 2008. Evo-devo: variations on ancestral themes. *Cell* 132: 185–195.

Gilbert, Scott F., 2006. *Developmental Biology*, 8th ed. Sunderland, Mass: Sinauer Associates.

Evolving Eyes

Jékely, Gáspár, Julien Colombelli, Harald Hausen, et al., 2008. Mechanism of phototaxis in marine zooplankton. *Nature* 456: 395–399.

Jonasova, Kristyna, and Zbynek Kozmik, 2008. Eye evolution: lens and cornea as an upgrade of animal visual system. *Seminars in Cell & Developmental Biology* 19: 71–81.

Lamb, Trevor D., Shaun P. Collin, and Edward N. Pugh, 2007. Evolution of the vertebrate eye: opsins, photoreceptors, retina and eye cup. *Nature Reviews Neuroscience* 8: 960–976. Lamb, Trevor, Edward Pugh, and Shaun Collin, 2008. The origin of the vertebrate eye. *Evolution: Education and Outreach* 1:415–426.

Oakley, Todd, and M. Pankey, 2008. Opening the "black box": The genetic and biochemical basis of eye evolution. *Evolution: Education and Outreach* 1: 390–402.

Convergent Evolution

Colosimo, Pamela F., Kim E. Hosemann, Sarita Balabhadra, et al., 2005. Widespread parallel evolution in sticklebacks by repeated fixation of ectodysplasin alleles. *Science* 307: 1928–1933.

Friedman, Matt, 2008. The evolutionary origin of flatfish asymmetry. *Nature* 454: 209–212.

Kozmik, Zbynek, Shivalingappa K. Swamynathan, Jana Ruzickova, et al., Cubozoan crystallins: evidence for convergent evolution of pax regulatory sequences. *Evolution & Development* 10: 52–61.

Wagner, Günter P., 2007. The developmental genetics of homology. *Nature Reviews Genetics* 8: 473–479.

Box: How *Not* To Study Evolution

Miller, Kenneth R., 2008. *Only a Theory: Evolution and the Battle for America's Soul*. New York: Viking Penguin.

Pennock, Robert T., 1999. *Tower of Babel: The Evidence Against the New Creationism*. Cambridge, Mass: MIT Press.

Scott, Eugenie Carol, 2008. *Evolution vs. Creationism: An Introduction*. 2nd ed. Westport, Conn.: Greenwood Press.

CHAPTER 9: THE ORIGIN OF SPECIES

General references

Coyne, Jerry A., and H. Allen Orr, 2004. *Speciation*. Sunderland, Mass: Sinauer Associates.

Schilthuizen, Menno, 2001. *Frogs, Flies, and Dandelions: Speciation—the Evolution of New Species*. Oxford: Oxford University Press.

Zimmer, Carl, 2008. What is a species? *Scientific American* 298: 72–79.

Detailed references

Introduction

Brown, David M., Rick A. Brenneman, Klaus-Peter Koepfli, et al., 2007. Extensive population genetic structure in the giraffe. *BMC Biology* 5: 57.

Good Barriers Make Good Species

Hurd, L. E., and Robert M. Eisenberg, 1975. Divergent selection for geotactic response and evolution of reproductive isolation in sympatric and allopatric populations of houseflies. *The American Naturalist* 109: 353.

Knowlton, N., and L. A. Weigt, 1998. New dates and new rates for divergence across the Isthmus of Panama. *Proceedings of the Royal Society B: Biological Sciences* 265: 2257.

Rings of Species

Irwin, Darren E., 2002. Phylogeographic breaks without geographic barriers to gene flow. *Evolution* 56: 2383–2394.

Irwin, Darren E., Staffan Bensch, Jessica H. Irwin, and Trevor D. Price. 2005. Speciation by distance in a ring species. *Science* 307: 414–416.

Species Side by Side

Barluenga, Marta, Kai N. Stölting, Walter Salzburger, Moritz Muschick, and Axel Meyer, 2006. Sympatric speciation in Nicaraguan crater lake cichlid fish. *Nature* 439: 719–723.

Seehausen, Ole, Yohey Terai, Isabel S, Magalhaes, et al., 2008. Speciation through sensory drive in cichlid fish. *Nature* 455: 620–626.

The Speed of Speciation

Coyne, J. A., and H. A. Orr. 1998. The evolutionary genetics of speciation. *Philosophical Transactions of the Royal Society of London. Series B, Biological Sciences* 353: 287–305.

Hegarty, Matthew J., and Simon J. Hiscock, 2008. Genomic clues to the evolutionary success of polyploid plants. *Current Biology*:18: R435–444.

Hendry, Andrew P., Peter R. Grant, B. Rosemary Grant, et al., 2006. Possible human impacts on adaptive radiation: beak size bimodality in Darwin's finches. *Proceedings of the Royal Society of London Series B—Biological Sciences* 273, no. 1596 (August 7): 1887–1894.

Uncovering Hidden Species

Agapow, Paul-Michael, Olaf R. Bininda-Emonds, Keith A. Crandall, et al., 2004. The impact of species concept on biodiversity studies. *The Quarterly Review of Biology* 79: 161–179.

Brown, David M., Rick A. Brenneman, Klaus-Peter Koepfli, et al., 2007. Extensive population genetic structure in the giraffe. *BMC Biology* 5: 57.

Species Beyond Barriers

Cohan, Frederick M., and Elizabeth B. Perry, 2007. A systematics for discovering the fundamental units of bacterial diversity. *Current Biology*17: R373–386.

The Origin of Our Own Species

Cox, Murray P., Fernando L. Mendez, Tatiana M. Karafet, et al., 2008. Testing for archaic hominin admixture on the X chromosome: model likelihoods for the modern human RRM2P4 region from summaries of genealogical topology under the structured coalescent. *Genetics* 178: 427–437.

Evans, Patrick D., Nitzan Mekel-Bobrov, Eric J. Vallender, Richard R. Hudson, and Bruce T. Lahn, 2006. Evidence that the adaptive allele of the brain size gene microcephalin introgressed into *Homo sapiens* from an archaic *Homo* lineage. *Proceedings of the National Academy of Sciences of the United States of America* 103: 18178–18183.

Fagundes, Nelson J. R., Nicolas Ray, Mark Beaumont, et al., 2007. Statistical evaluation of alternative models of human evolution. *Proceedings of the National Academy of Sciences of the United States of America* 104: 17614–17619.

Green, Richard E., Johannes Krause, Susan E. Ptak, et al., 2006. Analysis of one million base pairs of Neanderthal DNA. *Nature* 444: 330–336.

Krings, M., A. Stone, R. W. Schmitz, et al., 1997. Neandertal DNA sequences and the origin of modern humans. *Cell* 90: 19–30.

Noonan, James P., Graham Coop, Sridhar Kudaravalli, et al., 2006. Sequencing and analysis of Neanderthal genomic DNA. *Science* 314: 1113–1118.

Stoneking, Mark, 2008. Human origins. The molecular perspective. *EMBO Reports* 9, Suppl. 1 (July): S46–S50.

Wang, Jun, Wei Wang, Ruiqiang Li, et al., 2008. The diploid genome sequence of an Asian individual. *Nature* 456: 60–65.

CHAPTER 10: RADIATIONS AND EXTINCTIONS: BIODIVERSITY THROUGH THE AGES

Introduction

Alroy, John, 2008. Colloquium paper: dynamics of origination and extinction in the marine fossil record. *Proceedings of the National Academy of Sciences of the United States of America* 105: 11536–11542.

Alroy, John, Martin Aberhan, David J. Bottjer, et al., 2008. Phanerozoic trends in the global diversity of marine invertebrates. *Science* 321: 97–100.

Wilson, Edward O., 1999. *The Diversity of Life*. New York: W. W. Norton.

Riding the Continents

Beck, Robin M. D., Henk Godthelp, Vera Weisbecker, Michael Archer, and Suzanne J. Hand, 2008. Australia's oldest marsupial fossils and their biogeographical implications. *PLoS ONE* 3: e1858.

Boyer, Sarah L., Ronald M. Clouse, Ligia R. Benavides, et al., 2007. Biogeography of the world: a case study from cyphophthalmid Opiliones, a globally distributed group of arachnids. *Journal of Biogeography* 34: 2070–2085.

Lomolino, Mark V., 2006. *Biogeography*, 3rd ed. Sunderland, Mass: Sinauer Associates.

Shapiro, Beth, Dean Sibthorpe, Andrew Rambaut, et al., 2002. Flight of the dodo. *Science* 295: 1683.

Evolutionary Fits and Starts

Benton, M., and P. Pearson, 2001. Speciation in the fossil record. *Trends in Ecology & Evolution* 16: 405–411.

Gould, Stephen Jay, 2007. *Punctuated Equilibrium*. Cambridge, Mass: Belknap Press of Harvard University Press.

Hunt, Gene, 2007. The relative importance of directional change, random walks, and stasis in the evolution of fossil lineages. *Proceedings of the National Academy of Sciences of the United States of America* 104: 18404–18408.

The Lifetime of a Species

Foote, Michael, James S. Crampton, Alan G. Beu, et al., 2007. Rise and fall of species occupancy in Cenozoic fossil mollusks. *Science* 318: 1131–1134.

Cradles of Diversity

Jablonski, David, Kaustuv Roy, and James W. Valentine, 2006. Out of the tropics: evolutionary dynamics of the latitudinal diversity gradient. *Science* 314: 102–106.

Radiations

Barnosky, Anthony D., and Brian P. Kraatz, 2007. The role of climatic change in the evolution of mammals. *BioScience* 57: 523–532 .

Schluter, Dolph, 2000. *The Ecology of Adaptive Radiation*. Oxford: Oxford University Press.

Seehausen, Ole, 2006. African cichlid fish: a model system in adaptive radiation research. *Proceedings of the Royal Society of London, Series B—Biological Sciences*, 273 1987–1998.

A Gift for Diversity

Mayhew, Peter J., 2007. Why are there so many insect species? Perspectives from fossils and phylogenies. *Biological Reviews of the Cambridge Philosophical Society* 82: 425–454.

Lighting the Cambrian Fuse

Erwin, Douglas H., 2007. Disparity: morphological pattern and developmental context. *Palaeontology* 50: 57–73

Gould, Stephen Jay, 1989. *Wonderful Life: The Burgess Shale and the Nature of History*. New York: W.W. Norton.

Marshall, Charles R., 2006. Explaining the Cambrian "explosion" of animals. *Annual Review of Earth and Planetary Sciences* 34: 355–384.

Shen, Yanan, Tonggang Zhang, and Paul F. Hoffman, 2008. On the coevolution of Ediacaran oceans and animals. *Proceedings of the National Academy of Sciences of the United States of America* 105: 7376–7381.

Driven to Extinction

Bambach, Richard K., 2006. Phanerozoic biodiversity mass extinctions. *Annual Review of Earth and Planetary Sciences* 34 : 127–155.

Mayhew, Peter J., Gareth B. Jenkins, and Timothy G. Benton, 2008. A long-term association between global temperature and biodiversity, origination and extinction in the fossil record. *Proceedings of The Royal Society of London, Series B—Biological Sciences* 275: 47–53.

Payne, Jonathan L., and Seth Finnegan, 2007. The effect of geographic range on extinction risk during background and mass extinction. *Proceedings of the National Academy of Sciences of the United States of America* 104: 10506–10511.

Peters, Shanan E., 2008. Environmental determinants of extinction selectivity in the fossil record. *Nature* 454: 626–629.

When Life Nearly Died

Alvarez, Walter, 1997. *T. Rex and the Crater of Doom*. Princeton, N.J.: Princeton University Press.

Benton, M. J., 2003. *When Life Nearly Died: The Greatest Mass Extinction of All Time*. New York: Thames & Hudson.

Powers, Catherine M., and David J. Bottjer, 2007. Bryozoan paleoecology indicates mid-Phanerozoic extinctions were the product of long-term environmental stress. *Geology* 35: 995–998.

The New Die-off

Colwell, Robert K., Gunnar Brehm, Catherine L. Cardelús, Alex C. Gilman, and John T. Longino, 2008. Global warming, elevational range shifts, and lowland biotic attrition in the wet tropics. *Science* 322: 258–261.

Ehrlich, Paul R,, and Robert M, Pringle, 2008. Colloquium paper: where does biodiversity go from here? A grim business-as-usual forecast and a hopeful portfolio of partial solutions. *Proceedings of the National Academy of Sciences of the United States of America* 105, Suppl. 1 : 11579–11586.

Moritz, Craig, James L. Patton, Chris J. Conroy, et al., 2008. Impact of a century of climate change on small-mammal communities in Yosemite National Park, USA. *Science* 322: 261–264.

Pimm, Stuart, Peter Raven, Alan Peterson, Cagan H. Sekercioglu, and Paul R. Ehrlich, 2006. Human impacts on the rates of recent, present, and future bird extinctions. *Proceedings of the National Academy of Sciences of the United States of America* 103: 10941–10946.

Smith, Stephen A., David C. Tank, Lori-Ann Boulanger, et al., 2008. Bioactive endophytes warrant intensified exploration and conservation. *PLoS ONE* 3: e3052.

Zeebe, Richard E., James C. Zachos, Ken Caldeira, and Toby Tyrrell, 2008. Oceans. Carbon emissions and acidification. *Science* 321: 51–52.

CHAPTER 11: INTIMATE PARTNERSHIPS: HOW SPECIES ADAPT TO EACH OTHER

Introduction

Gaskett, A. C., C. G. Winnick, and M. E. Herberstein, 2008. Orchid sexual deceit provokes ejaculation. *The American Naturalist* 171: E206–E212.

Friends and Enemies

Zimmer, Carl, 2000. *Parasite Rex: Inside the Bizarre World of Nature's Most Dangerous Creatures*. New York: Free Press.

Locks and Keys

Brouat, C., N. Garcia, C. Andary, and D. McKey, 2001. Plant lock and ant key: pairwise coevolution of an exclusion filter in an ant–plant mutualism. *Proceedings of The Royal Society of London, Series B—Biological Sciences* 268: 2131–2141.

Mirror Trees

McCutcheon, John P., and Nancy A. Moran. 2007. Parallel genomic evolution and metabolic interdependence in an ancient symbiosis. *Proceedings of the National Academy of Sciences of the United States of America* 104: 19392–19397.

Natural Arms Races

Agrawal, Anurag A., and Mark Fishbein, 2008. Phylogenetic escalation and decline of plant defense strategies. *Proceedings of the National Academy of Sciences of the United States of America* 105: 10057–10060.

Gaskett, A. C., C. G. Winnick, and M. E. Herberstein, 2008. Orchid sexual deceit provokes ejaculation. *The American Naturalist* 171: E206–E212.

Geffeney, Shana, Edmund D. Brodie III, Peter C. Ruben, and Edmund D. Brodie Jr., 2002. Mechanisms of adaptation in a predator–prey arms race: TTX-resistant sodium channels. *Science* 297: 1336–1339.

The Tragedy of the Coevolutionary Commons

Kiers, E. Toby, Robert A. Rousseau, Stuart A. West, and R. Ford Denison, 2003. Host sanctions and the legume–rhizobium mutualism. *Nature* 425: 78–81.

A Geographic Mosaic

Anderson, Bruce, and Steven D. Johnson, 2008. The geographical mosaic of coevolution in a plant–pollinator mutualism. *Evolution; International Journal of Organic Evolution* 62: 220–225.

Hanifin, Charles T., Edmund D. Brodie Jr., and Edmund D. Brodie III, 2008. Phenotypic mismatches reveal escape from arms-race coevolution. *PLoS Biology* 6: e60.

Siepielski, Adam M., and Craig W. Benkman, 2008. A seed predator drives the evolution of a seed dispersal mutualism. *Proceedings of The Royal Society of London, Series B—Biological Sciences* 275: 1917–1925.

Thompson, John N., 2005. *The Geographic Mosaic of Coevolution* Chicago: University of Chicago Press.

Engines of Diversity

Barlow, Connie C., 2000. *The Ghosts of Evolution: Nonsensical Fruit, Missing Partners, and Other Ecological Anachronisms*. New York: Basic Books.

Godsoe, William, Jeremy B. Yoder, Christopher Irwin Smith, and Olle Pellmyr, 2008. Coevolution and divergence in the Joshua tree/yucca moth mutualism. *The American Naturalist* 171: 816–823. Lengyel, Szabolcs, Aaron D. Gove, Andrew M. Latimer, Jonathan D. Majer, and Robert R. Dunn, 2009. Ants sow the seeds of global diversification in flowering plants. *PloS ONE* 4:e5480.

Memmott, Jane, Paul G. Craze, Nickolas M. Waser, and Mary V. Price, 2007. Global warming and the disruption of plant–pollinator interactions. *Ecology Letters* 10: 710–717.

Two Species Become One

Gould, Sven B., Ross F. Waller, and Geoffrey I. McFadden, 2008. Plastid evolution. *Annual Review of Plant Biology* 59: 491–517.

Sapp, Jan, 1994. *Evolution by Association: A History of Symbiosis*. New York: Oxford University Press.

Tovar, Jorge, Gloria León-Avila, Lidya B. Sánchez, et al., 2003. Mitochondrial remnant organelles of *Giardia* function in iron-sulphur protein maturation. *Nature* 426: 172–176.

Invasion of the Genomic Parasites

Dewannieux, Marie, Francis Harper, Aurélien Richaud, et al., 2006. Identification of an infectious progenitor for the multiple-copy HERV-K human endogenous retroelements. *Genome Research* 16: 1548–1556.

Feschotte, Cédric, and Ellen J. Pritham, 2007. DNA transposons and the evolution of eukaryotic genomes. *Annual Review of Genetics* 41: 331–368.

Kapitonov, Vladimir V., and Jerzy Jurka, 2005. RAG1 core and V(D)J recombination signal sequences were derived from Transib transposons. *PLoS Biology* 3: e181.

CHAPTER 12: SEX AND FAMILY

General references

Alcock, John, 2009. *Animal Behavior: An Evolutionary Approach*. 9th ed. Sunderland, Mass: Sinauer Associates.

Andersson, Malte, 1994. *Sexual Selection*. Princeton, N.J: Princeton University Press.

Judson, Olivia, 2002. *Dr. Tatiana's Sex Advice to All Creation*. New York: Metropolitan Books.

Detailed references

Introduction

Zimmer, Carl, 2007. In ducks, war of the sexes plays out in the evolution of genitalia. *The New York Times*, May 1, sec. Science. http://www.nytimes.com/2007/05/01/science/01duck.html.

Why Sex?

Cooper, Tim F., 2007. Recombination speeds adaptation by reducing competition between beneficial mutations in populations of *Escherichia coli*. *PLoS Biology* 5: e225.

Hadany, Lilach, and Josep M. Comeron, 2008. Why are sex and recombination so common? *Annals of the New York Academy of Sciences* 1133: 26–43.

Running in Place

Decaestecker, Ellen, Sabrina Gaba, Joost A. M. Raeymaekers, et al., 2007. Host–parasite 'Red Queen' dynamics archived in pond sediment. *Nature* 450: 870–873.

Cheap Sperm and Expensive Eggs

Fabiani, Anna, Filippo Galimberti, Simona Sanvito, and A. Rus Hoelzel, 2004. Extreme polygyny among southern elephant seals on Sea Lion Island, Falkland Islands. *Behavioral Ecology* 15: 961–969.

Songs and Dances

Chapman, Tracey, Andrew Pomiankowski, and Kevin Fowler, 2005. Stalk-eyed flies. *Current Biology* 15: R533–535.

Forsman, Anders, and Mattias Hagman, 2006. Calling is an honest indicator of paternal genetic quality in poison frogs. *Evolution* 60: 2148–2157.

Lewis, Sara M., and Christopher K. Cratsley, 2008. Flash signal evolution, mate choice, and predation in fireflies. *Annual Review of Entomology* 53: 293–321.

Pryke, S. R., S. Andersson, and M. J. Lawes, 2001. Sexual selection of multiple handicaps in the red-collared widowbird: female choice of tail length but not carotenoid display. *Evolution:* 55: 1452–1463.

Rodd, F. Helen, Kimberly A. Hughes, Gregory F. Grether, and Colette T. Baril, 2002. A possible non-sexual origin of mate preference: Are male guppies mimicking fruit? *Proceedings: Biological Sciences* 269: 475-481.

Mating Games

Double, M., and A. Cockburn, 2000. Pre-dawn infidelity: females control extra-pair mating in superb fairy-wrens. *Proceedings of the Royal Society of London, Series B—Biological Sciences* 267: 465–570.

Sperm Wars

Birkhead, T. R., 2000. *Promiscuity: An Evolutionary History of Sperm Competition*. Cambridge, Mass: Harvard University Press.

Ramm, Steven A., Geoffrey A. Parker, and Paula Stockley, 2005. Sperm competition and the evolution of male reproductive anatomy in rodents. *Proceedings of the Royal Society of London, Series B—Biological Sciences* 272: 949–955.

Sexual Conflict

Arnqvist, Göran, 2005. *Sexual Conflict*. Monographs in behavior and ecology. Princeton, N.J.: Princeton University Press.

Brennan, Patricia L. R., Richard O. Prum, Kevin G. McCracken, et al., 2007. Coevolution of male and female genital morphology in waterfowl. *PLoS ONE* 2: e418.

Cornwallis, Charlie K., and Tim R. Birkhead, 2007. Experimental evidence that female ornamentation increases the acquisition of sperm and signals fecundity. *Proceedings of the Royal Society of London, Series B—Biological Sciences.* 274: 583–590.

Rice, William R., Andrew D. Stewart, Edward H. Morrow, et al., 2006. Assessing sexual conflict in the *Drosophila melanogaster* laboratory model system. *Philosophical Transactions of the Royal Society of London, Series B: Biological Sciences* 361: 287–299.

Evolving Parents

Klug, Hope, and Michael B. Bonsall, 2007. When to care for, abandon, or eat your offspring: the evolution of parental care and filial cannibalism. *The American Naturalist* 170: 886–901.

Reznick, David N., Cameron K. Ghalambor, and Kevin Crooks, 2008. Experimental studies of evolution in guppies: a model for understanding the evolutionary consequences of predator removal in natural communities. *Molecular Ecology* 17: 97–107.

Strassmann, Beverly I., and Brenda Gillespie, 2002. Life-history theory, fertility and reproductive success in humans. *Proceedings of the Royal Society of London, Series B—Biological Sciences* 269: 553–562.

Williams, George C., 1966. *Adaptation and Natural Selection; a Critique of Some Current Evolutionary Thought*. Princeton, N.J.: Princeton University Press.

Tipping Sex Ratios

Trivers, R. L., and D. E. Willard, 1973. Natural selection of parental ability to vary the sex ratio of offspring. *Science* 179: 90–92.

Uller, Tobias, Ido Pen, Erik Wapstra, Leo W. Beukeboom, and Jan Komdeur, 2007. The evolution of sex ratios and sex-determining systems. *Trends in Ecology & Evolution* 22: 292–297.

Parents in Conflict

Pogány, Akos, István Szentirmai, Jan Komdeur, and Tamás Székely, 2008. Sexual conflict and consistency of offspring desertion in Eurasian penduline tit *Remiz pendulinus*. *BMC Evolutionary Biology* 8: 242.

Conflict within the Genome

Burt, Austin, 2006. *Genes in Conflict: The Biology of Selfish Genetic Elements*. Cambridge, Mass: Belknap Press of Harvard University Press.

Wilkins, Jon F., and David Haig, 2003. What good is genomic imprinting: the function of parent-specific gene expression. *Nature Reviews Genetics* 4: 359–368.

CHAPTER 13: EVOLUTIONARY MEDICINE

General references

Here are two of the best introductions to evolutionary medicine. Nesse and Williams effectively founded the discipline, and Stearns and Koella have edited a detailed summary of where it stands today.

Nesse, Randolph M., and G. Williams, 1994. *Why We Get Sick: The New Science of Darwinian Medicine*. New York: Times Books.

Stearns, Stephen C., and Jacob C. Koella, eds., 2008. *Evolution in Health and Disease*, 2nd ed. Oxford: Oxford University Press.

Detailed references

Introduction

Maynard, Sharon, Franklin H. Epstein, and S. Ananth Karumanchi, 2008. Preeclampsia and angiogenic imbalance. *Annual Review of Medicine* 59: 61–78.

Evolving Parasites

Drosten, Christian, Wolfgang Preiser, Stephan Günther, Herbert Schmitz, and Hans Wilhelm Doerr, 2003. Severe acute respiratory syndrome: identification of the etiological agent. *Trends in Molecular Medicine* 9: 325–327.

Human Petri Dishes

Ebert, Dieter, 2008. Host-parasite coevolution: insights from the *Daphnia*–parasite model system. *Current Opinion in Microbiology* 11: 290–301.

Roberts, Anjeanette, Damon Deming, Christopher D. Paddock, et al., 2007. A mouse-adapted SARS-coronavirus causes disease and mortality in BALB/c mice. *PLoS Pathogens* 3: e5.

Shankarappa, Raj, Joseph B. Margolick, Stephen J. Gange, et al., 1999. Consistent viral evolutionary changes associated with the progression of human immunodeficiency virus type 1 infection. *Journal of Virology* 73): 10489–10502.

Molded by Parasites

Klein, Robyn S., 2008. A moving target: the multiple roles of CCR5 in infectious diseases. *The Journal of Infectious Diseases* 197: 183–186.

Sabeti, Pardis C., Emily Walsh, Steve F. Schaffner, et al., 2005. The case for selection at CCR5-Delta32. *PLoS Biology* 3: e378.

Evolution's Drug War

Peschel, Andreas, and Hans-Georg Sahl, 2006. The co-evolution of host cationic antimicrobial peptides and microbial resistance. *Nature Reviews Microbiology* 4: 529–536.

Zasloff, M., 1992. Antibiotic peptides as mediators of innate immunity. *Current Opinion in Immunology* 4: 3–7.

Silver Bullets

Mwangi, Michael M., Shang Wei Wu, Yanjiao Zhou, et al., 2007. Tracking the in vivo evolution of multidrug resistance in *Staphylococcus aureus* by whole-genome sequencing. *Proceedings of the National Academy of Sciences of the United States of America* 104: 9451–9456.

Perron, Gabriel G., Michael Zasloff, and Graham Bell, 2006. Experimental evolution of resistance to an antimicrobial peptide. *Proceedings of the Royal Society of London, Series B—Biological Sciences* 273: 251–256.

Zimmer, Carl, 2008. *Microcosm:* E. coli *and the New Science of Life*. New York: Pantheon Books.

Genetic Drift and Disease

Carr, R. E., N. E. Morton, and I. M. Siegel, 1971. Achromatopsia in Pingelap Islanders. Study of a genetic isolate. *American Journal of Ophthalmology* 72: 746–756.

McKusick, Victor A., 2000. Ellis–van Creveld syndrome and the Amish. *Nature Genetics* 24: 203–204.

Sacks, Oliver W., 1997. *The Island of the Colorblind ; and, Cycad Island*. New York: Alfred A. Knopf.

Tishkoff, S., A, A Goldman, F. Calafell, et al., 1998. A global haplotype analysis of the myotonic dystrophy locus: implications for the evolution of modern humans and for the origin of myotonic dystrophy mutations. *American Journal of Human Genetics* 62: 1389–1402.

Maladapted

Nesse, Randolph M., 2005. Maladaptation and natural selection. *The Quarterly Review of Biology* 80: 62–70.

Shubin, Neil, 2008. *Your Inner Fish: A Journey into the 3.5-Billion-Year History of the Human Body*. New York: Pantheon Books.

The Natural Selection of Cancer

Greaves, Mel, 2007. Darwinian medicine: a case for cancer. *Nature Reviews Cancer* 7, no. 3 (March): 213–221.

Merlo, Lauren M.F., John W. Pepper, Brian J. Reid, and Carlo C. Maley, 2006. Cancer as an evolutionary and ecological process. *Nature Reviews Cancer* 6: 924–935.

Zimmer, Carl, 2007. Evolved for cancer? *Scientific American* 296: 68–74.

Defending Against Our Own Imperfection

Komarova, Natalia L., 2005. Cancer, aging and the optimal tissue design. *Seminars in Cancer Biology* 15: 494–505.

Old Age: Evolution's Side Effect

Jenkins, Nicole L., Gawain McColl, and Gordon J. Lithgow, 2004. Fitness cost of extended lifespan in *Caenorhabditis elegans*. *Proceedings of the Royal Society of London, Series B—Biological Sciences* 271: 2523–2526..

Kirkwood, T. B. L., 2008. Understanding ageing from an evolutionary perspective. *Journal of Internal Medicine* 263: 117–127.

Oberdoerffer, Philipp, Shaday Michan, Michael McVay, et al., 2008. SIRT1 Redistribution on chromatin promotes genomic stability but alters gene expression during aging. *Cell* 135: 907–918.

Stearns, S. C., M. Ackermann, M. Doebeli, and M. Kaiser, 2000. Experimental evolution of aging, growth, and reproduction in fruitflies. *Proceedings of the National Academy of Sciences of the United States of America* 97: 3309–3313.

Sick from Sexual Conflict

Amor, David J., and Jane Halliday, 2008. A review of known imprinting syndromes and their association with assisted reproduction technologies. *Human Reproduction* 23: 2826–2834.

Bowdin, Sarah, Cathy Allen, Gail Kirby, et al., 2007. A survey of assisted reproductive technology births and imprinting disorders. *Human Reproduction* 22: 3237–3240.

Crespi, Bernard, 2008. Genomic imprinting in the development and evolution of psychotic spectrum conditions. *Biological Reviews of the Cambridge Philosophical Society* 83: 441–493.

Maher, Eamonn R., 2005. Imprinting and assisted reproductive technology. *Human Molecular Genetics* 14: R133–138.

Rodier, Francis, Judith Campisi, and Dipa Bhaumik, 2007. Two faces of p53: aging and tumor suppression. *Nucleic Acids Research* 35: 7475–7484.

Mismatched with Modern Life

Chen, Yu, and Martin J. Blaser, 2008. *Helicobacter pylori* colonization is inversely associated with childhood asthma. *The Journal of Infectious Diseases* 198: 553–560.

Gluckman, Peter D., 2006. *Mismatch: Why Our World No Longer Fits Our Bodies.* Oxford: Oxford University Press.

Palsdottir, Astridur, Agnar Helgason, Snaebjorn Palsson, et al., 2008. A drastic reduction in the life span of cystatin C L68Q carriers due to life-style changes during the last two centuries. *PLoS Genetics* 4: e1000099.

Sachs, Jessica Snyder, 2007. *Good Germs, Bad Germs: Health and Survival in a Bacterial World.* New York: Hill & Wang.

Strachan, D. P., 1989. Hay fever, hygiene, and household size. *BMJ* (Clinical Research Ed.) 299: 1259–1260.

Evolutionary Medicine: Gloom and Hope

Fry, Bryan G., Holger Scheib, Louise van der Weerd, et al., 2008. Evolution of an arsenal: structural and functional diversification of the venom system in the advanced snakes (Caenophidia). *Molecular & Cellular Proteomics* 7: 215–246.

Oda, Masako, Yoko Satta, Osamu Takenaka, and Naoyuki Takahata, 2002. Loss of urate oxidase activity in hominoids and its evolutionary implications. *Molecular Biology and Evolution* 19: 640–653.

Ramazzina, Ileana, Claudia Folli, Andrea Secchi, Rodolfo Berni, and Riccardo Percudani, 2006. Completing the uric acid degradation pathway through phylogenetic comparison of whole genomes. *Nature Chemical Biology* 2: 144–148.

CHAPTER 14: MINDS AND MICROBES: THE EVOLUTION OF BEHAVIOR

General references

For textbooks covering the topics in this chapter, see Alcock and Cartwright. Pinker's book is intended for the general public.

Alcock, John, 2009. *Animal Behavior: An Evolutionary Approach.* 9th ed. Sunderland, Mass: Sinauer Associates.

Cartwright, John, 2008. *Evolution and Human Behavior: Darwinian Perspectives on Human Nature.* 2nd ed. Cambridge, Mass.: MIT Press.

Pinker, Steven, 1997. *How the Mind Works.* New York: Norton.

Detailed references

Introduction

Krause, Johannes, Carles Lalueza-Fox, Ludovic Orlando, et al., 2007. The derived FOXP2 variant of modern humans was shared with Neandertals. *Current Biology*: 17: 1908–1912.

Behavior Evolves

Hillesland, Kristina L., Gregory J. Velicer, and Richard E Lenski, 2009. Experimental evolution of a microbial predator's ability to find prey. *Proceedings of the Royal Society of London, Series B—Biological Sciences* 276: 459-467.

The Origin of Nerves

King, Nicole, M. Jody Westbrook, Susan L. Young, et al., 2008. The genome of the choanoflagellate *Monosiga brevicollis* and the origin of metazoans. *Nature* 451: 783–788.

Manning, Gerard, Susan L. Young, W. Todd Miller, and Yufeng Zhai, 2008. The protist, *Monosiga brevicollis*, has a tyrosine kinase signaling network more elaborate and diverse than found in any known metazoan. *Proceedings of the National Academy of Sciences of the United States of America* 105: 9674–9679.

Richards, Gemma S., Elena Simionato, Muriel Perron, et al., 2008. Sponge genes provide new insight into the evolutionary origin of the neurogenic circuit. *Current Biology* 18: 1156–1161.

Sakarya, Onur, Kathryn A. Armstrong, Maja Adamska, et al., 2007. A post-synaptic scaffold at the origin of the animal kingdom. *PLoS ONE* 2: e506.

Tissir, Fadel, and Andre M. Goffinet, 2003. Reelin and brain development. *Nature Reviews Neuroscience*: 496–505.

Learning New Tricks

Arnold, Stevan J., 1977. Polymorphism and geographic variation in the feeding behavior of the garter snake *Thamnophis elegans*. *Science* 197: 676–678.

Dukas, Reuven, 2008. Evolutionary biology of insect learning. *Annual Review of Entomology* 53: 145–160.

Mery, Frederic, and Tadeusz J. Kawecki, 2003. A fitness cost of learning ability in *Drosophila melanogaster*. *Proceedings of the Royal Society of London, Series B—Biological Sciences* 270: 2465–2469.

The Vertebrate Brain: Not an All-purpose Computer

Catania, Kenneth C., and Erin C. Henry, 2006. Touching on somatosensory specializations in mammals. *Current Opinion in Neurobiology* 16: 467–473.

Janvier, P., 2008. The brain in the early fossil jawless vertebrates: evolutionary information from an empty nutshell. *Brain Research Bulletin* 75: 314–318.

Krubitzer, Leah, 2007. The magnificent compromise: cortical field evolution in mammals. *Neuron* 56: 201–208.

Marcus, Gary F., 2008. *Kluge: The Haphazard Construction of the Human Mind*. Boston: Houghton Mifflin.

Shu, D.-G., S. Conway Morris, J. Han, et al., 2003. Head and backbone of the Early Cambrian vertebrate *Haikouichthys*. *Nature* 421: 526–529.

The Origin of Emotions

Domes, Gregor, Markus Heinrichs, Andre Michel, Christoph Berger, and Sabine C. Herpertz, 2007. Oxytocin improves "mind-reading" in humans. *Biological Psychiatry* 61: 731–733.

Feldman, Ruth, Aron Weller, Orna Zagoory-Sharon, and Ari Levine, 2007. Evidence for a neuroendocrinological foundation of human affiliation: plasma oxytocin levels across pregnancy and the postpartum period predict mother–infant bonding. *Psychological Science* 18: 965–970.

Fiorillo, Christopher D., William T. Newsome, and Wolfram Schultz, 2008. The temporal precision of reward prediction in dopamine neurons. *Nature Neuroscience* 11: 966–973.

LeDoux, Joseph E., 1996. *The Emotional Brain: The Mysterious Underpinnings of Emotional Life*. New York: Simon & Schuster.

Leng, Gareth, Simone L. Meddle, and Alison J. Douglas, 2008. Oxytocin and the maternal brain. *Current Opinion in Pharmacology* 8: 731–734.

Panksepp, Jaak, 2007. Neuroevolutionary sources of laughter and social joy: modeling primal human laughter in laboratory rats. *Behavioural Brain Research* 182: 231–244.

Primates: Eyes, Tools, Societies

Byrne, Richard W., and Lucy A. Bates, 2007. Sociality, Evolution and Cognition. *Current Biology* 17: R714–R723.

Byrne, Richard W., 2007. Culture in great apes: using intricate complexity in feeding skills to trace the evolutionary origin of human technical prowess. *Philosophical Transactions of the Royal Society of London, Series B—Biological Sciences* 362: 577–585.

Duffy, Kimberly G., Richard W. Wrangham, and Joan B. Silk, 2007. Male chimpanzees exchange political support for mating opportunities. *Current Biology*: 17: R586–587.

Dunbar, R. I. M., and Susanne Shultz, 2007. Evolution in the social brain. *Science* 317: 1344–1347.

Gilad, Yoav, Victor Wiebe, Molly Przeworski, Doron Lancet, and Svante Pääbo, 2004. Loss of olfactory receptor genes coincides with the acquisition of full trichromatic vision in primates. *PLoS Biology* 2: E5.

Martin, Robert D., Christophe Soligo, and Simon Tavaré, 2007. Primate origins: implications of a Cretaceous ancestry. *Folia Primatologica: International Journal of Primatology* 78,: 277–296.

Reader, Simon M., and Kevin N. Laland, 2002. Social intelligence, innovation, and enhanced brain size in primates. *Proceedings of the National Academy of Sciences of the United States of America* 99: 4436–4441.

Silk, Joan B., 2007. The adaptive value of sociality in mammalian groups. *Philosophical Transactions of the Royal Society of London, Series B—Biological Sciences* 362: 539–559.

Silk, Joan B., 2007. Social components of fitness in primate groups. *Science* 317: 1347–1351.

Sol, Daniel, Sven Bacher, Simon M. Reader, and Louis Lefebvre, 183. Brain size predicts the success of mammal species introduced into novel environments. *The American Naturalist* 172: S63–S71.

Tavaré, Simon, Charles R. Marshall, Oliver Will, Christophe Soligo, and Robert D. Martin, 2002. Using the fossil record to estimate the age of the last common ancestor of extant primates. *Nature* 416: 726–729.

Uniquely Human (or at Least, Uniquely Hominid)

Gazzaniga, Michael S., 2008. *Human: The Science Behind What Makes Us Unique*. New York: Ecco.

Gil-da-Costa, Ricardo, Alex Martin, Marco A. Lopes, et al., 2006. Species-specific calls activate homologs of Broca's and Wernicke's areas in the macaque. *Nature Neuroscience* 9: 1064–1070.

Herrmann, Esther, Josep Call, Maráa Victoria Hernàndez-Lloreda, Brian Hare, and Michael Tomasello, 2007. Humans have evolved specialized skills of social cognition: the cultural intelligence hypothesis. *Science* 317: 1360–1366.

Kenneally, Christine, 2007. *The First Word: The Search for the Origins of Language*. New York: Viking.

Krause, Johannes, Carles Lalueza-Fox, Ludovic Orlando, et al., 2007. The derived FOXP2 variant of modern humans was shared with Neandertals. *Current Biology* 17: 1908–1912.

Liégeois, Frédérique, Torsten Baldeweg, Alan Connelly, et al., 2003. Language fMRI abnormalities associated with FOXP2 gene mutation. *Nature Neuroscience* 6: 1230–1237.

Pollick, Amy S., and Frans B. M. de Waal, 2007. Ape gestures and language evolution. *Proceedings of the National Academy of Sciences of the United States of America* 104: 8184–8189.

Rilling, James K., Matthew F. Glasser, Todd M. Preuss, et al., 2008. The evolution of the arcuate fasciculus revealed with comparative DTI. *Nature Neuroscience* 11: 426–428.

Sherwood, Chet C., Francys Subiaul, and Tadeusz W. Zawidzki, 2008. A natural history of the human mind: tracing evolutionary changes in brain and cognition. *Journal of Anatomy* 212: 426–454.

Slocombe, Katie E., and Klaus Zuberbühler, 2007. Chimpanzees modify recruitment screams as a function of audience composition. *Proceedings of the National Academy of Sciences of the United States of America* 104: 17228–17233.

Taglialatela, Jared P., Jamie L. Russell, Jennifer A. Schaeffer, and William D. Hopkins, 2008. Communicative Signaling Activates 'Broca's' Homolog in Chimpanzees. *Current Biology* 18: 343–348.

Tomasello, Michael, 2008. *Origins of Human Communication*. The Jean Nicod lectures 2008. Cambridge, Mass.: The MIT Press.

Credits

CHAPTER 1

Page 2: Ambulocetus, Carl Buell. Page 3: Hans Thewissen, photo courtesy of Hans Thewissen. Page 7: Neanderthal skull © John Reader/Photo Researchers, Inc. Page 8: Dwarf minke whales © Adrian Baddeley/iStockphoto. Page 9: Skeleton of Ambulocetus, photo courtesy of Hans Thewissen. Page 11, Figure 1.1: Whale tree evogram, Carl Buell and Echo Medical Media. Page 13: Dolphins © David Schrader/iStockphoto. Page 15: Whale hunting, Nathenial Currier © MPI/Getty Images.

CHAPTER 2

Page 16: Galapagos marine iguana © AllCanadaPhotos/Photoshot. Page 17: Charles Darwin watercolor portrait painted by George Richmond from *Origins*, Richard Leakey and Roger Lewin. Page 19: Carl Linnaeus painted by Alexander Roslin. Page 20: Paul Turner © Michael Marsland/Yale University. Page 21: (left) Nicholas Steno original painting by J. Trap; (right) Shark teeth illustration from Steno's 1667 paper; (bottom) Georges Buffon original painting by François-Hubert Drouais. Page 23: Mastodon fossil © The Academy of Natural Sciences, Ewell Sale Stewart Library. Page 24: (left) William Smith portrait by Hugues Fourau; (right) Map of Great Britain published by William Smith in 1815. Page 25: (top) Buckland's diagram of Earth's crust © The Natural History Museum, London; (bottom) Jean-Baptiste Lamarck © Photo by Hulton Archive/Getty Images. Page 27: HMS Beagle engraving © World Illustrated/Photoshot. Page 28: (collage, row 1, left to right) Warbler finch © Photograph by P.R. Grant; Woodpecker finch © David Hosking/Photo Researchers, Inc.; Small ground finch © David Hosking/FLPA; (row 2, left to right) Warbler finch © Photograph by P.R. Grant; Medium ground finch © David Hosking/FLPA; Medium tree finch © David Hosking/FLPA. Page 29: Darwin's notebook from the American Museum of Natural History. Page 30, Figure 2.1: Limbs, Lineworks. Page 31, Figure 2.2: Gill arches, Echo Medical Media. Page 33: (top) Beetle with horns © photo by D. Emlen and O. Helmy, reprinted with permission from PNAS; (bottom) Figure 2.3: Natural selection, Lineworks. Page 34: Charles Darwin photo by J. Cameron.

CHAPTER 3

Page 36: (top) Stromatolites © Georgette Douwma/Photo Researchers, Inc.; (bottom) Stromatolite fossil, courtesy of Abigail Allwood. Page 37: Abigail Allwood, courtesy of Abigail Allwood. Page 41, Figure 3.1: Radioactive clock, Echo Medical Media. Page 43, Figure 3.2: Dating a fossil, Lineworks. Page 44, Figure 3.3: Fossil formation, Lineworks. Page 46: (left) Zircon, courtesy of Alexander Nemchin; (right) Biomarkers, Echo Medical Media. Page 49: Bangiomorpha, Nicholas J. Butterfield. Page 50: (top) Doushantuo embryos, Shuhai Xiao at Virginia Tech; (bottom) Ediacaran fauna © Chase Studio/Photo Researchers, Inc. Page 51: (top, left) Opabinia, Echo Medical Media; (top, right) Trilobite © James L. Amos/Photo Researchers, Inc.; (bottom, left) Haikouichthys, Carl Buell; (bottom, right) Dunkleosteus, Carl Buell. Page 52: Wattieza © Nature article *Giant cladoxylopsid trees resolve the enigma of the Earth's earliest forest stumps at Gilboa*; (reconstruction, left) Frank Mannolini, New York State Museum, Albany NY; (photograph, right) South Mountain Trunk, William Stein, State University of New York at Binghamton, NY. Page 53: (top, left) Millipede fossil from the National Museums Scotland, Edinburgh; (top, right) Millipede drawing, Carl Buell; (bottom) Silvanerpeton, Carl Buell. Page 54: (top) Dimetrodon, Carl Buell; (bottom) Parasaurolophus, Carl Buell. Page 55: (top) Amber insect © George Bernard/Photo Researchers, Inc.; (bottom left) Sahelanthropus, Carl Buell; (bottom right) Sahelanthropus fossil © AFP/Getty Images. Page 56: Homo sapiens cranium (Omo 2) © The Natural History Museum, London.

CHAPTER 4

Page 58: Tiktaalik, Carl Buell. Page 59: (top) Ted Daeschler and Neil Shubin, courtesy of Neil Shubin; (bottom) Paleontologists at Tiktaalik site, Jason Downs. Page 60: Tiktaalik skeleton/reconstruction © Ted Daeschler/VIREO. Page 61, Figure 4.1: Vertebrate trait tree, Echo Medical Media. Page 62: Isaac Newton © Steven Wynn/iStockphoto. Page 64: Coelacanth © Tom & Therisa Stack/NHPA/Photoshot. Page 65: Coelacanth pectoral fins illustration, Carl Buell. Page 66, Figure 4.2: Fish/tetrapod limb, Carl Buell. Page 67, Figure 4.3: Tetrapod evogram, Carl Buell and Echo Medical Media. Page 69, Figure 4.4: Mammal ear, Echo Medical Media. Page 70, Figure 4.5: Ear evogram, Carl Buell and Echo Medical Media. Page 72: (left) Archaeopteryx fossil © Jason Edwards/Getty Images; (right) Archaeopteryx illustration, Carl Buell. Page 73, Figure 4.6: Bird evogram, Carl Buell and Echo Medical Media. Page 74 (left) Sinornithosaurus fossil © Mick Ellison/AMNH; (right) Sinornithosaurus illustration, Carl Buell. Page 75: (left) Feather nodes © Mick Ellison *Feather Quill Knobs in the Dinosaur Velociraptor*, by Alan H. Turner, Peter J. Makovicky, Mark A. Norell, Science 21 Sept. 2007, Vol. 317 no. 5845, Page 1721. Reprinted with permission from AAAS; (top, right) Dinosaur nest fossil by Finn/Ellison © American Museum of Natural History; (bottom, right) Bird nesting © Martin Ruegner/Getty Images; Page 76, Figure 4.7: Coccyx, Echo Medical Media. Page 78, Figure 4.8: Hominid evogram, Carl Buell and Echo Medical Media. Page 79: Orrorin bone, courtesy of Brian G. Richmond. Page 80: (left) Australopithecus afarensis skeleton cast from Museum national d'histoire naturelle, Paris; (right) Australopithecus afarensis illustration, Carl Buell. Page 81: Laetoli footprints © Kenneth Garrett/Getty Images. Page 82: (left) Homo erectus skeleton, courtesy of Alan Walker; (right) Homo erectus illustration, Carl Buell.

CHAPTER 5
Page 84: Xia and Bao Xishun © photo by Tao Qi/ChinaFotoPress/Getty Images. Page 85: Joel Hirschhorn, courtesy of Joel Hirschhorn. Page 88, Figure 5.1: Proteins, Echo Medical Media. Page 90, Figure 5.2: DNA, Echo Medical Media. Page 91, Figure 5.3: Transcription, Echo Medical Media. Page 93, Figure 5.4: Asexual horizontal gene transfer, Echo Medical Media. Page 94, Figure 5.5: Sexual reproduction, Lineworks. Page 95: Gregor Mendel, "Mendel's principles of heredity: A Defence," by William Bateson. Page 97, Figure 5.6: Mendel's peas, Lineworks. Page 98: (photo) Francis Galton, courtesy of galton.org; Figure 5.7: Galton height chart, Lineworks. Page 100, Figure 5.8: Height heritability diagram, Lineworks.

CHAPTER 6
Page 102: Pod Mracu lizard, courtesy of Anthony Herrel. Page 103: Duncan Irschick, courtesy of Duncan Irschick. Page 106, Figure 6.1: Mutation chart, Lineworks. Page 110, Figure 6.2: Hardy-Weinberg model, Lineworks. Page 112, Figure 6.3: Sickle cell Hardy-Weinberg model, Lineworks. Page 113, Figure 6.4: (top) bw75 drift graph, Lineworks; (drosophila photo) © Tomasz Zachariasz/iStockphoto; Figure 6.5: (bottom) Jelly bean drift diagram, Lineworks. Page 114, Figure 6.6: Drift simulation from Hartl, Lineworks. Page 116, Figure 6.7: Fitness diagram, Lineworks. Page 117, Figure 6.8: Large population simulation, Lineworks. Page 118, Figure 6.9: Lenski experiment protocal, Lineworks. Page 119, Figure 6.10: Lenski growth chart, Lineworks; (photo) E. coli © Luis M. Molina/iStockphoto. Page 120, Figure 6.11: Sickle cell fitness graph, Lineworks. Page 121, Figure 6.12: Strength of selection graph, Lineworks. Page 122, Figure 6.13: Evolutionary response graph, Lineworks. Page 123, Figure 6.14: Adaptive landscape, Lineworks. Page 124: Daphne Major, courtesy of Kenneth Petren. Page 125, Figure 6.15: Grant beak graph, Echo Medical Media. Page 127, Figure 6.16: Linkage disequilibrium chart, Lineworks. Page 129, Figure 6.17: Scarlet kingsnake ranges, Echo Medical Media; (photos) Scarlet kingsnake, Wayne Vandevender; Scarlet kingsnake mimicking, Wayne Vandevender; Eastern coral snake , John D. Wilson.

CHAPTER 7
Page 132: (collage, row 1, left to right) Man © Shelly Perry/iStockphoto; Woman with glasses © Will Rennick/iStockphoto; Man © Lee Pettet/iStockphoto; Woman © Cliff Parnell/iStockphoto; (row 2, left to right) Woman © Lukasz Laska/iStockphoto; Man © Michael Koehl/iStockphoto; Woman © Achirangshu De/iStockphoto; Man © Dan Brandenburg/iStockphoto; (row 3, left to right) Man © Sergey Kashkin/iStockphoto; Woman © Eduardo Jose Bernardino/iStockphoto; Man © Vikram Raghuvanshi/iStockphoto; Woman © iofoto/iStockphoto; (row 4, left to right) Woman © digitalskillet/iStockphoto; Man © Stephanie Swartz/iStockphoto; Woman © Vikram Raghuvanshi/iStockphoto; Man © Eduardo Jose Berardino/iStockphoto. Page 133: Sarah Tishkoff, courtesy of Sarah Tishkoff. Page 135, Figure 7.1: T7 virus tree, Lineworks. Page 137, Figure 7.2: Human, chimp, gorilla gene tree, Echo Medical Media. Page 138, Figure 7.3: Gene tree (Ya5AH137), Lineworks. Page 140, Figure 7.4: Tetrapod tree, Echo Medical Media. Page 141, Figure 7.5: Tishkoff human tree, Lineworks. Page 142, Figure 7.6: Finch tree, Echo Medical Media. Page 144, Figure 7.7: HIV tree, Lineworks; (photo) HIV molecule © Russel Knightley Media. Page 146, Figure 7.8: DNA mutations, Lineworks. Page 147, Figure 7.9: HIV clock, Lineworks. Page 149, Figure 7.10: Selection diagram, Lineworks. Page 150, Figure 7.11: FOXP2 tree, Echo Medical Media.

CHAPTER 8
Page 154: Snake with bared fangs © Pete Oxford/Nature Picture Library. Page 155: Brian Fry, courtesy of Brian Fry. Page 158, Figure 8.1: Bacteria adaptation graph, Lineworks; (photo) E.coli © Luis M. Molina/iStockphoto. Page 160, Figure 8.2: Gene duplication, Lineworks. Page 161, Figure 8.3: Crotamine venom tree, Echo Medical Media. Page 163, Figure 8.4: Venom gene recruitment, Echo Medical Media. Page 164, Figure 8.5: Venom evolution, Echo Medical Media. Page 165, Figure 8.6: Development circuit, Echo Medical Media. Page 166, Figure 8.7: Finch beak, Echo Medical Media. Page 167: Peter and Rosemary Grant, photograph by P.R. Grant. Page168, Figure 8.8: Feather diagram, Echo Medical Media. Page 170, Figure 8.9: Fly vs. Mouse diagram, Echo Medical Media. Page 171, Figure 8.10: Dorsal ventral diagram, Echo Medical Media; (photos) Lobster © Soyka/Shutterstock; Cat © Eric Iselée/Shutterstock. Page 173, Figure 8.11: Eye diversity diagram, Echo Medical Media; (photos) Jellyfish © ANT Photo Library/Photo Researchers, Inc.; Octopus © Kerry L. Werry/Shutterstock; Fly © Stana/Shutterstock; Human eye © Bplucinski/Shutterstock. Page 176, Figure 8.12: Opsin origin, Lineworks. Page 177, Figure 8.13: Opsin evolution, Echo Medical Media. Page 178, Figure 8.14: Crystallins, Echo Medical Media. Page 179, Figure 8.15: Vertebrate eye evolution, Echo Medical Media. Page 182, Figure 8.16: Convergence evolution, Carl Buell. Page 183, Figure 8.17: Flatfish evolution tree, Echo Medical Media; (photo) Flatfish © Undersea Discoveries/Shutterstock.

CHAPTER 9
Page 186: (collage, clockwise left to right) Rothschild giraffe © Klaas Lingbeek-van Kranen/iStockphoto; Angolan giraffe © Dirk Ruter/iStockphoto; West African giraffe © Jean-Patrick Suraud; Reticulated giraffe © Liz Leyden/iStockphoto; South African giraffe © Ricardo De Mattos/iStockphoto; Masai giraffe © Brian Raisbeck/iStockphoto. Page 187: Rick Brenneman, courtesy of Rick Brennemnan. Page 190: (top)

Sea eagles © Enrique Aguirre/Getty Images; (bottom) Figure 9.1 Evolution of species, Lineworks. Page 193, Figure 9.2: Panama speciation, Echo Medical Media. Page 194: (top) Greenish warblers, courtesy of Jessica H. Irwin; (bottom) Figure 9.3 Greenish warbler ranges, Lineworks. Page 196: Lake Apoyo and cichlid fish photos, courtesy of Axel Meyer. Page 197, Figure 9.4: Reproductive isolation, Lineworks; (photo) Drosophila © Tomasz Zachariasz/iStockphoto. Page 198, Figure 9.5: Allopolyploidy, Lineworks; (photo) Sunflower, courtesy of Jason Rick. Page 200, Figure 9.6: Giraffe geographical ranges, Lineworks. Page 201, Figure 9.7: Three domain tree, Lineworks. Page 203, Figure 9.8: Bacterial mosaics, Echo Medical Media; Figure 9.9 Bacteria web of life, black and white tree skeleton, Tal Dagan and William Martin, Düsseldorf University; Colored tree, Dagan T, Artzy-Randrup Y, Martin W (2008) *Modular networks and cumulative impact of lateral transfer in prokaryote genome evolution*, PNAS USA 105:10039-10044. Page 204, Figure 9.10: Ecovar species, Echo Medical Media. Page 207, Figure 9.11: Neanderthal hybridization, Carl Buell and Lineworks.

CHAPTER 10

Page 210: Beetle mosaic © Christopher Marley/Form and Pheromone. Page 212, Figure 10.1: Tropic diversity map, Echo Medical Media © ML Design after W. Barthlott, 1997, from *The Seventy Great Mysteries of the Natural World* edited by Michael J. Benton, Thames & Hudson, London and New York, 2008; (photo) Flower © Elaine Davis/Shutterstock. Page 213, Figure 10.2: Species diversity curve, Lineworks. Page 214, Figure 10.3: Mite harvestman map, Echo Medical Media; (photo) Mite harvestman, courtesy of Gonzalo Giribet. Page 215, Figure 10.4: Marsupial timeline, Echo Medical Media. Page 217, Figure 10.5: Diversification rates diagram, Lineworks. Page 219, Figure 10.6: Jackson punctuated equilibrium, Lineworks; (drawings) Metrarabdotos and Rhizosolenia, Echo Medical Media. Page 220: (left) Dodo, Carl Buell; (right) Carolina parakeet, *Birds of America* by John James Audubon. Page 221, Figure 10.7: Foote species lifetime graph, Lineworks © Dmitry P. Filippenko & Mikhail O. Son *The New Zealand mud snail Potamopyrgus antipodarum (Gray, 1843) is colonising the artificial lakes of Kaliningrad City, Russia (Baltic Sea Coast)*, Aquatic Invasions (2008) Volume 3, Issue 3: 345-347. Page 223, Figure 10.8: Mountain beaver tree, Carl Buell and Echo Medical Media. Page 224: (collage, left to right) Blue/black cichlid © Andreas Gradin/Shutterstock; White cichlid © Andreas Gradin/Shutterstock; Blue cichlid © Andreas Gradin/Shutterstock; Yellow cichlid © Andreas Gradin/Shutterstock. Page 225, Figure 10.9: Insect diversity graph, Lineworks; (photo) Beetle © Evgeniy Ayupov/Shutterstock. Page 227, Figure 10.10: Cambrian phylum phylogeny, Echo Medical Media. Page 228, Figure 10.11: Marshall Cambrian diagram, Echo Medical Media. Page 230, Figure 10.12: (left) Cambrian bioturbation, Artist: Peter Trusler, Melbourne, from *The Rise of Animals. Evolution and Diversification of the Kingdom Animalia* by M. Fedonkin, J. Gehling, K. Grey, G. Narbonne & P. Vickers-Rich, Johns Hopkins University Press, 2007; (right) Cloudina fossils, Bengtson, Stefan and Yue Zhao, 1992. *Predatorial borings in Late Precambrian mineralized exoskeletons.* Science, 257: 367–369. Page 231, Figure 10.13: Alroy origin and extinction graph, Linworks. Page 233: (left) Chixculub impact, Echo Medical Media; (right) Chixculub crater, Virgil L. Sharpton, University of Alaska, Fairbanks. Page 235, Figure 10.14: Deforestation, Echo Medical Media; (photo) Deforestation, Jamil Dwyer. Page 236, Figure 10.15: Modern extinction graph, Lineworks. Page 237, Figure 10.16: Modern CO2 graph, Lineworks. Page 238: Coral reef, National Oceanic and Atmospheric Administration.

CHAPTER 11

Page 240: Dupe wasp and tongue orchid © Esther Beaton Wild Pictures. Page 241: Anne Gaskett with dupe wasps © Esther Beaton Wild Pictures. Page 243, Figure 11.1: Coevolution chart, Echo Medical Media. Page 246: Sharpshooter bacteriomes, courtesy of Roman Rakitov and Daniela M. Takiya. Page 247, Figure 11.2: Cospeciation trees, Lineworks. Page 248, Figure 11.3: Sulcia trees, Lineworks; (photo) Micrograph of Sulcia and Baumannia, *Symbiosis as an adaptive process and source of phenotypic complexity*, by Nancy A. Moran. PNAS May 15, 2007 vol. 104 no. Suppl. 1 8627-8633 Fig. 3 (the two symbionts, Sulcia and Baumannia) © 2007 National Academy of Sciences, U.S.A. Page 253, Figure 11.4: Rough-skinned newt and snake map, Echo Medical Media; (photo) Garter snake eating rough-skinned newt, courtesy of Edmund D. Brodie III. Page 256, Figure 11.5: Mitochondria, Echo Medical Media. Page 257, Figure 11.6: Plant diagram, Lineworks.

CHAPTER 12

Page 262: Widowbird © Fritz Polking/FLPA. Page 263: Patricia Brennan, courtesy of Bernard J. Brennan. Page 266, Figure 12.1: Sex benefit diagram, Lineworks. Page 267, Figure 12.2: E. coli sex results, Lineworks; (photo) E. coli © Dennis Kunkel Microscopy, Inc. Page 269: Red Queen, "Through the Looking-Glass," by Lewis Carroll. Page 270: Sperm fertilizing egg © James Steidl/Shutterstock. Page 271: (top) Fiddler crab © Rod Porteous/Getty Images; (bottom) Southern elephant seals © Peter Bassett/naturepl.com. Page 272, (top) Figure 12.3: Firefly graph, Lineworks; (photo) Firefly © Cathy Keifer/Shutterstock; (bottom) Stalk-eyed flies, courtesy of Jerry Wilkinson. Page 273: Orange-colored fruit and guppies, courtesy of Gregory F. Grether. Page 277: Damselfly scraping © Doug Lemke/Shutterstock. Page 278, Figure 12.4: Graph of duck results, Lineworks; (photo) Duck phallus, courtesy of Patricia Brennan. Page 280, Figure 12.5: Dogon family graph, Lineworks; (photo) Dogon women © Beverly I. Strassmann. Page 281, Figure 12.6: Reznick guppies graph, Lineworks; (photos) Guppies, Paul Bentzen; River, Andrew Hendry. Page 283: Seychelles warblers, courtesy of David S. Richardson. Page 284, Figure 12.7: Inclusive fitness diagram, Lineworks.

Page 288, Figure 12.8: Imprinting diagram, Lineworks.

CHAPTER 13

Page 290: Immune cell attacking bacteria © Dennis Kunkel Microscopy, Inc. Page 291: S. Ananth Karumanchi, courtesy of S. Ananth Karumanchi. Page 293, Figure 13.1: Mother-fetal blood pressure system, Echo Medical Media. Page 295, Figure 13.2: Human herpesvirus 5 tree, Lineworks; (photo) HHV-5, Dr. Haraszti/CDC. Page 296: SARS panic in Hong Kong © Eightfish/Getty Images. Page 297, Figure 13.3: SARS tree, Lineworks; (photos) Human © Ronen/Shutterstock; Bat, Nevada Bureau of Land Management. Page 298, Figure 13.4: HIV patient tree, Lineworks; (photo) HIV virus © Russel Knight Media. Page 299: Plasmodium micrograph © Eye of Science/Photo Researchers, Inc. Page 300, Figure 13.5: Water flea virulence, Lineworks; (photo) Water flea © Paul Hebert. Page 305, Figure 13.6: Antibiotic JH timeline, Lineworks; (photo) Staphylococcus aureus, Janice Haney Car/CDC. Page 308, Figure 13.7: Ellis-van Creveld syndrome, Lineworks; (photo) Amish child with Ellis-van Creveld syndrome, courtesy of the Alan Mason Chesney Medical Archives. Page 310, Figure 13.8: Cancer diagram, Echo Medical Media. Page 312, Figure 13.9: Aging graph, Lineworks; (photo) Usain Bolt © Hamish Blair/Getty Images. Page 316, Figure 13.10: Infectious disease graph, Lineworks. Page 318: World at night satellite photo, NASA/Goddard Space Flight Center Scientific Visualization Studio. Page 320, Figure 13.11: Uric acid pathway, Echo Medical Media.

CHAPTER 14

Page 324: Neanderthals, Carl Buell. Page 325: El Sidron excavation site, courtesy of Carles Lalueza-Fox. Page 328: (left) Myxococcus xanthus © Juergen Berger/Photo Researchers, Inc.; (right) Myxococcus mound, Trance Gemini. Page 331, Figure 14.1: Choanoflagellates, Echo Medical Media; (photos) Choanoflagellates, courtesy of Monika Abedin; Sponges © Marsha Goldenberg/Shutterstock. Page 332, Figure 14.2: Baby gull pecking diagram, Lineworks. Page 336: (top) Haikouichthys, Carl Buell; (bottom) Figure 14.3 Vertebrate brains, Echo Medical Media. Page 337, Figure 14.4: Mammal cortex evolution, Echo Medical Media. Page 338, Figure 14.5: Human brain diagram, Echo Medical Media. Page 339, Figure 14.6: (top) Rodent touch maps, Carl Buell; (bottom) Human homunculus © The National History Museum, London. Page 345, Figure 14.7: Dunbar brain size chart, Lineworks; (photo) Primate brain, Image by Todd Preuss, Yerkes Primate Research Center *Molecular Insights into Human Brain Evolution*, Bradbury J PLoS Biology 3/3/2005. Page 347, Figure 14.8: Foxp2, Echo Medical Media. Page 348, Figure 14.9: Arcuate fasciculus, Echo Medical Media. Page 349: (top) Chimpanzee gesturing © Mary Beth Angelo/Photo Researchers, Inc.; (bottom) Stone tools © John Reader/Photo Researchers, Inc. Page 350, Figure 14.10: Hominid brain graph, courtesy of Nicholas Matzke; (photos, left) Australopithecus afarensis reconstruction, Museum of Man, San Diego, California; (right) Homo ergaster skull reconstruction, Museum of Man, San Diego, California. Page 351: Lascoux Cave Painting © Ralph Morse/Time & Life Pictures/Getty Images.

Index